PHYSICAL AND MATERIALS CONSTANTS

Name	Symbol	Magnitude	Units
Electron charge	q	1.6×10^{-19}	C
Boltzmann's constant	k	1.38×10^{-23}	J/K
Room temperature	T	300	K (27°C)
Thermal voltage	kT/q	0.026	V (at 27°C)
Dielectric constant of vacuum	ϵ_0	8.85×10^{-14}	F/cm
Dielectric constant of silicon	ϵ_{si}	$11.7\epsilon_0$	F/cm
Dielectric constant of SiO$_2$	ϵ_{ox}	$3.97\epsilon_0$	F/cm
Intrinsic carrier concentration of silicon (squared)	n_i^2	$1.5 \times 10^{32} T^3 e^{\frac{-1.15q}{kT}}$	(carriers/cm^8)^2
Room temperature value	n_i^2	2.1×10^{20}	(carriers/cm^8)^2
Nominal surface mobilities:			
Electrons	μ_n	580	cm^2/V·s
Holes	μ_p	230	cm^2/V·s

ANALYSIS AND DESIGN OF
DIGITAL INTEGRATED CIRCUITS

ANALYSIS AND DESIGN OF DIGITAL INTEGRATED CIRCUITS

Second Edition

David A. Hodges

Horace G. Jackson

Department of Electrical Engineering and Computer Sciences
University of California, Berkeley

McGraw-Hill Book Company

New York   St. Louis   San Francisco   Auckland   Bogotá   Caracas   Colorado Springs
Hamburg   Lisbon   London   Madrid   Mexico   Milan   Montreal   New Delhi
Oklahoma City   Panama   Paris   San Juan   São Paulo   Singapore   Sydney   Tokyo   Toronto

To Hazel and Susie

This book was set in Times Roman by Publication Services.
The editors were Alar E. Elken and James W. Bradley;
the cover was designed by Rafael Hernandez;
the production supervisor was Leroy A. Young.
Arcata Graphics/Halliday was printer and binder.

ANALYSIS AND DESIGN OF DIGITAL INTEGRATED CIRCUITS 2/e

2 3 4 5 6 7 8 9 0 HAL HAL 8 9 3 2 1 0 9 8

ISBN 0-07-029158-6

Hodges, David A., (date).
    Analysis and design of digital integrated circuits.
    1. Includes bibliographies and index.
    1. Digital electronics.   2. Integrated circuits.
I. Jackson, Horace G.   II. Title
TK7868.D5H63   1988   621.381173   87-31066
ISBN 0-07-029158-6

ABOUT THE AUTHORS

David A. Hodges earned his B.E.E. degree at Cornell University and his M.S. and Ph.D. degrees in Electrical Engineering at the University of California, Berkeley. From 1966 to 1970 he worked at Bell Telephone Laboratories, first in the components area at Murray Hill, then as Head of the System Elements Research Department at Holmdel. Now he is Professor of Electrical Engineering and Computer Sciences at the University of California, Berkeley, where he has been a member of the faculty since 1970. He is active in teaching and research on microelectronics technology and design, on communications and computer systems, and on computer-integrated manufacturing systems. In 1983 he was elected to membership in the National Academy of Engineering. He is now Editor of the new *IEEE Transactions on Semiconductor Manufacturing*, to be published beginning early in 1988. He is a former Editor of the *IEEE Journal of Solid-State Circuits* and a past Chairman of the International Solid-State Circuits Conference. Professor Hodges serves as a consultant to Cyclotomics, Inc., a Kodak Company.

Horace G. Jackson was born and educated in England. In 1947 he went to Canada, where until 1956 he was engaged in nuclear physics research at the Chalk River Laboratories of Atomic Energy of Canada. In 1956 he joined the Lawrence Berkeley Laboratory of the University of California, where he is presently a Senior Staff Scientist working on the development of instrumentation for nuclear science research. Since 1968 he has been Resident Lecturer in the Department of Electrical Engineering and Computer Sciences at the University of California, Berkeley. He has published widely in nuclear science and electronic engineering journals and is coauthor of two textbooks: "Introduction to Integrated Circuits" (McGraw-Hill, 1975) and "Analysis and Design of Digital Integrated Circuits" (McGraw-Hill, 1983). His research interests are in high-frequency analog and high-speed digital integrated circuits.

CONTENTS

*Denotes material that may be omitted without loss of continuity.

PREFACE TO SECOND EDITION

Digital integrated circuit technology has continued to evolve rapidly in the five years since publication of the first edition of this book. Much of the progress has taken the form of steady reductions in the internal dimensions of transistors and other devices, together with increases in the maximum practical size of chips. The maximum feasible complexity of a single microelectronic chip, whether measured in terms of transistors, logic gates, or memory cells, continues a long-term trend of doubling every one to two years. The term *VLSI* (very large scale integration) is now commonplace in referring to chips with 100,000 or more components, and we have the new term *ULSI* (ultra large scale integration) denoting chips with a million components or more.

Despite the rapid growth in component density and chip complexity, circuit design techniques for optimum peformance have changed relatively slowly. Perhaps the strongest trend has been toward greater use of CMOS (complementary metal-oxide-semiconductor) technology. Bipolar transistor technology has shown more capacity for improvement than some expected and is still an important factor. Another trend is the gradual introduction of GaAs (gallium arsenide) digital integrated circuit technology for selected applications. While the relative volume of usage of the various technologies will change with time, it seems clear that no single technology will ever meet all the various needs for digital circuits. We continue to believe that engineers working with microelectronics should have knowledge of the comparative characteristics of all the major technologies.

The second edition reflects the above-noted trends with a number of revisions and additions. Specific changes include the following. Chapter 1 has been reordered and revised. Material on logic functions and logic design has been moved to Appendix A. Most students now cover this material in other courses. The material on fabrication technology has been expanded somewhat, and a section on layout fundamentals has been added. Chapter 3 has been expanded by addition of material on the new CMOS logic families. Chapters 4 through 8 have undergone minor changes. Chapters 9 and 10 have been expanded to include new material on memory sense amplifiers and output buffers for VLSI circuits. Additional problems are included with most of these chapters.

Chapter 11, on gallium arsenide digital integrated circuits, is entirely new. We believe this to be the first textbook treatment of this advanced technology for high-peformance digital circuits. Transistor properties, the SPICE model, and the three principal GaAs circuit families are covered. Included are comparisons among these families and comparison with conventional silicon bipolar digital circuits. Special thanks go to Professors Stephen Long and Robert Meyer for their helpful critiques of this material.

Many other helpful corrections, comments, and suggestions on the first edition have come from our own students and from students and colleagues around the world. We acknowledge these with thanks as well as: Stephen Director, Carnegie-Mellon and McGraw-Hill Consulting Editor; Clarence Joh, Tufts University; Gerold W. Neudeck, Purdue University; Frank L. Raposa, Boston University; Mani Soma, University of Washington; and Andrezej J. Strojwas, Carnegie-Mellon University.

David A. Hodges
Horace G. Jackson

PREFACE TO FIRST EDITION

This textbook deals with the analysis and design of digital integrated circuits (ICs). Although a large part of the book is concerned with the internal design of digital ICs, we also want it to be helpful to the user of digital ICs. In practice we find the design and use of digital ICs to be closely linked, and a knowledge of both is important to the designer and the user. There are, by far, many more users than designers, but it is our experience that a working knowledge of IC design is a great advantage to the IC user. This is particularly true when the user must choose from a number of competing designs to satisfy a particular requirement. An understanding of the IC structure is important in evaluating the relative merits of different designs in the presence of electrical noise or variations in supply voltage. The user who understands the internal operation of integrated circuits is better able to interpret manufacturers' data sheets. He or she is also better prepared to anticipate the likely significance of progress in integrated circuit technology.

The book contains many worked-out examples. These are used to illustrate the principles of analysis and design, and also to impart some practical knowledge of digital ICs. Within most chapters, at the conclusion of main sections, there are exercise problems (with answers at the back of the book). Thus students are able to assure themselves that they comprehend the section before proceeding to the next. At the end of each chapter there are a number of problems covering the subject matter of the whole chapter. Solutions for these problems are included in a *Solutions Manual* available from the publisher.

A summary is given at the end of each chapter. It is intended to help the student review the material and to focus attention on the essential concepts developed in the chapter.

At the end of Chapters 2 through 8 two or three demonstrations or laboratory experiments are described. These are intended to stimulate understanding and retention of the material, and to illustrate the quality of agreement between design theory and experimental reality. In a small class these demonstrations may be incorporated into the lecture period. At Berkeley, students perform these experiments in weekly 3-hour laboratory sessions which are a required part of our course. We find that students profit from the experience of performing these experiments themselves, preferably following the coverage of the subject material in lectures. We believe that in engineering, theory alone is only half a loaf.

We expect that the third- and fourth-year electrical engineering students using this text will already have had an introductory course in electronic circuits and will have been introduced to the basic elements of logic design. Students who have completed a course on semiconductor devices will find that they can cover Chapters 2, 4, and 5 rather quickly. However, this material should not be omitted because the important emphasis here is on directly measurable electrical characteristics, circuit properties of devices, and device model parameters for circuit simulation. Many semiconductor device courses, in contrast, place principal emphasis on semiconductor band structure and carrier transport phenomena.

For a 15-week semester course, with 3 hours of lecture and a 3-hour laboratory period each week, the text may be covered at the rate of about one chapter each week, with the exception that two weeks are required to cover each of Chapters 3 and 7. In a 10-week quarter course, only the first 8 chapters can be covered thoroughly.

A chapter-by-chapter outline of the topics follows.

Chapter 1
INTRODUCTION TO DIGITAL ELECTRONICS

In this first chapter we briefly review important concepts of logic functions and Boolean identities. The essentials of integrated circuit fabrication technology are briefly described. Definitions of the various complexity levels up to very large scale integration (VLSI) are presented. We introduce the basic properties of a digital circuit and describe the ideal logic element in terms of its static input-output characteristics. We also define important characteristics of digital circuits such as noise margin and propagation delay time. The growing role of computer tools in analysis and design is noted. As a practical example of digital integrated circuit design, we introduce the programmable logic array.

Chapter 2
METAL-OXIDE-SEMICONDUCTOR (MOS) TRANSISTOR

The tremendous impact of the MOSFET in digital ICs is reflected in our early introduction of the subject. At the start of this chapter we briefly discuss some of the physical properties of the MOSFET, as well as the fabrication process for the device. We then give a detailed analysis of the static and then the dynamic characteristics of the MOS transistor. The SPICE model for the MOS transistor is described, and means for measuring model parameters on a given device are presented.

Chapter 3
MOS INVERTERS AND GATE CIRCUITS

This is the core chapter on MOSFET digital circuits. We first describe the static properties of a simple MOSFET inverter, which is mainly concerned with obtaining the voltage transfer characteristics (V_{out} versus V_{in}) of the various inverter-load connections of the NMOS transistor. The analysis of dynamic properties, that is, the switching time, is then presented for NMOS inverters. Next the static and dynamic properties of a CMOS inverter are developed. We then describe the analysis and design of simple NMOS and CMOS gate circuits in terms of their static and dynamic properties. Circuit modeling and simulation using program SPICE, dynamic logic techniques, and the important topic of scaling in MOS circuits are also presented in this chapter.

Chapter 4
SEMICONDUCTOR DIODES

The topic now changes from unipolar to bipolar circuits. At the start of this chapter we briefly discuss some of the physical properties of the *pn* junction diode, including the equilibrium barrier potential and depletion region charge. The *I-V* characteristics describing the operation of the device with forward and reverse bias are then derived. The effect of temperature on these equations is also discussed. The dynamic characteristics of the *pn* junction are covered by an analysis of the diode switching times. For practical application we describe the properties of various diode configurations possible with integrated circuits. The Schottky-barrier diode is introduced and its static and dynamic properties described. The SPICE model of the diode and methods for measurement of the model parameters are included. Finally, the effect of voltage breakdown in the *pn* junction is briefly discussed.

Chapter 5
BIPOLAR JUNCTION TRANSISTOR

Following on from Chapter 4, in this chapter, after a brief description of transistor operation, we derive the basic static equations for the bipolar transistor. The various modes of operation of the transistor are then described and simpler equations derived. The SPICE model of the bipolar transistor and methods for measurement of the model parameters are presented.

Chapter 6
BIPOLAR TRANSISTOR INVERTER

The core material for the bipolar digital ICs is found in Chapters 6 and 7. We start this chapter with a development of the static characteristics of a simple bipolar inverter, namely, the voltage transfer characteristic. From this we obtain the noise margins and derive equations for the fan-out. The dynamic characteristic of the inverter are described in terms of the charge-control model. Simpler forms of the charge-control equations are then derived for each of the operating modes of the transistor. These equations are then used in an illustrative example to compute the switching times of the bipolar inverter. In a similar manner the static and dynamic properties of a Schottky-clamped inverter are covered. A comparison of the results of hand analysis and computer simulation using SPICE is also included.

Chapter 7
BIPOLAR DIGITAL GATE CIRCUITS

In this chapter we present a detailed study of the major types of IC digital gates, namely, RTL, DTL, TTL, ECL, and I^2L. Especially emphasized are the latest developments in TTL, ECI, and I^2L. Both static and dynamic characteristics are covered in detail.

Chapter 8
REGENERATIVE LOGIC CIRCUITS

From strictly combinational circuits, in Chapter 8 we move our attention to sequential circuits. After describing the basic operation of a simple bistable circuit, we describe the properties of SR latch and the JK and D flip-flops with the aid of logic diagrams. Examples are then given of the implementation of these regenerative circuits in both MOS and bipolar technologies, specifically NMOS, CMOS, TTL, ECL and I^2L. Also included in this chapter are descriptions of the Schmitt trigger circuit as well as monostable and astable multivibrator circuits. Examples of each of these circuits implemented with CMOS and bipolar technologies are presented.

Chapter 9
SEMICONDUCTOR MEMORIES

In Chapter 9 we enter the world of large-scale integration (LSI). Read-only memories (ROMs) are described in both MOS and bipolar technologies. Details of the cells in the array, as well as the peripheral circuits, are presented. The use of MOSFET and bipolar circuits in static read-write memories (SRAMs) are also described. Three-transistor (3-T) and one-transistor (1-T) cells, widely used in dynamic read-write memories (DRAMs), are explained, and information on application of standard dynamic RAMs is included. The chapter concludes with a short section on bucket-brigade and charge-coupled device (CCD) serial memories.

Chapter 10
CIRCUIT DESIGN FOR LSI AND VLSI

Several more advanced topics are covered in Chapter 10. Advantages and drawbacks of gate arrays, which are popular in the design of semicustom digital ICs, are described for CMOS and bipolar technologies. More complex circuits of this form of design, such as standard cells and programmable logic arrays, are also described. The final section of this chapter is concerned with specialized examples of circuit design for VLSI.

We acknowledge with thanks the many comments and suggestions of students and colleagues that aided us in preparing this textbook. Professor Robert Dutton and Lanny Lewyn of Stanford University gave us very helpful, detailed suggestions on several chapters. We especially wish to thank Professor Donald O. Pederson of the University of California, Berkeley. He greatly influenced our planning of the organization and content of this text, and has been a continuous source of constructive criticism and encouragement. Our appreciation goes also to Ms. Bettye Fuller and Ms. Doris Simpson, who typed portions of the manuscript.

David A. Hodges
Horace G. Jackson

ANALYSIS AND DESIGN OF DIGITAL INTEGRATED CIRCUITS

CHAPTER

1

INTRODUCTION TO DIGITAL ELECTRONICS

1.0 INTRODUCTION

The design of modern digital systems requires contributions from several engineering specialists. First, a *system designer*, or *system architect*, determines the desired characteristics for the final system and prepares a detailed specification that should define all inputs, outputs, environmental conditions, operating speeds, etc. The *logic designer* translates the system specification into a logic design that can meet the functional requirements. Some basic principles of logic design are briefly reviewed in Appendix A.

The main subject for study in this book is digital circuit design. The task of the *circuit engineer* is to design circuits that provide the required logic functions. Whenever many copies of the desired system are to be manufactured, it is important to achieve high reliability of operation and a good balance among cost and performance characteristics. The chapters that follow address in depth the issues of electronic design that determine these characteristics. Hence in Secs. 1.1 and 1.2 we introduce some useful properties of digital circuits and characterize an ideal logic element. Some technical terms that describe the electrical performance of digital circuits are defined in Secs. 1.3 and 1.4.

Computer aids to design (CAD) are essential in analysis and design of digital integrated circuits. Section 1.5 presents a short introduction to this important subject and describes the role these modern tools will have in our study of digital circuits. Manual analysis is used only for quick approximate calculations to compare different configurations.

Today, virtually all digital systems are based on integrated circuit (IC) technology. Various design options and trade-offs exist. Choices must be made of circuit family, level of integration (the number of circuits on a chip), and programmable versus fixed-function ("hard-wired") circuits.

The various integrated circuit technologies have widely differing characteristics. Integrated circuit *process and device engineers* continue to make improvements in these technologies. Some understanding of integrated circuit fabrication techniques is required to understand the relative characteristics of different circuit families, such as TTL, ECL, NMOS, and CMOS. An appreciation for the direction and rate of change in fabrication technology is important if product designs are to provide good possibilities for evolutionary improvements. Basic techniques for fabrication of integrated circuits are described briefly in Sec. 1.6. The influence of fabrication technology on design is nowhere more important than in the layout of the circuit for the making of the photomasks used in the fabrication of the integrated circuit. Some fundamental guidelines in layout design are given in Sec. 1.7.

Good system design requires that design decisions result in a good balance among system characteristics, logic design, circuit design, layout design, and fabrication technology. Since compromises must usually be made and alternatives evaluated, it is important that the various specialists mentioned above have some knowledge of the related fields.

1.1   PROPERTIES OF DIGITAL CIRCUITS

Certain important characteristics are desired of electronic circuits for processing digital information.

1. The binary output signal must be a prescribed function of the binary input or inputs. This is termed the *logic function* of the circuit.
2. Quantization of amplitudes within the normal range of operating voltage is required, as illustrated in Fig. 1.1*a*. This implies strong nonlinearities in circuit operation. Amplitudes within the boxed regions in Fig. 1.1*a* and *b* represent each of the two binary states. At the circuit input, the uncertain region between the two boxed regions should be as small as possible.
3. Amplitude levels should be regenerated in passing from the input to the output of a digital circuit, as illustrated in Fig. 1.1*b*. This requirement dictates a voltage transfer characteristic in the general shape shown in Fig. 1.1*c*. Voltage gain should be greater than unity somewhere between the logic states. The two nominal output levels are denoted V_{OH} and V_{OL}, as in Fig. 1.1*c*. The input voltages V_{IL} and V_{IH} are defined by the points at which the magnitude of the slope of the voltage transfer characteristic is unity, as seen in Fig. 1.1*c*.
4. Directivity is required for a useful logic circuit. Changes in an output level should not appear at any unchanging input of the same circuit; that is, there must be an explicit, unilateral cause-effect relationship between input(s) and output(s).

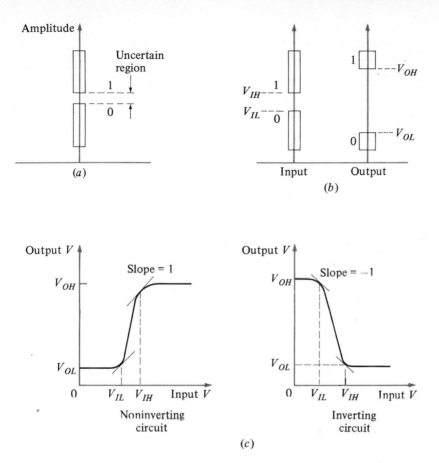

FIGURE 1.1
Amplitude and voltage transfer characteristics of digital circuits.

5. The output of one circuit must be capable of driving more than one input of similar circuits. The number that can be driven is termed the *fan-out* of the circuit. Similarly, for general use digital circuits must be capable of accepting more than one input. The number of independent input nodes is termed the *fan-in*.

With the above concepts in mind we can define an ideal digital circuit. This will be a basis for comparison of alternative practical realizations of digital circuits, most of which fall short of ideal in significant ways.

1.2 THE IDEAL DIGITAL LOGIC ELEMENT

Figure 1.2*a* shows an "ideal" logic gate. It operates from a single power source, from which it draws a minimum of power (ideally zero, of course). The two binary

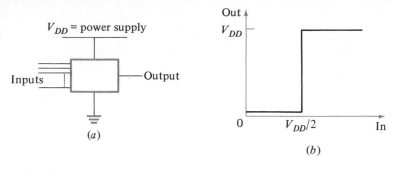

FIGURE 1.2
Ideal digital circuit.

output levels are at zero and at the supply voltage V_{DD}. The output impedance is low so that large currents may be driven into external resistive or capacitive loads without altering the output voltage levels. The transition between states at the output occurs abruptly for an input level equal to one-half the supply voltage, as in Fig. 1.2b. There is negligible time delay between the input transition and the consequent output transition. Virtually any number of inputs are available; the input impedance is high so that the circuit imposes little loading on the driving signal. We will see in the following chapters that all practical logic elements fall short of the ideal performance defined here, but depending upon the application, some are better than others. Thus there are opportunities for good engineers to make useful compromises as a part of circuit design.

1.3   DEFINITION OF NOISE MARGIN

The word "noise" in the context of digital circuits and systems means unwanted variations of voltages or currents at logic nodes. If the magnitude of noise is too great, it will cause logic errors. However, if noise amplitude at the input of any logic circuit is smaller than a specified critical magnitude known as the *noise margin* of that circuit, the noise amplitude will be attenuated as the desired signal passes from input to output. In properly functioning digital systems, noise is attenuated in passing through every circuit while the desired logic signals are restored to full amplitude without error. Noise does not accumulate from one logic stage to the next as it does in analog systems. Digital systems have an important advantage in this respect.

   The term *noise* refers to signals originating outside the circuit under study. Noise is transferred to logic nodes or interconnecting lines by unwanted capacitive or inductive coupling, as illustrated in Fig. 1.3a. Series inductance and resistance in ground and power supply lines shared by many logic elements are a common source of noise problems.

   Digital circuits typically exhibit variations in logic levels, the voltages that represent logic signals. It is desirable to minimize these variations so that output logic levels are held within two narrow voltage ranges, as illustrated in Fig. 1.3b.

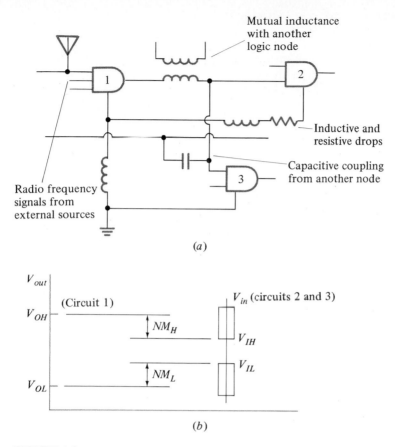

FIGURE 1.3
(a) Sources of noise in digital systems. (b) Definition of noise margins.

These output levels may vary due to circuit manufacturing tolerances, temperature changes, power supply variations, and electrical loading at the output node. The worst combinations of all these effects are used to define the worst-case output voltage range.

At the input of a digital circuit, the range of voltages accepted as representing each of the two binary logic values should be as wide as possible without the two ranges overlapping and without the possibility of a valid output level coming from another circuit being interpreted erroneously. The boundaries with the uncertain region are of course most critical; the upper boundary on the 1 level and the lower boundary on the 0 level, as long as these are outside the range of valid output levels, are not important. Note that worst-case analysis of input levels narrows the valid range and widens the uncertain region.

Figure 1.3b shows the definition of noise margins for digital circuits. In general, noise margins are different in high and low logic states. They are denoted, respectively, as $NM_H = V_{OH} - V_{IH}$ and $NM_L = V_{IL} - V_{OL}$. Frequent use will be made of the terms and concepts discussed above in the remainder of this book.

1.4 DEFINITION OF TRANSIENT CHARACTERISTICS

Specific definitions of pulse transition times and propagation delay times are needed for a description of the dynamic characteristics of logic circuits. Once such definitions are established, calculations of these times can be made.

Standard definitions of digital circuit delay times are illustrated in Fig. 1.4. Rise and fall times t_r and t_f are defined between the 10 and 90% points of the total voltage transition at the input of an inverter or gate. The total voltage range at both input and output is taken to be V_{OL} to V_{OH}, because this is the nominal situation when identical inverters or gates are cascaded. Input transition voltages V_{IL} and V_{IH} are not normally used in specification of transient performance.

High-low and low-high transition times at the output of a gate are defined as t_{HL} and t_{LH}, again between the 10 and 90% points, as seen in Fig. 1.4a. Propagation delay times from input to output, denoted t_{PHL} and t_{PLH}, are defined between the 50% points of the input and output pulse waveforms. Cycle time t_{cyc} is the time between identical points of successive cycles in the signal waveform as seen at any single node, as in Fig. 1.4a. Often cycle time is specified in terms of its reciprocal, clock frequency f_{clk}. Practical digital systems operate with cycle times 20 to 50 times the propagation delay of a single gate circuit.

In hand calculations, it is difficult to take into account the finite rise and fall times of signals at inverter or gate inputs. Consequently, it is common to approximate the real situation by an ideal pulse input with zero rise and fall times, as in Fig. 1.4b. This ideal input signal is positioned with its edges at the 50% points of an actual input signal, as illustrated in Fig. 1.4b. Propagation delay times are then approximated by the time from the edge of the ideal input pulse to the 50% point of output voltage transitions. Computer circuit simulation should be used if more accurate results are needed.

1.5 COMPUTER-AIDED DESIGN OF DIGITAL CIRCUITS

Prior to 1970, electronic circuits were analyzed and designed almost exclusively by hand, a situation that is reflected in the content of textbooks from that era. Rapid growth in the feasible complexity of integrated circuits has made computer aids essential to the design process today.

Most of the computer aids in use today do not really do any *design*. Instead, they can perform rapid analyses of a given design, varying parameters as specified by the engineer. Computer aids can also cope with masses of data that would overwhelm the unaided engineer. Computer-based *layout design systems* are extremely useful in preparing and modifying the geometric patterns required for integrated circuit masks.

Description of all the types of computer-aided design (CAD) tools is beyond the scope of this book. However, we will make extensive use of one type of CAD tool, the circuit simulator. The program used for the examples in this

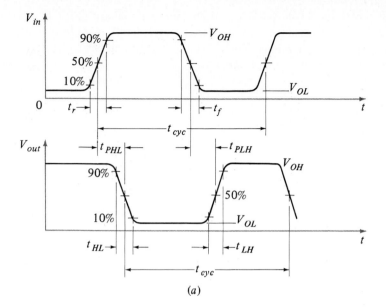

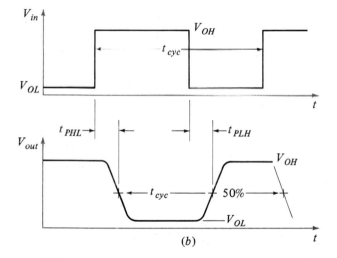

FIGURE 1.4
Definitions of transition and delay times. (*a*) Actual transient voltages. (*b*) Idealized transient voltages for hand calculations.

book is SPICE,* probably the most widely used circuit simulator. SPICE can perform nonlinear dc, large-signal time domain (transient), small-signal frequency domain, and other types of simulations. The dc and transient analysis capabilities are of greatest interest for digital circuit studies.

The inputs to SPICE are a listing of all circuit elements including parameter values (voltage sources, resistance values, transistor parameters, etc.) and all connections. The outputs desired are specified by the user. These may include tables or plots showing dc operating point, dc voltage transfer characteristics, and node voltages and loop currents as functions of time.

The internal numerical accuracy of programs such as SPICE is high; errors seldom exceed 1%. The accuracy of SPICE simulation in predicting circuit performance depends entirely upon how completely and correctly the input data describe the real physical circuit. There are two aspects to this issue. First, we must have a mathematical or numerical *model* that adequately represents each physical device. Usually the model for a device is stated as equations that relate voltages and currents. For instance, the model equation for a single linear resistor is $i(t) = v(t)/R$, and for a single linear capacitor, $i(t) = C\, dV/dt$. Model equations for transistors are much more complex, because these devices exhibit nonlinear resistive and capacitive characteristics. Second, we must have practical means for determining numerical values applying to each device, for instance, R and C for the example just cited. Measurement of capacitive parameters of IC devices is often impractical, so calculations must be used to obtain values for such parameters.

The next four chapters devote much attention to the models used to represent integrated circuit devices, including the simplifications and approximations that are a practical necessity in the development of these models. Also described in those chapters are means of measuring or calculating device parameters in forms suitable for hand analysis and computer simulation. By working carefully, one can usually obtain simulation results that are within 0.1 V of measurements for dc analyses and within ±20% of measurements for propagation delays and other transient characteristics.

Shortly after they learn to use a computer circuit simulator, many students develop the mistaken idea that skill in hand analysis is no longer necessary. Computer simulation appears to be much quicker and more accurate than hand analysis, especially for complex circuits with many nonlinearities. However, even when a fully defined circuit is under study, hand analysis before starting computer simulation is helpful. For instance, hand analysis is the best means for determining an appropriate simulation time interval, as well as driving pulse duration and rise time, before first simulation of a given circuit. It makes no sense to use a driving pulse 10 ns wide for simulation of a circuit that has a 100-ns rise time! Hand analysis also helps focus attention on possible model limitations and parasitic elements that may require special care in the parameter determination phase before circuit simulation.

---

*SPICE is an acronym for *S*imulation *P*rogram, *I*ntegrated *C*ircuit *E*mphasis.

More importantly, only through considerable experience with hand analysis do most engineers develop the imagination and insight needed for skill in *design,* which requires the specification of many interdependent device parameters, and for *synthesis,* in which a choice must be made among competing technologies or interconnections of elements *(configurations)* to meet specific requirements. While it is extremely helpful in making final adjustments of design parameters, computer simulation is of limited usefulness in design and synthesis of digital ICs until the range of alternatives to be investigated is reduced to a small number.

Throughout this text, hand analysis will be used for quick first-order approximate analyses in which allowable errors may be as much as $\pm 50\%$. One important aim is to develop in the student some skill at making good engineering approximations. Choices of circuit configuration and initial values of circuit parameters will be determined by hand. A scientific calculator is an essential tool in the hand analysis phase.

Simulation will be used to improve on the accuracy of hand analyses and to refine the choice of device parameters in design work. The emphasis in our study of simulation techniques is on developing familiarity with commonly used device models and on determining of the model parameters needed for analysis and design of modern digital integrated circuits.

1.6   INTEGRATED CIRCUIT FABRICATION TECHNOLOGY

Silicon transistors and integrated circuits are manufactured starting with pure slices, or *wafers,* of single-crystal silicon, 100 to 150 millimeters (mm) in diameter and about 0.2 mm thick. This thickness is determined by the need to provide enough mechanical strength so that the wafer is not easily broken. The thickness necessary to meet the electronic requirements is 10 micrometers (μm) or less. The wafers are polished to a mirror finish by abrasive lapping with finer and finer grits, followed by a chemical etch which leaves the surface virtually free of scratches and imperfections. At this point the polished wafer is worth a few dollars.

A layer of silicon dioxide (SiO_2) is grown on the surface of the wafer to protect the silicon surface from damage or contamination. A sequence of 100 to 300 distinct operations such as cleaning, selectively etching, doping with *p-* or *n*-type impurities, and material deposition (further described below) is required to produce the complete integrated circuits. After this processing, each wafer is sawed into hundreds or thousands of identical rectangular chips, typically between 1 and 10 mm in size on each edge, as depicted in Fig. 1.5. Integrated circuit chips may contain as few as 10 devices (transistors, resistors, diodes, etc.) to 1 million or more. Plan views of a single metal-oxide-semiconductor (MOS) transistor and a single bipolar junction transistor (BJT) are shown in Fig. 1.5c and d. The processing sequence that forms the devices and circuits comprises a sequence of pattern definition steps interspersed with other processes such as oxidation, etching, controlled introduction of desired elements *(doping),*

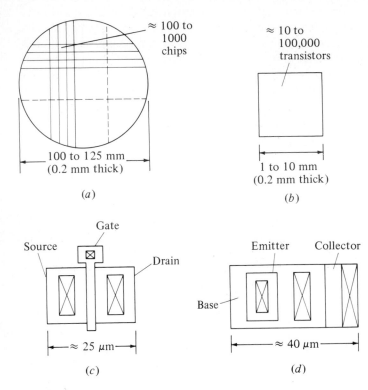

FIGURE 1.5
Simplified views of integrated circuits: (*a*) wafer, (*b*) chip, (*c*) single MOS transistor, (*d*) single bipolar transistor.

and material deposition. Simple examples of such sequences will be described shortly.

The process of pattern transfer and pattern definition is repeated 4 to 12 times during the fabrication of an integrated circuit wafer. Each of these so-called *masking steps* requires that the wafer be coated with a photosensitive emulsion known as *photoresist,* then optically exposed in desired geometric patterns using a previously prepared photographic plate. After development of the photoresist (which removes the photoresist in the selected areas), a specified process such as etching or doping is carried out in the patterned areas of the wafer. This entire pattern transfer process is known as *photolithography,* or *optical lithography.*

Steady improvements in optical lithography have made it possible to reduce the smallest surface dimensions on an IC chip from about 25 μm (or microns) in 1960 to about 2 μm today. (A hair from your head is about 50 μm in diameter!) The cost per logic gate or memory cell is reduced as more devices and circuits are formed per unit chip area. Furthermore, smaller devices have smaller capacitances and hence can switch faster, leading to better circuit performance. Minimum feature sizes less than 1 μm are feasible from the standpoint of device operation but cannot be achieved with optical lithography because the wavelength of light is about 0.4 μm. To overcome this limitation of optical lithography, electron beam lithography or x-ray lithography is now coming into use where very fine line geometries are called for.

1.6.1   Bipolar IC Fabrication Process

A typical process sequence for bipolar transistor integrated circuit fabrication is illustrated in Fig. 1.6. Included in the figure is the cross section of a typical *npn* transistor at various stages of the process. Of course, many devices can be formed at once over the surface area of the wafer if appropriate patterns are provided.

MASK 1. *Select buried-layer locations.* A thin layer (about 1 μm thick) of silicon dioxide (SiO_2) is formed on all surfaces of a *p*-type silicon wafer by exposing it to oxygen or water vapor in an electric furnace at a temperature of approximately 1000°C. The first masking step defines the area for a heavily doped *n*-type (n^+)* *subcollector*, or *buried layer*, that will provide a low-resistance connection between the active base-collector area (under the center of the completed device) and the collector contact area on the top surface. The SiO_2 is removed in these areas by chemical etching, opening "windows" through which impurities can be deposited on the silicon surface. Where no windows exist, the oxide prevents the impurities from reaching the silicon. The wafer is now placed in a diffusion furnace at a temperature of about 1000°C, and a gas containing an *n*-type impurity such as arsenic is passed over the surface of the wafer. The gas decomposes and the impurity atoms are deposited on the surface, and because of the elevated temperature they diffuse into the silicon to form the desired heavily doped *n*-type buried-layer region.

The SiO_2 masking layer is removed, exposing the entire silicon wafer surface. A high-temperature chemical vapor deposition (CVD) process known as *epitaxy* forms a layer of *n*-type single-crystal silicon 2 to 5 μm thick over the entire wafer surface. During the epitaxial process, the *n* type dopant previously introduced in the buried-layer areas diffuses in all directions. A new layer of SiO_2 is then grown.

MASK 2. *Select isolation areas.* A second masking step defines a border completely enclosing *n*-type islands of silicon that are to be the electrically isolated collectors of transistors. In the older junction-isolated bipolar process, a heavy *p*-type (typically boron) diffusion into the border areas is continued until the entire epitaxial layer has been penetrated, as illustrated in Fig. 1.6. Thus, islands of *n*-type silicon are bounded on all sides by *p*-type silicon. Electrical isolation is

---

*The notation n^+ implies doping $> 10^{18}$ atoms/cm³, n^- implies doping $< 10^{15}$ atoms/cm³, and *n* is not specific. The same convention applies for designations of *p*-type material.

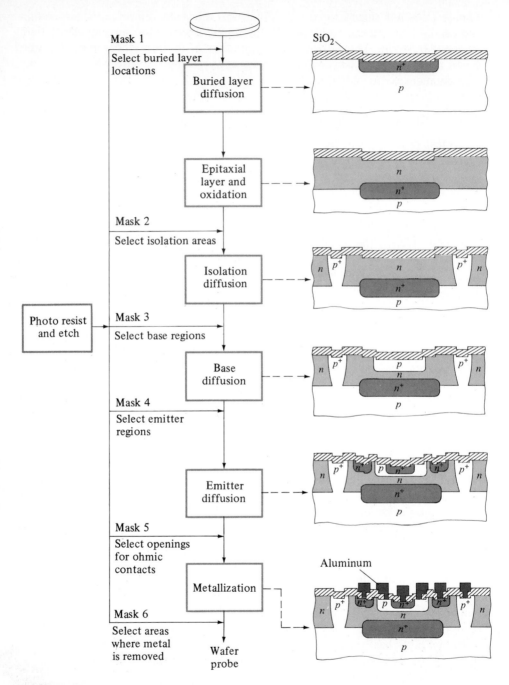

FIGURE 1.6
Process sequence for fabrication of integrated circuit *npn* transistor. [*After Hamilton and Howard (4).*]

achieved by applying voltages such that this *pn* junction is always reverse-biased. A new layer of thermal oxide is grown over the isolation areas.

Newer oxide-isolated bipolar processes oxidize the silicon in the border areas, resulting in an insulating boundary that completely surrounds the transistor. Then the base region can abut the isolation region without causing a short circuit, and transistor area and capacitances are significantly reduced.

MASK 3. *Select base regions.* The third masking step defines base regions of *npn* transistors. Patterns for resistors are formed simultaneously in separate isolated *n*-type regions. Boron is diffused or implanted to form bases and resistors. The *n*-type collector material is converted to *p*-type when the density of *p*-type impurities exceeds that of *n*-type impurities, a situation known as *overcompensation*.

MASK 4. *Select emitter regions.* The fourth masking step defines *n*-type transistor emitters and *n*-type regions for low-resistance contact to collector regions. Again, conversion of *p*-type base material to *n*-type requires overcompensation. Inevitably, each succeeding diffused layer must be more heavily doped than the one it must overcompensate.

MASK 5. *Select openings for ohmic contacts.* The fifth mask defines windows in the oxide where connection is to be made to the collector, base, and emitter of the transistor. A thin layer of aluminum is then evaporated over the entire surface of the wafer.

MASK 6. *Select areas for metal removal.* The aluminum is then etched in a pattern defined by a sixth mask to form the desired interconnections. A protective passivating layer (often termed *scratch protection*, or simply *scratch)* is deposited over the entire surface. A final masking step removes this insulating layer over the pads where contacts will be made. Circuits are tested using needlelike probes on the contact pads. Defective units are marked with a dot of ink; then the wafer is sawed into individual chips. Good chips are packaged and undergo a final test.

The process sequence just described is the simplest one that can produce high-performance bipolar transistor integrated circuits. More complex processes with more masking steps are widely used to achieve advantages in performance, density, etc. By using *ion implanters* the location and concentration of the impurity atoms can be more closely controlled, leading to higher-performance circuits. Considerable improvement in circuit density is possible if the *p*-type isolation region is replaced with SiO_2, using a selective *local oxidation* process. A frequent addition to the bipolar IC process is a second layer of interconnecting metallization.

We note here that the cross section of the transistor in Fig. 1.6 is not to scale. Often the drawings are distorted in the horizontal or vertical dimension to give the particular emphasis required. To give the reader some feeling for the aspect ratio involved between the surface dimensions and the junction depths, an exact-scale drawing of a typical *npn* transistor fabricated by the process just described is shown in Fig. 1.7. In this layout the design rules, or minimum feature size, are 5 μm. The drawing does not include the metallization.

1.6.2 MOS IC Fabrication Process

The process sequence for fabrication of n-channel MOS integrated circuits is illustrated in Fig. 1.8.

MASK 1. *Define transistor areas.* A CVD process deposits a thin layer of silicon nitride (Si_3N_4) on the entire wafer surface. The first photolithographic step defines areas where transistors are to be formed. The silicon nitride is removed outside the transistor areas by chemical etching. Boron (p-type) is implanted in the exposed regions to suppress unwanted conduction between transistor sites. Such implants are colloquially known as *channel stoppers*. Next, a layer of silicon dioxide (SiO_2) about 1 μm thick is grown thermally in these inactive, or *field*, regions by exposing the wafer to oxygen in an electric furnace. This is known as a *selective*, or *local, oxidation* process. The Si_3N_4 is impervious to oxygen and thus inhibits growth of the thick oxide in the transistor regions.

The Si_3N_4 is next removed by an etchant that does not attack SiO_2. A clean thermal oxide about 0.1 μm thick is grown in the transistor areas, again by exposure to oxygen in a furnace. This is the *thin*, or *gate, oxide*. Another CVD process deposits a layer of polycrystalline silicon *(poly)* over the entire wafer.

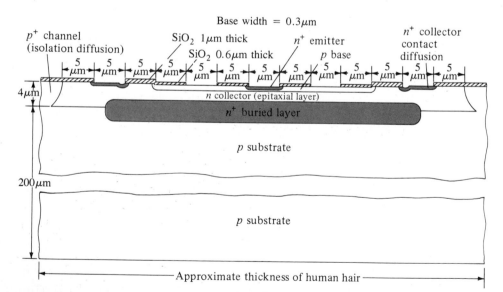

FIGURE 1.7
Cross section of an integrated circuit transistor shown to correct scale. [*After Hamilton and Howard (4)*.]

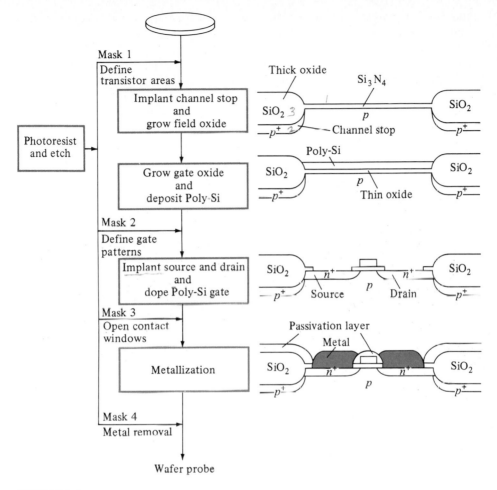

FIGURE 1.8
Process sequence for fabrication of n-channel MOS integrated circuit.

MASK 2. *Define gate patterns.* The second photolithographic step defines the desired patterns for gate electrodes. Undesired poly and the underlying thin oxide are removed by chemical or plasma (reactive gas) etching. An n-type dopant (phosphorus or arsenic) is introduced into the regions that will become the transistor source and drain. Either thermal diffusion or ion implantation may be used for this doping process. The thick field oxide and the poly gate are barriers to the dopant, but in this process, the poly becomes heavily n-type. Another CVD process deposits an insulating layer, often SiO_2, over the entire wafer.

MASK 3. *Define contact openings.* The third masking step defines the areas in which contacts to the transistors are to be made. Chemical or plasma etching selectively exposes bare silicon or poly in the contact areas. Aluminum (Al) is then deposited over the entire wafer by evaporation from a hot crucible in a vacuum evaporator.

MASK 4. *Define areas for metal removal.* The fourth masking step patterns the Al as desired for circuit connections. The final steps of the process are identical to those described above for bipolar transistor ICs. A pictorial representation of the entire NMOS IC manufacturing process is shown in Fig. 1.9.

This is the simplest process for forming NMOS circuits. More advanced NMOS and CMOS processes require 7 to 12 masking steps.

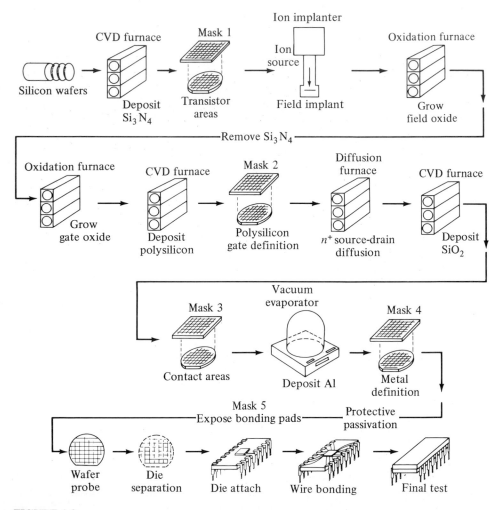

FIGURE 1.9
Manufacturing process for NMOS silicon-gate integrated circuits. (*Integrated Circuit Engineering Corporation.*)

1.6.3   Passive Components: Diffused Resistors

In this section we describe the process for making resistors in the IC technology. Capacitors and inductors, while feasible in the monolithic technology, have not proven to be advantageous in digital ICs.

The resistance of a diffused layer of uniform resistivity that is L cm long, W cm wide, and t cm thick, as shown in Fig. 1.10, is as for any conducting bar and is given by

$$R = \frac{\rho L}{tW} \; \Omega$$

where ρ = resistivity of the material in $\Omega \cdot$cm.

Now with $L = W$, the surface dimension is a square, and we have

$$R_S = \frac{\rho}{t}$$

where R_S is termed the *sheet resistance* and has the units of *ohms per square* (Ω per square). The sheet resistance of any diffused layer can readily be measured. Then the resistance of the layer is simply

$$R = \frac{L}{W} R_S$$

Or for a desired resistance R, the diffused layer is made R/R_S squares long.

Diffused resistors are usually formed during the base diffusion, though for very small valued resistors the emitter diffusion is used. The plan view and cross section of a base-diffused resistor is illustrated in Fig. 1.11. Notice from the schematic drawing of the resistor that there is a parasitic pn junction diode formed by the p-type base diffusion and the n-type epitaxial layer. This is usually of no consequence provided the diode is always reversed-biased by connecting the epitaxial layer to a voltage that is always more positive than that of the resistor.

The base sheet resistance is typically in the range 100 to 200 Ω per square, so resistances 50 Ω to 10 kΩ are practical. For the emitter layer, the sheet resistance is typically 5 to 50 Ω per square. To conserve die area, high-valued resistors, i.e., long resistors, generally take on a meander pattern, as shown in Fig. 1.12.

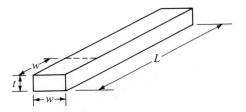

FIGURE 1.10
Calculation of the sheet resistance of a diffused layer.

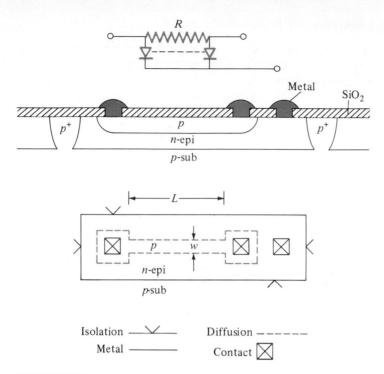

FIGURE 1.11
Plan view and cross section of base-diffused resistor.

When still higher values of resistance are needed, a special masking step and ion implant may be used to produce sheet resistances up to about 10 kΩ per square. The geometries and parasitics are similar to those of diffused resistors. In MOS technologies, the polysilicon layer may be doped to provide a wide range of resistor values, from 50 Ω to 50 MΩ being feasible. The geometries are similar to those for diffused resistors, but there are no parasitic diodes. The only parasitic effect is the capacitance to the layer below and (possibly) above.

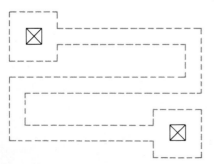

FIGURE 1.12
Meander pattern for a long resistor.

1.7 LAYOUT FUNDAMENTALS

An important step in the design of any IC is the layout of the various masks used in the fabrication of the circuit. Layout design is as much an art as it is a science, but there are some fundamental guidelines that must be adhered to if a successful product is to be produced.

1. While adhering to the technology-imposed limits of minimum feature sizes and the circuit-imposed minimum transistor sizes needed to meet current-handling requirements, minimize the area to obtain low parasitic capacitances and high yield.
2. Group components together wherever possible to minimize the number of isolated regions and number of long interconnections.
3. Large-area junctions should be avoided because of the associated large parasitic capacitances and leakage currents.
4. Circuit topology should be defined to minimize the number of crossing interconnection lines. Such *crossovers* can be costly in terms of parasitic resistance and/or capacitance as well as production yield.
5. No conducting region should be allowed to *float;* that is, every region should have its potential set to a known value by an ohmic connection.
6. The layout must be arranged so that the bonding pads for external connections all come to the periphery of the chip.
7. The layout must be arranged so that bonding wires from bonding pads to header pins do not cross.

These are general guidelines. More specific layout rules including minimum (and sometimes maximum) sizes of features on all mask layers accompany every IC manufacturing process. There is continuing progress in integrated circuit fabrication technology. With a new technology comes new process sequences and new design rules. The details of new fabrication technologies, including the layout rules, are generally proprietary and are rarely disclosed by manufacturers.

1.8 SCALE OF INTEGRATION

In practice, many gates are manufactured on a single integrated circuit chip. Although there are no universally accepted definitions for levels of complexity, when between 1 and 10 gates are included on a chip, the usual term referring to this level of complexity is *small-scale integration (SSI)*.

Medium-scale integrated (MSI) circuits are generally considered to include 10 to 100 gates on a chip, while *large-scale integration (LSI)* refers to complexities in the range 100 to approximately 10,000 gates or bits of memory per chip. The term *very large scale integration (VLSI)* is commonly used for integrated circuit chips containing more than 10,000 equivalent gates; the first commercial digital circuits at this level of complexity became available about 1980. Ultimately it may be possible to incorporate a million or more gates on a chip; a new identifier such as *ultra large scale integration (ULSI)* may come into common usage.

Two distinct classifications for integrated circuits are as *standard parts* (components used by many system manufacturers) or *custom circuits* (components designed and manufactured for one customer.) Custom circuits are used when suitable standard parts are not available, or to reduce costs by providing exactly the function needed for a specific application. A third category, known as *semi-custom circuits,* includes *gate array* and *standard cell* ICs. Semicustom circuits have some standardized patterns or masking layers that are used in many different final ICs and some patterns or layers that are designed to meet a particular user's requirements.

1.9   SUMMARY

This chapter provides a condensed presentation of basic concepts that underlie the subject of digital integrated circuits.

- Five important properties of digital circuits are defined, and the characteristics of an ideal circuit realization are illustrated.
- The important concept of *noise* in digital systems is described. Standard definitions of logic levels and noise margins are presented. Terms used in the description of transient characteristics, such as propagation delay and cycle time, are defined.
- The role of *computer-aided design (CAD)* relative to older methods for analysis and design of digital integrated circuits is stated. The advantages of CAD in analysis, and its relative limitations for synthesis and design, are noted.
- Fabrication sequences for integrated circuits employing bipolar transistor and MOS transistor technology are explained briefly, and the fundamentals of good layout design are presented.
- Finally, the general concepts of small-scale integration (SSI), medium-scale integration (MSI), and large-scale integration (LSI), as well as ultra large-scale integration (ULSI), are given. Also, the general classifications of standard, semicustom, and custom parts are explained.

REFERENCES

1. V. H. Grinich and H. G. Jackson, *Introduction to Integrated Circuits,* McGraw-Hill, New York, 1975.
2. H. Taub and D. Schilling, *Digital Integrated Electronics,* McGraw-Hill, New York, 1977.
3. M. M. Mano, *Computer Logic Design,* Prentice-Hall, Englewood Cliffs, N.J., 1972.
4. D. J. Hamilton and W. G. Howard, *Basic Integrated Circuit Engineering,* McGraw-Hill, New York, 1975.
5. R. A. Colclaser, *Microelectronics: Processing and Device Design,* Wiley, New York, 1980.
6. D. K. Reinhard, *Introduction to Integrated Circuit Engineering,* Houghton Mifflin, Boston, 1987.

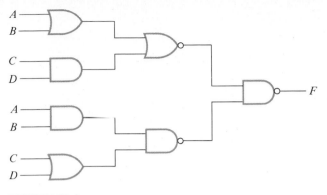

FIGURE P1.2

7. W. Maly, *An Atlas of IC Technologies,* Benjamin/Cummings, Menlo Park, Calif., 1987.
8. W. S. Ruska, *Microelectronic Processing,* McGraw-Hill, New York, 1987.

PROBLEMS

These problems also cover the text material found in Appendix A.

P1.1. Simplify the following functions and express each (*a*) as a sum of products and (*b*) as a product of sums:

$$F = CB\overline{A} + C\overline{B}A + \overline{C}BA + CBA$$

$$F = DC\overline{A} + DC\overline{A} + \overline{D}\overline{C}A + DCA$$

P1.2. Analyze the circuit shown in Fig. P1.2 and write the minimal expression as (*a*) a sum of products and (*b*) a product of sums.

P1.3. A light in a long hallway is to be controlled by three switches *A*, *B*, and *C*. If the switch is up, its logic value is 1. If an odd number of switches are up, the light is on; if an even number, off. Design the logic circuit for *F* (*F* = 1 for light on) from *A*, *B*, and *C* (*a*) as an all-NAND circuit and (*b*) as an all-NOR circuit.

P1.4. A logic circuit has four inputs that represent decimal numbers in BCD ($A - 2^0$, $B = 2^1$, $C = 2^2$, $D = 2^3$). Design an all-NAND circuit that gives a 1 output if the BCD value is ≥ 5 and gives a 0 if the value is < 5. The four true inputs are the only ones available.

P1.5. This problem is similar to P1.4, but this time design an all-NOR circuit that gives a 1 output if the BCD value is a multiple of 3.

P1.6. In Fig. A.14*a*, $V^+ = +4$ V, $V^- = -4$ V, $R_1 = 1$ kΩ, and $R_2 = 2$ kΩ. Plot V_F versus V_A as V_B takes on values of $-4, -2, 0, +2$, and $+4$ V. Let V_A range from -4 to $+4$ V. Assume a diode characteristic similar to Fig. A.13*b*, but $V_{D(on)} = 0.8$ V.

P1.7. Repeat P1.6 for the circuit shown in Fig. A.15*a*.

P1.8. In Fig. A.14*a*, $V^- = -5$ V and $R_1 = 5$ kΩ. Solve for V^+ and R_2 such that the limits at V_F are $+4$ and 0 V. Assume $V_{D(on)} = 0.8$ V.

P1.9. Repeat P1.8 for the circuit shown in Fig. A.15*a* but with $V^+ = +5$ V and $R_1 = 5$ kΩ. Solve for V^- and R_2.

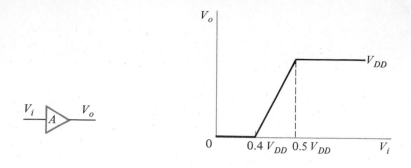

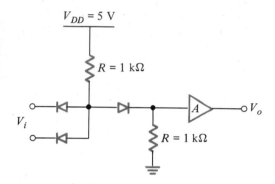

FIGURE P1.12

P1.10. Consider a cascade connection of two circuits of the type described in P1.6 such that the output of the first drives both inputs of the second circuit. Sketch the VTC for V_F versus $V_A = V_B$. What conclusion can you draw about the usefulness of cascaded diode logic circuits?

P1.11. Use SPICE to establish the VTCs for P1.6 and P1.7 using the built-in diode model with $I_s = 10^{-16}$ A. Compare the results with your hand analysis.

P1.12. A buffer amplifier that has the VTC shown in Fig. P1.12 is available. A diode AND circuit is to be used with this amplifier, as shown. Assume $V_{D(on)} = 0.8$ V. Plot V_{R2} versus V_i.

2

THE METAL-OXIDE-SEMICONDUCTOR (MOS) TRANSISTOR

2.0 INTRODUCTION

Small-scale digital integrated circuits based on complementary MOS technology have been in use for many years. MOS technology is the basis for most of the large-scale integrated (LSI) digital memory and microprocessor circuits. The most important advantage of MOS circuits over bipolar circuits for LSI is that more transistors and more circuit functions may be successfully fabricated on a single chip with MOS technology. The reasons for this are threefold. First, an individual MOS transistor occupies less chip area. Second, the MOS fabrication process involves fewer steps and as a result achieves fewer critical defects per unit chip area than in bipolar circuit fabrication. This makes feasible somewhat larger chips in MOS technology. Third, dynamic circuit techniques that require fewer transistors to realize a given circuit function are practical in MOS technology but not in bipolar technology. The result of these differences is that MOSLSI circuits are significantly cheaper to manufacture than bipolar circuits of equivalent function. Consequently, MOSLSI circuits are making up a steadily increasing fraction of the total market for digital LSI.

Many students embarking upon a study of this text will have had a course in semiconductor devices, including the MOS transistor. Even these students should study this chapter, however, because here the emphasis is on the specific dc and transient characteristics of MOS devices that are critical to LSI digital circuit design.

2.1 ALTERNATIVE MOS PROCESSES

The first MOS circuits, made in metal-gate p-channel (PMOS) technology, required special supply voltages (for example, -9 V, -12 V) and functioned only at very low digital data rates (for example, 200 kb/s to 1 Mb/s). The change to n-channel (NMOS) silicon-gate technology and other improvements have resulted in LSI circuits that require only a single standard $+5$-V supply and operate at digital data rates up to 40 Mb/s.

A significant feature of MOS circuits is that reductions in the internal dimensions of individual devices result in very sharp improvements in circuit speed. In contrast, bipolar circuit speed improves only gradually as internal dimensions are similarly reduced. With the steady improvements in pattern definition capability, the difference in performance between bipolar and MOS circuits has steadily become smaller. For LSI circuits, the speed difference has shrunk from a factor of 10 or more in 1970 to less than a factor of 2 in 1987.

The most important limitation of MOS circuits is in driving high currents and high voltages. Standard MOSLSI processes are limited to 5 or 10 V maximum operating voltage and are inefficient when driving more than about 20 mA into a load. The latter is an important speed-limiting factor in digital systems employing data busses that exhibit high capacitance. Use of terminated transmission-line busses, which typically have characteristic impedances of 50 to 100 Ω, is impractical in systems employing MOSLSI due to the limited current-driving capability of these circuits. Because bipolar digital circuits, particularly emitter-coupled logic (ECL), can drive highly capacitive loads and terminated transmission lines at high speeds, such circuits continue to be preferred for use in systems that must operate at off-chip data rates of 20 Mb/s and higher. Typically these are systems such as large computers and high-speed signal processors.

A great many variations of MOS technology have been and continue to be used. Furthermore, for commercial reasons, manufacturers of components often coin their own name or acronym for what may be a process technology used by others under a different name.

The most prevalent version of MOS technology today is self-aligned silicon-gate NMOS. Modern versions of this process employ a technique known as local oxidation to increase circuit density and performance. The transistor structure that results from this process is described in the next section. When such a process employs gate dielectrics thinner than 1000 Å and surface dimensions smaller than 5 micrometers (μm), designations such as HMOS (for high-speed MOS), SMOS (scaled MOS), or XMOS are used by various manufacturers. A frequent addition to the process, particularly for important memory applications, is a second layer of polysilicon.

Metal-gate PMOS and NMOS are older versions of the MOS process that are not much used for new designs. As device geometries get smaller, possibly metal-gate techniques will come into use again in the future to overcome certain limitations of silicon-gate processes. If so, doubtless a new name will be found.

Complementary MOS (CMOS) technologies provide both n- and p-channel devices in one chip, at the expense of some increase in fabrication complexity and chip area compared to basic NMOS. As will be shown, the great advantage of CMOS digital circuits is that they may be designed for essentially zero power consumption in steady-state condition for both logic states. Power is consumed only when circuits switch between logic states; average power consumption is usually much smaller than for NMOS circuits. CMOS is widely used for digital wristwatches and other battery-powered equipment, and it is coming into wider use in computers and communication equipment.

Metal-gate CMOS is still widely used because it is the simplest, least expensive form. Density and performance are adequate for high-volume applications such as digital watches and simple calculators and for SSI and MSI digital circuits such as the 4000 and 74C logic families.

Silicon-gate CMOS is considerably more complex to manufacture, but it offers significantly higher circuit density and better high-speed performance when used in LSI. CMOS memories and microprocessors usually employ a silicon-gate process.

A number of additional forms of the MOS process have been used to a more limited extent than the standard NMOS and CMOS processes discussed above. Among the variations of MOS in this category are DMOS (double-diffused MOS) and its equivalent, DSA (diffusion self-aligned), VMOS (V-groove MOS), and SOS (silicon-on-sapphire) MOS. All of these claim performance and/or density advantages over standard processes. However, in all cases, the advantages gained from a change in device structure have proven to be relatively small compared to the improvements achieved by reducing internal device dimensions in standard processes. Furthermore, many of the unusual device structures proved to be much harder to manufacture in LSI form. For these reasons, these additional forms of MOS technology will not be considered further in this book.

2.2 STRUCTURE AND OPERATION OF THE MOS TRANSISTOR

The analysis and design of integrated circuits depends heavily on the use of suitable models for the integrated circuit components. This is true in hand analysis, where fairly simple models are generally used, and in computer analysis, where much more complex models are utilized.

Since any analysis or design is only as accurate as the models used, it is essential that the circuit designer have a thorough understanding of the models commonly used and the degree of approximation involved in each. For these reasons, this chapter as well as Chaps. 4 and 5 are devoted to studies of the internal structure and conduction processes in transistors and diodes and of the equations or models used to approximate these devices in the analysis and design of digital circuits.

A perspective view of and the schematic symbol for an *n*-channel silicon-gate MOS transistor are shown in Fig. 2.1*a*. More details are shown in the cross-sectional view of Fig. 2.1*b*. In simplest terms, a voltage applied to the gate electrode is used to set up an electric field that controls conduction between heavily doped *n*-type (n^+)* source and drain regions. Because of this use of an electric field, the device is one form of field-effect transistor (FET). Note that the gate is completely insulated from the other electrodes. This fact leads to the designation insulated-gate field-effect transistor (IGFET). Still another name, much older, is unipolar transistor. This name arises from the fact that only a single type of charge carrier (electrons in NMOS) is necessary for device operation. The mobile holes in the *p*-type substrate of an NMOS transistor are not involved in normal transistor operation. In contrast to a MOS or other unipolar transistor, an *npn* or *pnp* bipolar transistor must involve both electrons and holes in its operation.

Unfortunately, no standardized terminology is widely accepted, and the terms MOS, MOST, MOSFET, FET, and IGFET are used by different organizations and people to designate the device shown in Fig. 2.1.

The device structure shown in Fig. 2.1*a* is formed by a complex sequence of steps including oxidation, pattern definition, diffusion, ion implantation, and material deposition and removal processes. A brief description of the process sequence is presented in Chap. 1. The final structure has a number of features deserving mention.

The *body*, or *substrate*, is a single-crystal silicon wafer which is the starting material for circuit fabrication and provides physical support for the final circuit. The *p*-type doping density is a factor in device electrical behavior, as shown in the next section. The top surface of the body comprises *active*, or *transistor*, regions and *passive*, or *field* (or sometimes, *isolation*), regions. The main requirement on the field regions is that they should never permit conduction between separate active regions. In NMOS, all conduction is via electrons, so the field region fulfills its purpose if electrons can never pass through it. Extra *p*-type doping is used in the field regions to achieve this result. A thick layer (0.5 to 1 μm) of silicon dioxide over the field regions is used to minimize unwanted capacitance from interconnecting metal and polysilicon to the body.

The transistor regions in the body comprise n^+ source and drain regions and a *p*-type channel region. Typical devices are symmetrical, just as the one shown here; source and drain are interchangeable. Strictly speaking, source and drain

---

*The notation n^+ implies doping $> 10^{18}$/cm^3, n^- implies doping $< 10^{15}$/cm^3, and n is not specific. The same convention applies for designations of *p*-type material.

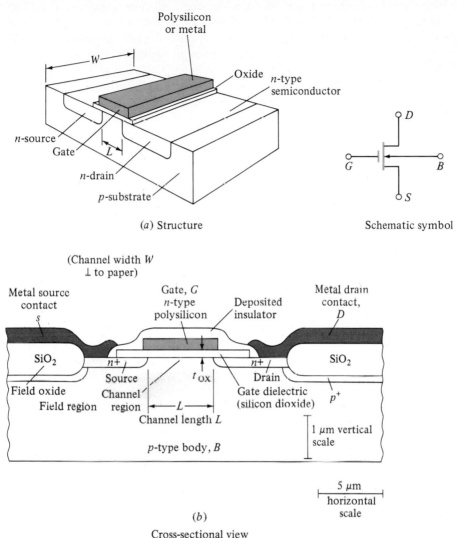

(*a*) Structure

Schematic symbol

(*b*)

Cross-sectional view

FIGURE 2.1
NMOS transistor.

can be identified only after the polarities of applied voltages are established. In NMOS, the more positive electrode is defined as the drain.

With no bias voltages applied, the path from drain to source has two *pn* junction diodes in series, back to back, with the body as a *p* region common to them. The only current that can flow is diode reverse leakage. Under these conditions, no conducting channel is present.

Now consider the result when source, drain, and body are all tied to ground and a positive voltage is applied to the gate. From simple ideas of electrostatics, a positive gate voltage will tend to draw electrons into the channel region. The n^+ source and drain regions provide nearby copious sources of electrons. Once electrons are present in the channel region, a conducting path is present between drain and source. Current will flow from drain to source if there is a voltage difference between them.

The conducting channel does not form for very small positive gate voltages. First the electrostatic potential at the surface of the p-type channel region must be made positive by application of $+0.5$ to $+2$ V of gate-source voltage. The gate voltage needed to initiate formation of a conducting channel is termed the *threshold voltage* V_T. This important device parameter is analyzed in the next section.

In many practical MOS devices, the channel doping concentration is modified by ion implantation to alter the threshold voltage. A p-type implant will make the threshold voltage more positive. On the other hand, an n-type implant makes the threshold voltage more negative. A larger n-type implant can be used to make the channel of an NMOS transistor n-type with zero gate-source voltage. Negative gate bias then can be used to reduce electron concentration in the channel, or even eliminate the conducting channel. The threshold voltage in this case is negative.

NMOS transistors that have no conducting channel at zero gate-source voltage are termed *normally off* devices. A common alternative term is *enhancement-mode* device, meaning that gate-source voltage of the same polarity as drain source voltage (positive for NMOS, negative for PMOS) is required to initiate conduction. A device which is *normally on* (at zero gate-source bias) is termed a *depletion-mode* device. For an NMOS depletion device, negative gate-source bias depletes the conducting channel. Threshold voltages of enhancement and depletion devices (both are often used in the same circuit) are denoted V_{TE} and V_{TD}, respectively.

PMOS devices are symmetrical but opposite to NMOS devices in terms of voltage polarities. The above discussion applies to PMOS devices, provided the polarity designations n-type, p-type and positive, negative are interchanged wherever they appear. In CMOS circuits, device threshold voltages are typically denoted V_{TN} and V_{TP}. Depletion-mode devices are rarely used in CMOS circuits, so all devices are assumed to be enhancement-mode. Figure 2.2 shows a cross-sectional view of a silicon-gate CMOS structure and two widely used schematic notations for devices in CMOS circuits. The polarity conventions for voltages and currents defined in Fig. 2.2*b* will be used throughout this text.

Figures 2.1*a* and 2.2*a* are drawn approximately true to scale. In the fabrication process, horizontal device dimensions are made as small as possible with the available fabrication technology in order to maximize both circuit density and high-speed performance. The most important horizontal dimension is channel length L, shown in the figure. Typical values of L are in the range of 2 to 6 μm. Perpendicular to the plane of the figure is the channel width W, typically 3 to 500 μm, depending on the desired current-handling capability. Gate oxide thickness t_{ox}, the most important vertical dimension, is typically 40 to 100 nm

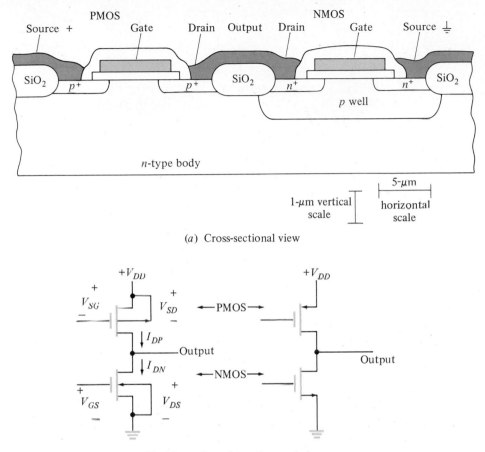

(a) Cross-sectional view

(b) Alternative schematic sysmbols

FIGURE 2.2
CMOS transistors.

(400 to 1000 Å). As we shall see in Sec. 2.4, gate length and width and gate oxide thickness are the major parameters determining the electrical characteristics of the MOS transistor.

2.3   THRESHOLD VOLTAGE OF THE MOS TRANSISTOR

The derivation of the transfer characteristics of the enhancement-mode NMOS device shown in simplified form in Fig. 2.3 begins by noting that with $V_{GS} = 0$ V, the source and drain regions are separated by back-to-back pn junctions. These junctions are formed between the n-type source and drain regions and the p-type substrate or body and result in an extremely high resistance (thousands of megohms) between source and drain when the device is off.

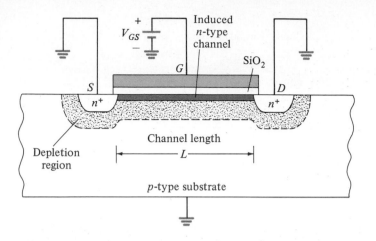

FIGURE 2.3
Idealized NMOS device cross section with positive V_{GS} applied showing depletion regions and the induced channel.

Now consider body, source, and drain to be grounded and a voltage V_{GS} (initially zero) applied to the gate as shown in Fig. 2.3. Conventionally the equilibrium electrostatic potential ϕ_F in a semiconductor is defined as

$$\phi_F = \frac{kT}{q} \ln \frac{n_i}{p} \qquad (p\text{-type semiconductor}) \qquad (2.1a)$$

or

$$\phi_F = \frac{kT}{q} \ln \frac{n}{n_i} \qquad (n\text{-type semiconductor}) \qquad (2.1b)$$

where p or n is the equilibrium majority mobile carrier concentration (assumed equal to the doping concentration N_A or N_D) and n_i is the intrinsic carrier concentration of the semiconductor.* (Refer to the front inside cover for values of constants.) In this text, we employ the convention that ϕ_F is negative for p-type semiconductor material and positive for n-type material. This choice gives the expected results when device operation is explained using simple electrostatic principles.[†]

The gate and body of the MOS transistor form the plates of a capacitor with the SiO_2 as a dielectric. The gate oxide capacitance per unit area is defined as C_{ox}. As V_{GS} changes from zero to a positive value, positive charge accumulates on

---

*A semiconductor is said to be intrinsic when it is free of dopant atoms. The only mobile carriers are those caused by thermal excitation, a process that creates equal numbers of holes and electrons; that is, $p = n = n_i$. The intrinsic carrier concentration n_i is a strong function of temperature. For silicon, at 300 K, $n_i = 1.45 \times 10^{10}$ cm^{-3}.

[†]Texts that explain device operation using semiconductor band theory employ the opposite sign convention.

the gate and negative charge in the substrate under the gate. Initially, the negative charge in the p-type body is manifested by creation of a *depletion region* in which mobile holes are pushed down under the gate, leaving behind negatively charged immobile acceptor ions. (This is the same depletion region that appears in a *pn* junction, described in Sec. 4.3.)

A simple analysis can be applied to find the thickness X_d of the depletion region in the silicon below the gate as a function of the surface electrostatic potential ϕ_s just inside the depletion layer at the oxide-silicon interface. The doping density in the p-type substrate is denoted and N_A, and ϵ_{si} is the permittivity of the silicon. To create the shaded depletion region shown in Fig. 2.3, mobile holes must be pushed back. The mobile charge dQ of the holes originally contained in an infinitesimal horizontal layer of p-type material is given by

$$dQ = q(-N_A)\, dX_d$$

The change in surface potential $d\phi_s$ needed to displace the mobile charge dQ is

$$d\phi = -X_d\, dE = -\frac{X_d\, dQ}{\epsilon_{si}} = \frac{qN_A X_d\, dX_d}{\epsilon_{si}}$$

Integrating both sides and evaluating the constant of integration as equal to the equilibrium potential ϕ_F of the p material, we obtain

$$\phi_s = \frac{qN_A X_d^2}{2\epsilon_{si}} + \phi_F$$

When this is solved for X_d (with $|\phi_s - \phi_F| \geq 0$), we finally get

$$X_d = \left(\frac{2\epsilon_{si}|\phi_s - \phi_F|}{qN_A}\right)^{1/2} \tag{2.2}$$

The immobile charge per unit area due to acceptor ions that have been stripped of their mobile holes is

$$Q = -qN_A X_d = -\sqrt{2qN_A\epsilon_{si}|\phi_s - \phi_F|} \qquad |\phi - \phi_F| \geq 0 \tag{2.3}$$

As V_{GS} is increased in the positive direction, the potential ϕ at the silicon surface increases from its original (negative) equilibrium value of ϕ_F, through zero, until $\phi_s = -\phi_F$. Under this condition the density of mobile electrons at the surface is equal to the density of mobile holes in the original substrate or body, and the surface potential has changed by $-2\phi_F$. The value of V_{GS} needed to cause this change in surface potential is (somewhat arbitrarily) defined as the threshold voltage V_T for a MOS transistor. Note that $-2\phi_F$ is positive for n-channel devices and negative for p-channel devices. Since practical devices employ carrier concentrations in the body ($\simeq$ body doping) in the range of 10^{15} cm^{-3} (see Example 2.1),

$$2|\phi_F| = 2(0.026) \ln \frac{10^{15}}{1.45 \times 10^{10}} \simeq 0.6 \text{ V}$$

The phenomenon known as *strong inversion* is conventionally defined to occur at $\phi_s = -\phi_F$. In essence, the semiconductor surface becomes *n*-type. Further increases in gate voltage produce only slight additional changes in ϕ_s and in the depletion-layer width. Instead, the additional charge induced in this region appears as additional electrons in this thin inversion layer. The electrons are drawn into the inversion layer from the strongly *n*-type source, with the positive surface potential as the attractive force. The additional gate voltage increases the electric field in the gate oxide and increases the electron concentration in the channel.

As long as there is only a small voltage difference between drain and source, the induced layer of electrons extends continuously from source to drain, producing a continuous region with mobile electrons present throughout. The conductivity of the *conducting channel* thus formed can be increased or decreased (*modulated*) by increasing or decreasing the gate voltage. In the presence of an inversion layer, and with no body bias ($V_{SB} = 0$), the depletion region then contains a fixed negative charge that may be found by using Eq. (2.3) together with the fact that $\phi_s = -\phi_F$.

$$Q_{B0} = -\sqrt{2qN_A\epsilon_{si}| - 2\phi_F|} \qquad (2.4a)$$

If there is a voltage V_{SB} between source and body (V_{SB} is normally positive for *n*-channel and negative for *p*-channel devices), the surface potential required to produce inversion becomes $|-2\phi_F + V_{SB}|$ and the charge stored in the depletion region in this case is

$$Q_B = -\sqrt{2qN_A\epsilon_{si}| - 2\phi_F + V_{SB}|} \qquad (2.4b)$$

Example 2.1. A *p*-type silicon substrate has $N_A = 10^{15}$ cm^{-3}. Find the limiting value of depletion-layer width and the total charge contained in the depleted region. From Eq. (2.1),

$$2|\phi_F| = \frac{2kT}{q} \ln \frac{n_i}{p} = 0.58 \text{ V}$$

From Eq. (2.2),

$$X_d = 0.87 \times 10^{-4} \text{ cm} = 0.87 \text{ } \mu\text{m}$$

From Eq. (2.4b),

$$Q_{B0} = -1.39 \times 10^{-8} \text{ C/cm}^2$$

The value of gate voltage V_{GS} required to produce strong inversion, called the *threshold voltage* V_T, can now be calculated. It consists of several components:

1. A gate voltage $(-2\phi_F - Q_B/C_{ox})$ is needed to change the surface potential and to offset the depletion-layer charge Q_B.

2. A voltage term ϕ_{GC} representing the difference in work functions between the gate (G) material and the bulk silicon in the channel (C) region must be added. For silicon-gate devices, $\phi_{GC} = \phi_{F(substrate)} - \phi_{F(gate)}$. For aluminum-gate (metal-gate) devices, $\phi_{GC} = \phi_{F(substrate)} - \phi_M$, where $\phi_M = +0.6$ V.

3. There is always an undesired positive charge Q_{ox} (often denoted Q_{ss}) present at the interface between the oxide and the bulk silicon. This charge is due to impurities and/or imperfections at the interface. It makes a negative contribution to the threshold voltage of an amount $-Q_{ox}/C_{ox}$.

Thus we have a threshold voltage V_T:

$$V_T = \phi_{GC} - 2\phi_F - \frac{Q_B}{C_{ox}} - \frac{Q_{ox}}{C_{ox}} \tag{2.5a}$$

$$= \phi_{GC} - 2\phi_F - \frac{Q_{B0}}{C_{ox}} - \frac{Q_{ox}}{C_{ox}} - \frac{Q_B - Q_{B0}}{C_{ox}} \tag{2.5b}$$

$$= V_{T0} + \gamma \left(\sqrt{|-2\phi_F + V_{SB}|} - \sqrt{2|\phi_F|} \right) \tag{2.6}$$

where Eqs. (2.4) have been used and V_{T0} is the threshold voltage with $V_{SB} = 0$. The parameter γ (gamma) is termed the *body-effect coefficient*, or *body factor*. Comparing Eqs. (2.4) to (2.6), we see that γ is given by

$$\gamma = \frac{1}{C_{ox}} \sqrt{2q\epsilon_{si}N_A} \tag{2.7}$$

Gate oxide capacitance per unit area is defined by

$$C_{ox} = \frac{\epsilon_{ox}}{t_{ox}} \tag{2.8}$$

where ϵ_{ox} and t_{ox} are the permittivity and thickness of the gate dielectric. Typical values for $t_{ox} = 0.1 \ \mu m$ and $N_A = 5 \times 10^{14} \ cm^{-3}$ are $\gamma = 0.37 \ V^{1/2}$ and $C_{ox} = 34.5 \times 10^{-9} \ F/cm^2$.

It is easy to become confused about the signs of the various terms in the threshold voltage equations. The above equations give correct results for NMOS and PMOS if the signs shown in Table 2.1 are used.

TABLE 2.1
Signs in threshold voltage equation

Parameter	NMOS	PMOS
Substrate	p type	n type
ϕ_{GC}:		
Metal gate	−	−
n^+ Si gate	−	−
p^+ Si gate	+	+
ϕ_F	−	+
Q_{B0}, Q_B	−	+
Q_{ox}	+	+
γ	+	−
X_d, C_{ox}	+	+
Source-to-body voltage V_{SB}	+	−

Example 2.2. Calculate threshold voltage (for $V_{SB} = 0$) and body-effect coefficient for a NMOS silicon-gate transistor that has substrate doping $N_A = 10^{15}$ cm^{-3}, gate doping $N_D = 10^{20}$ cm^{-3}, gate oxide thickness $t_{ox} = 700$ Å, and 2×10^{10} cm^{-2} singly charged positive ions at the oxide-silicon interface.

$$\phi_{F(sub)} = \frac{kT}{q} \ln \frac{n_i}{N_A} = -0.026 \ln \frac{10^{15}}{1.4 \times 10^{10}} = -0.29 \text{ V}$$

$$\phi_{GC} = \phi_{F(sub)} - \phi_{F(gate)} = -0.29 - \frac{kT}{q} \ln \frac{10^{20}}{1.4 \times 10^{10}} = -0.88 \text{ V}$$

$$\epsilon_{ox} = 3.9\epsilon_0 = 3.5 \times 10^{-13} \text{ F/cm}$$

$$C_{ox} = \frac{\epsilon_{ox}}{7 \times 10^{-6}} = 49 \times 10^{-9} \text{ F/cm}^2$$

$$Q_{B0} = -(2 \times 1.6 \times 10^{-19} \times 10^{15} \times 1.04 \times 10^{-12} \times | -0.58|)^{1/2}$$
$$= -1.4 \times 10^{-8} \text{ C/cm}^2$$

$$\frac{Q_{B0}}{C_{ox}} = -\frac{1.4 \times 10^{-8}}{49 \times 10^{-9}} = -0.28 \text{ V}$$

$$\frac{Q_{ox}}{C_{ox}} = \frac{2 \times 10^{10} \times 1.6 \times 10^{-19}}{49 \times 10^{-9}} = 0.065 \text{ V}$$

$$V_T = -0.88 - (-0.58) - (-0.28) - 0.07 = -0.09 \text{ V}$$

$$\gamma = \frac{1}{49 \times 10^{-9}} \sqrt{2 \times 1.6 \times 10^{-19} \times 1.04 \times 10^{-12} \times 10^{15}} = 0.37 \text{ V}^{1/2}$$

The small negative value of threshold voltage calculated above is not desirable for use in digital circuits. Although in principal the threshold voltage may be set to any value by proper choice of doping concentrations and oxide capacitance, considerations such as breakdown voltage and junction capacitance frequently dictate the desirable specifications for these variables. Therefore, in practice the value of V_{T0} is determined during circuit fabrication by ion implanting dopant atoms into the substrate in the channel region.

Extra p-type impurities are implanted to make $V_{T0} = +0.5$ to $+1.5$ V for n-channel enhancement devices. By implanting n-type impurities in the channel region, a strongly conducting channel can be obtained even with $V_{GS} = 0$. The resulting NMOS transistor is termed a *depletion device* and typically has V_{T0} in the range -1 to -4 V. If Q_C is the charge density per unit area in the channel region due to the implant, then the threshold voltage V_{T0} given by Eqs. (2.5) and (2.6) is shifted by approximately Q_C/C_{ox}. It is assumed that all implanted ions are electrically active. The threshold voltage is shifted positively for p-type implants and negatively for n-type implants.

Example 2.3. Calculate the ion implant doses N_I needed to achieve threshold voltages of $+1.0$ and -3.0 V for the process defined in Example 2.2. Assume all implanted ions are electrically active.

For $V_T = 1.0$ V:

$$N_I = \frac{Q_c}{q} = \frac{C_{ox}}{q}[1 - (-0.09)] = \frac{49 \times 10^{-9}}{1.6 \times 10^{-19}} \times 1.09$$

$$= 3.3 \times 10^{11} \text{ ions/cm}^2 \quad (p\text{-type})$$

For $V_T = -3.0$ V:

$$N_I = \frac{49 \times 10^{-9}}{1.6 \times 10^{-19}}[-3.0 - (-0.09)] = 8.9 \times 10^{11} \text{ ions/cm}^2 \quad (n\text{-type})$$

The above calculations of threshold voltage do not give exact quantitative results in practical cases. Reasons for this include the facts that body doping may vary near the oxide interface, oxide thickness and dielectric constant may vary due to process variations, and oxide charge is not exactly controlled. Calculations of threshold voltage are useful for predicting how V_T varies as a function of dopings and dimensions. As a practical matter for purposes of circuit design, nominal values and statistical variations of the threshold voltage and body-effect coefficient must be determined by direct measurements of actual devices, as described in Sec. 2.6.

Exercise 2.1. Calculate the threshold voltage and body-effect coefficient for a silicon-gate PMOS transistor that has substrate doping $N_D = 10^{16}$ cm^{-3}, gate doping (n-type) $N_D = 10^{20}$ cm^{-3}, $Q_{ox}/q = 4 \times 10^{10}$ cm^{-3}, and $t_{ox} = 0.10$ μm. Find the ion implant doses (ions/cm^2) needed to give $V_{T0} = -1.0$ and $+3.0$ V.

2.4   CURRENT-VOLTAGE CHARACTERISTICS

The equations derived above can now be used to calculate the large-signal characteristics of MOS transistors for dc or slowly changing applied signals. We assume an NMOS device with source grounded and bias voltages V_{GS}, V_{DS}, and V_{BS} applied as shown in Fig. 2.4. If V_{GS} is greater than V_T, a conducting channel is present and V_{DS} causes a drift current I_D to flow from drain to source. The voltage V_{DS} causes a larger reverse bias from drain to body than that present from source to body, and thus a wider depletion layer exists at the drain. However, for simplicity we assume that the voltage drop along the channel is small so that the threshold voltage and depletion-layer width are approximately constant along the channel.

At a distance y along the channel, the voltage with respect to the source is $V(y)$ and the gate-to-channel voltage at that point is $V_{GS} - V(y)$. We assume this voltage exceeds the threshold voltage V_T, and thus the induced charge per unit area at the point y in the channel is

$$Q_I(y) = C_{ox}[V_{GS} - V(y) - V_T] \qquad (2.9)$$

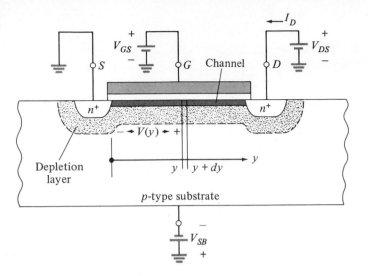

FIGURE 2.4
NMOS device with bias voltages applied.

The resistance dR of a length dy of the channel is

$$dR = \frac{dy}{W\mu_n Q_I(y)}$$

where W is the width of the device perpendicular to the plane of Fig. 2.4 and μ_n is the average mobility of electrons in the channel.

Mobility is the ratio of carrier (electron or hole) velocity to electric field causing carrier motion. Its dimensional units are cm/s over V/cm, or $cm^2/V{\cdot}s$. The physical details determining variation of mobility with local fields, crystal structure, etc., are complex. For the purposes of MOS digital circuit design, mobility is best used as an adjustable parameter to fit measured transistor I-V characteristics to the simple device equations derived below.

To continue with the analysis of MOS transistor conduction, the voltage drop dV along the length of channel dy is

$$dV = I_D\, dR = \frac{I_D}{W\mu_n Q_I(y)}\, dy$$

Substitution of Eq. (2.9) in the above and minor rearrangement gives

$$I_D\, dy = W\mu_n C_{ox}(V_{GS} - V - V_T)\, dV$$

For simplicity, we assume that V_T does not vary significantly along the length of the channel. Integrating the left side along the channel from $y=0$ to L and the right side in voltage from $V=0$ to V_{DS} and substituting *the process transconductance parameter* k' that is defined as

$$k' = \mu_n C_{ox} = \frac{\mu_n \epsilon_{ox}}{t_{ox}} \tag{2.10}$$

we obtain

$$I_D \int_0^L dy = Wk' \int_0^{V_{DS}} (V_{GS} - V - V_T)\, dV$$

$$I_D = k'\frac{W}{L}\left[(V_{GS} - V_T)V_{DS} - \frac{V_{DS}^2}{2}\right] \qquad V_{GS} \geq V_T, V_{DS} \leq V_{GS} - V_T \qquad (2.11a)$$

The *device transconductance parameter* is defined as $k = k'(W/L)$. Substituting this in the above yields

$$I_D = \frac{k}{2}\left[2(V_{GS} - V_T)V_{DS} - V_{DS}^2\right] \qquad V_{GS} \geq V_T V_{DS} \leq V_{GS} - V_T \qquad (2.11b)$$

for the so-called *linear* region of operation. This equation is important. It describes the current-voltage (*I-V*) characteristics of the MOS transistor, assuming a continuous channel is present from source to drain. Typical values of k' for $t_{ox} = 0.1\,\mu m$ are about 20 $\mu A/V^2$ for NMOS and 8 $\mu A/V^2$ for PMOS devices.

As the value of V_{DS} is increased, the induced conducting channel charge Q_I decreases near the drain. Equation (2.9) shows that Q_I at the drain end approaches zero as V_{DS} approaches $V_{GS} - V_T$. When V_{DS} equals or exceeds $V_{GS} - V_T$, the channel is said to be *pinched off*. Increases in V_{DS} above this critical voltage produce little change in I_D, and Eq. (2.11) no longer applies. The value of I_D in this region is obtained by substituting $V_{DS} = V_{GS} - V_T$ in Eq. (2.11b), giving

$$I_D = \frac{k}{2}(V_{GS} - V_T)^2 \qquad V_{GS} \geq V_T, V_{DS} \geq V_{GS} - V_T \qquad (2.12)$$

for the MOS transistor operating in this so-called *saturation* region.[*] The word "saturation" is used because I_D reaches a limit, or saturates, at the level given by Eq. (2.12). The choice of word is unfortunate because the same word has a different meaning in relation to bipolar transistor operation.

The drain current of an MOS transistor in the saturation region in fact is not completely independent of V_{DS}, because the depletion layer at the drain widens as V_{DS} increases, shortening the electrically effective value of L. Also, there is significant electrostatic coupling between the drain and the mobile charge in the channel, such that increasing the drain voltage increases Q_I above the value given by Eq. (2.9). Each of these effects acts to increase the drain current as drain voltage increases. An empirical approximation to the actual drain current is given by

$$I_D = \frac{k}{2}(V_{GS} - V_T)^2(1 + \lambda V_{DS}) \qquad (2.13)$$

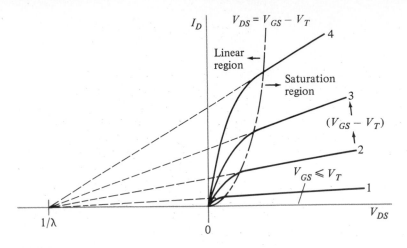

FIGURE 2.5
NMOS device I_D-V_{DS} characteristics. Channel-length modulation parameter here is much smaller than usual, leading to a steeper slope for I_D-V_{DS} in saturation than is usually observed.

where the *channel-length modulation parameter* λ (lambda) has typical values in the range 0.1 to 0.01 V^{-1} and represents the small influence of V_{DS} on I_D in saturation. To avoid discontinuities in the I_D-V_{DS} characteristic, the $1 + \lambda V_{DS}$ term may be included for both saturated and linear regions with negligible error. Usually the value of λ has little effect on the operating characteristics of digital MOS circuits. Unless otherwise stated, $\lambda = 0$ is assumed in this book.

A plot of I_D versus V_{DS} for an NMOS transistor is shown in Fig. 2.5. Below pinch-off the device behaves as a nonlinear voltage-controlled resistor. This is termed the *linear, resistance, triode,* or *nonsaturation* region of operation. Above pinch-off the device approximates a voltage-controlled current source. Note that for depletion-mode NMOS devices, V_T is negative and drain current can flow even for $V_{GS} = 0$. For PMOS devices, all polarities of voltages and current are reversed.

As we will see in the next chapter, MOS circuits are usually designed with MOS transistors as the only circuit elements. One or two values of threshold voltage V_T are usually available on a chip. Transistor width W and length L are typically the only parameters available to be specified by the circuit designer.

Exercise 2.2. A NMOS transistor with $k = 20$ $\mu A/V^2$ and $V_T = 1.5$ V is operated at $V_{GS} = 5$ V and $I_{DS} = 100$ μA. Determine whether Eq. (2.11) or (2.12) is appropriate, and find V_{DS}.

2.5   CAPACITANCES OF THE MOS TRANSISTOR

Since the MOS transistor is a field-effect device (majority carriers only), the switching speed of MOS digital circuits is limited only by the time required to charge and discharge the capacitances between device electrodes and from interconnecting lines to ground or other lines. Within LSI circuits all of these capacitances are so small (each in the approximate range 0.01 to 1.0 pF) that they are difficult or impossible to measure directly. For circuit analysis, these capacitances must be calculated from data on internal device dimensions and dielectric constants.

Figure 2.6 shows the significant capacitances between electrodes of an MOS transistor. The capacitances C_{sb} and C_{db} are n^+p junction capacitances that are readily calculated as shown in Sec. 4.3. For the present, we assume that measured values of junction capacitance are given. Concerning the capacitances from gate to other electrodes, to a first approximation, the sum C_g of C_{gs}, C_{gb}, and C_{gd} is constant, equal to

$$C_g = WLC_{ox} = WL\frac{\epsilon_{ox}}{t_{ox}} \tag{2.14}$$

where C_{ox} is the capacitance per unit area of the gate dielectric as defined in Eq. (2.8).

The division of C_g into its three elements is fairly complex and varies depending on whether the device is in the cutoff, linear, or saturation region. The SPICE circuit analysis program makes the necessary calculations from data on oxide thickness and device dimensions. Further details on modeling of transistor capacitances for hand analysis are presented in Sec. 3.3.2.

Probably the most significant limitation on logic delays and clock rate in LSI digital systems is the capacitances of off-chip connections. Capacitance on an off-chip connection includes device protection (see below), bonding pad, package, and printed wiring board capacitance and totals 5 to 10 pF per connection on each chip. One LSI circuit may have to drive a number of others, so total capacitance which must be driven can reach 100 pF or more.

The above comments relate to limits on the maximum operating speed of MOS transistor circuits. There is an important limitation on the minimum operating speed of certain MOS logic and memory circuits known as *dynamic circuits*. For proper operation, these circuits rely upon storage of logic information as charge on the small capacitance of a circuit node, as illustrated in Fig. 2.7.

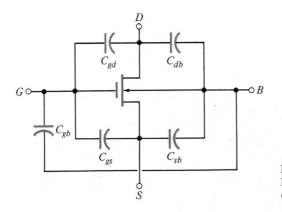

FIGURE 2.6
MOS transistor capacitances. $C_g = C_{gs} + C_{gb} + C_{gd}$.

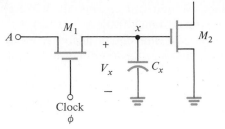

FIGURE 2.7
Essentials of dynamic MOS circuit operation. C_x is the total capacitance at node x.

The circuit of Fig. 2.7 operates as follows. With a logic 1 or 0 on node A, M_1 is turned on, then turned off, by clock signal ϕ. While M_1 is on, the high or low voltage at A is transferred to node x, charging the node capacitance C_x. If a logic 1 is present, M_2 is supposed to retain a conducting channel after M_1 turns off due to the high voltage stored on C_x. However, the charge on C_x leaks away through the n^+p junction of M_1. The rate of leakage determines how long the desired logic signal will be safely represented by V_x, and thus the maximum allowable repetition period for clock signal ϕ. A typical value for C_x is 0.1 pF; increasing this is not desirable because it would consume additional chip area. The reverse (or leakage) current for an n^+p junction in an MOS circuit at maximum temperature might range as high as 1 to 10 pA. This leads to leakage rates in the range 0.01 to 0.1 V/ms. A common specification for minimum clock rate for such circuits is 500 Hz, corresponding to a 2-ms clock period.

Exercise 2.3. Calculate approximate values of C_g, C_{sb}, and C_{db} for a silicon-gate MOS transistor for which

$$L = 5 \ \mu\text{m} \qquad W = 20 \ \mu\text{m}$$

$$\text{Source, drain surface dimensions} = 6 \times 24 \ \mu\text{m}$$

$$\text{Body doping} = 5 \times 10^{14} \ \text{cm}^{-3}$$

$$\text{Source, drain doping} = 10^{20} \ \text{cm}^{-3}$$

$$t_{ox} = 700 \ \text{Å}$$

Exercise 2.4. Calculate the capacitance of a conducting line 0.5 cm long and 6 μm wide. It is separated from the body by 5000 Å of silicon dioxide. Compare this with the device capacitances calculated in Exercise 2.3.

Exercise 2.5. Calculate the average current I_{av} needed to drive a 100-pF load through a 2-V change in 30 ns. (MOS transistors with very large W/L ratios are needed to drive such off-chip connections with acceptably small delays.)

2.6   MODELING THE MOS TRANSISTOR FOR CIRCUIT SIMULATION

Computer circuit simulators are essential tools in the analysis and design of MOSLSI circuits. The computer program SPICE is widely used for such work and is the basis for examples in this text. Other circuit simulators require similar modeling considerations.

2.6.1 SPICE MOSFET Models

SPICE provides three MOSFET device models which differ in the formulation of the *I-V* characteristic. Only the LEVEL 1 model, which is the simplest, is considered in this book. The LEVEL 1 model employs exactly the dc equations derived in Secs. 2.3 and 2.4. Simulation accuracy using LEVEL 1 is adequate for most MOS digital circuit analysis and design, provided experimentally determined dc parameters are used.

The dc characteristics of the MOSFET are defined by the device electrical parameters VTO, KP, LAMBDA, PHI, and GAMMA. (Correspondences between SPICE parameter names, which are restricted to use of the uppercase English alphabet, and the algebraic symbols used in the preceding pages are shown in Table 2.2.) Electrical parameters are computed by SPICE if process parameters (NSUB, TOX, . . .) are given, but user-specified values always override. VTO is positive (negative) for enhancement-mode and negative (positive) for depletion-mode *n*-channel (*p*-channel) devices.

Charge storage is modeled in several parts, defined so they may be calculated easily from actual circuit layouts. Three constant capacitors, CGSO, CGDO, and CGBO, represent gate-source, gate-drain, and gate-body overlap capacitances. The nonlinear gate-channel thin-oxide capacitance is calculated by the program as a function of applied voltages and distributed among the gate, source, drain, and body regions. The sum of all the above-mentioned capacitances is approximately equal to C_g of Sec. 2.5. The nonlinear depletion-layer capacitances, for both source-body and drain-body *pn* junctions, is divided into bottom and periphery, which vary as the MJ and MJSW power of junction voltage, respectively. These capacitances are determined by the parameters CBD, CBS, CJ, CJSW, MJ, MJSW, and PB.

There are two built-in models of the charge storage effects associated with the thin oxide. LEVEL 1 uses the piecewise-linear voltage-dependent capacitance model proposed by Meyer.* The thin-oxide charge storage effects are included only if TOX is specified in the input description.

There is some overlap among the parameters describing the junctions; for example, the reverse current can be input either as IS (in A) or as JS (in A/m²). Whereas the first is an absolute value, the second is multiplied by AD and AS to give the reverse current of the drain and source junctions, respectively. This flexibility has been provided so that junction characteristics can either be entered as absolute values on .MODEL cards or related to areas AD and AS entered on device cards. The same flexibility is available also for the zero-bias junction capacitances CBD and CBS (in F) on the one hand, and CJ (in F/m²) on the other. The parasitic drain and source series resistance can be expressed as either

---

*J. E. Meyer, "MOS Models and Circuit Simulation," *RCA Review,* Vol. 32, March 1971, pp. 42–63.

TABLE 2.2
SPICE MOS transistor model parameters

Symbol	Name	Parameter	Units	Default	Example
	LEVEL	Model index		1	
V_{T0}	VTO	Zero-bias threshold voltage	V	0.0	1.0
k'	KP	Transconductance parameter	A/V^2	2.0E−5	3.1E−5
γ	GAMMA	Bulk threshold parameter	$V^{1/2}$	0.0	0.37
$2|\phi_F|$	PHI	Surface potential	V	0.6	0.65
λ	LAMBDA	Channel-length modulation	V^{-1}	0.0	0.02
r_d	RD	Drain ohmic resistance	Ω	0.0	1.0
r_s	RS	Source ohmic resistance	Ω	0.0	1.0
C_{db}	CBD	Zero-bias B-D junction capacitance	F	0.0	2.0E−14
C_{sb}	CBS	Zero-bias B-S junction capacitance	F	0.0	2.0E−14
	IS	Bulk junction saturation current	A	1.0E−14	1.0E−15
ϕ_0	PB	Bulk junction potential	V	0.8	0.87
	CGSO	Gate-source overlap capacitance per meter channel width	F/m	0.0	4.0E−11
	CGDO	Gate-drain overlap capacitance per meter channel width	F/m	0.0	4.0E−11
	CGBO	Gate-bulk overlap capacitance per meter channel length	F/m	0.0	2.0E−10
	RSH	Drain and source diffusion sheet resistance	Ω/square	0.0	10.0
C_{j0}	CJ	Zero-bias bulk junction bottom capacitance per square meter of junction area	F/m^2	0.0	2.0E−4
m	MJ	Bulk junction bottom grading coefficient		0.5	0.5
	CSJW	Zero-bias bulk junction sidewall capacitance per meter of junction perimeter	F/m	0.0	1.0E−9
m	MJSW	Bulk junction sidewall grading coefficient		0.33	
	JS	Bulk junction saturation current per square meter of junction area	A/m^2		1.0E−8
t_{ox}	TOX	Oxide thickness	m	1.0E−7	1.0E−7
N_A or N_D	NSUB	Substrate doping	cm^{-3}	0.0	4.0E15
Q_{ss}/q	NSS	Surface state density	cm^{-2}	0.0	1.0E10
	NFS	Fast surface state density	cm^{-2}	0.0	1.0E10
	TPG	Type of gate material: +1 opposite to substrate −1 same as substrate 0 Al gate		1.0	
X_j	XJ	Metallurgical junction depth	m	0.0	1.0E−6
L_D	LD	Lateral diffusion	m	0.0	0.8E−6
μ	UO	Surface mobility	$cm^2/V \cdot s$	600	700

RD and RS (in Ω) or RSH (in Ω per square), the latter being multiplied by the number of squares NRD and NRS input on the device card.

The SPICE parameters used in LEVEL 1 are listed in Table 2.2 together with the corresponding symbols used in this text. The SPICE MOSFET model is schematically summarized in Fig. 2.8.

Over the past 20 years much effort has gone into analysis of the MOS transistor. Analytical equations much more sophisticated than those presented in Secs. 2.3 and 2.4 have been developed, and more sophisticated models are available as LEVEL 2 and 3 models in SPICE. Nevertheless, it is still impossible to accurately predict transistor current-voltage (*I-V*) characteristics based only on knowledge of device dimensions and the bulk properties of silicon and silicon dioxide.

To obtain satisfactory results in circuit analysis and design, measured data on samples of MOS transistors must be obtained. Satisfactory results will be obtained for most digital circuits by fitting measured data to Eqs. (2.6), (2.11), (2.12), and (2.13) over the intended operating range of voltages and currents. More sophisticated model equations are necessary for use at very low and very high voltages and currents—outside the normal ranges for most digital MOS circuits.

2.6.2 Measurement of Device Parameters

Methods of taking and reducing data to determine V_T, γ, $k = k'(W/L)$, and λ are illustrated in Fig. 2.9. Suggested values of test voltages for transistors intended

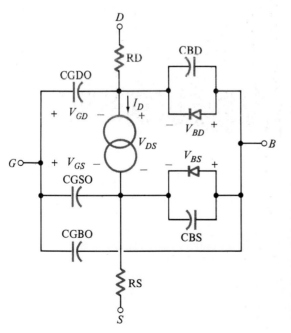

FIGURE 2.8
SPICE NMOS model, LEVEL 1. Drain current I_D as a function of voltages is given by Eqs. (2.6), (2.11), and (2.12). the three gate capacitances shown represent only the gate overlap outside the active channel region. SPICE calculates the voltage-dependent channel region gate capacitance and allocates it among source, body, and drain.

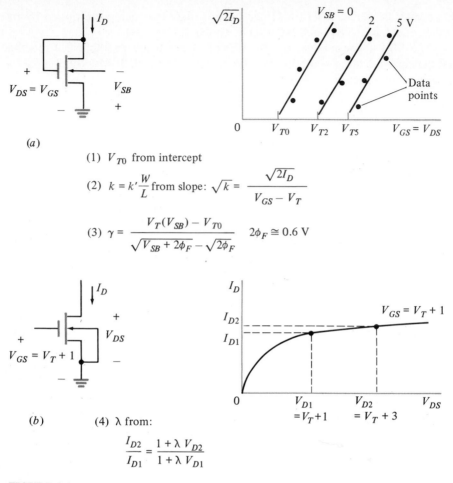

(1) V_{T0} from intercept

(2) $k = k'\dfrac{W}{L}$ from slope: $\sqrt{k} = \dfrac{\sqrt{2I_D}}{V_{GS} - V_T}$

(3) $\gamma = \dfrac{V_T(V_{SB}) - V_{T0}}{\sqrt{V_{SB} + 2\phi_F} - \sqrt{2\phi_F}}$ $2\phi_F \cong 0.6$ V

(4) λ from:

$$\dfrac{I_{D2}}{I_{D1}} = \dfrac{1 + \lambda V_{D2}}{1 + \lambda V_{D1}}$$

FIGURE 2.9
MOS transistor parameters.

for operation in 5-V digital circuits are shown. Threshold voltage V_T, device transconductance parameter k, and body-effect coefficient γ are all determined from the first set of data. The second group of measurements is used to determine channel-length modulation parameter λ, if needed. Demonstration D2.1 covers these measurement techniques.

The resulting parameters are the most important ones for the LEVEL 1 model in the SPICE circuit analysis program. Only minor errors will result from allowing other dc parameters to take on default values. Capacitance calculations for SPICE are described in Sec. 3.3.2.

The threshold voltage V_{T0} used in SPICE and in this chapter is sometimes called the *extrapolated threshold voltage* because it is obtained by extrapolating to zero drain current. It is not the threshold voltage parameter commonly measured in a production environment. The common measurement is of gate voltage necessary to achieve a specified drain current, usually 1 or 10 μA, for a device of specified W and L in saturation. The V_{T0} needed for use in hand analysis and for input to SPICE will usually be smaller than the value established from the production measurement.

2.7 LIMITATIONS ON THE MOS TRANSISTORS

In this section we describe some of the limitations on electrical characteristics of integrated circuit MOS devices.

2.7.1 Voltage Limitations

MOS transistors employ *pn* junctions and are subject to voltage limitations determined by avalanche breakdown, which typically occurs at 50 to 100 V for the drain-body junction. However, before reaching avalanche breakdown, most modern MOS devices with channels shorter than about 10 μm exhibit *punch-through,* in which the depletion region surrounding the drain extends through the channel to the source. Drain current rises sharply and is no longer under control of the gate voltage. A lower bound on the punch-through voltage may be calculated from Eq. (4.3.6). When the depletion layer extends out from the drain by a distance equal to the electrical channel length L, the device will punch through. Punch-through occurs at drain voltages in the range 15 to 20 V for modern devices. It does not cause permanent damage to the transistor unless enough power is dissipated to cause local melting.

Gate dielectric breakdown is another important limitation. Usually this is a destructive event, resulting in a short circuit between gate and source. A 1000-Å gate oxide is safe from breakdown up to perhaps 50 V; thinner oxides withstand proportionately less voltage.

None of the voltage breakdown effects is included in the SPICE model for MOS transistors.

Most inputs to a MOS integrated circuit are connected to gates of transistors, which of course have a nearly infinite input resistance. Total capacitance seen at an input to a circuit is a few picofarads. A very common problem with early MOS circuits was failure due to static electricity charging circuit inputs to the point of destructive breakdown. (In practice, even the discharge of 100 pC through a gate dielectric will destroy it.) Static charge is most commonly picked up when a person handling the circuit brushes it against a garment.

This problem has been overcome through incorporation of a simple protection circuit at each input to an MOS circuit. The protection device is designed to break down nondestructively at a voltage lower than the breakdown voltage of gate oxide in order to drain away the static charge. Figure 2.10 shows typical input protection circuits for NMOS and CMOS circuits.

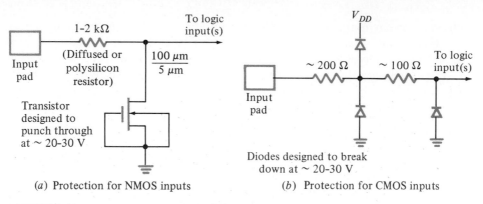

(a) Protection for NMOS inputs   (b) Protection for CMOS inputs

FIGURE 2.10
Input protection for MOS circuits.

2.7.2 Parasitic Bipolar Transistors and Latch-Up

All MOS transistor integrated circuits have undesired and potentially troublesome parasitic bipolar transistors which will conduct if one or more pn junctions become forward-biased. This may be understood by examining the NMOS transistor structure of Fig. 2.1. Although the study of bipolar transistors does not begin until we reach Chap. 5 in this text, we assume the reader is aware of the basic structure of a bipolar transistor. The device shown in Fig. 2.1 may be operated as an npn bipolar transistor by biasing the n^+ source negatively with respect to the p-type body. Electrons injected from the source into the body may be collected at the n^+ drain if it is positive with respect to the body, thereby achieving bipolar transistor action. In NMOS circuits, this type of bipolar transistor action can discharge desired positive voltage levels on floating or dynamic nodes such as node x in Fig. 2.7.

The potential consequences of bipolar transistor action are much more serious in CMOS circuits. Figure 2.2 makes it clear that a pnp transistor is possible with the n-type body as its base, while an npn transistor is possible with an n^+ source or drain electrode as its emitter, the p well as its base, and the n body as a collector.

With only rare exceptions, the undesired parasitic circuit shown schematically in Fig. 2.11a will be present in any CMOS circuit. The bipolar transistors originate as just described. The resistors R_1 and R_2 (also parasitic elements in the sense that they make no contribution to the desired MOS circuit function) originate in the bulk semiconductor material of the body and the p-type well. Low values of resistance are desirable in order to make it more difficult to forward-bias junctions.

The two-terminal current-voltage characteristics of this parasitic circuit are depicted in Fig. 2.11b. Above some critical voltage V_L (related to punch-through;

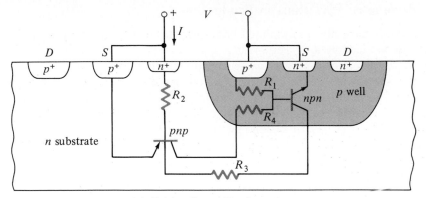

(a) Origin of parasitic elements

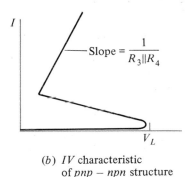

(b) *IV* characteristic
of *pnp* − *npn* structure

FIGURE 2.11
Latch-up in CMOS circuits.

10 to 20 V for most modern CMOS circuits) both bipolar transistors begin to conduct and the current rises sharply from leakage levels (under 1 μA) to a value limited by resistors R_3 and R_4, often many milliamperes. This phenomenon is known as *latch-up*. It can occur even at normal operating voltages if voltages applied to input or output pins cause forward-biasing of *pn* junctions within the chip. Circuits are often permanently damaged by the resulting high currents.

The obvious solution to latch-up problems is to prevent junctions from ever becoming forward-biased and to limit externally applied voltages at levels safely below V_L. In practice this proves to be virtually impossible due to the many spurious signals that occasionally occur in digital systems. Practical solutions to the latch-up problem involve special care in device and circuit design to reduce bipolar transistor current gain and reduce the values of R_1 and R_2. Circuits connected to input and output pins are most critical with respect to latch-up. When power is switched on and off, voltages applied to the pins of a chip frequently go outside the normal range. This can cause latch-up.

2.7.3   Parameter Variations in Production

MOS transistors have always exhibited broad variations in major device parameters among production lots. Particularly significant are variations in channel
length, threshold voltage, and gate oxide thickness. While these matters are progressively coming under better control, it is still quite common to find circuits
with a 2 to 1 (or larger) unit-to-unit variation in dc power consumption and speed
coming out of a single production facility. A common response to this situation
in industry is to use testing to sort the circuits according to speed and/or power
consumption. Typically the fastest circuits are sold at a higher price.

2.7.4   Temperature Effects

Parameters of MOS transistors display a temperature dependence which must often
be considered in circuit design and application. Testing circuits at high temperature is very costly, so means are usually sought to predict high-temperature performance from room temperature tests. Most important in dynamic circuits is the
fact that leakage current doubles with every 6 to 10°C temperature increase. This
rapidly reduces the length of time a desired charge can be held on a capacitive
logic node. Leakage current is rarely an issue in static circuits.

   Mobility of carriers in the channel of an MOS transistor is an inverse
function of absolute temperature according to the following empirical expression:

$$\mu(T) = \frac{\mu(300)}{(T/300)^a}$$

where temperature T is in kelvins (absolute) and the exponent a is between 1.0
and 1.5. Thus, for a 100° temperature increase, mobility may decrease by as
much as 40%. The result is a proportional decrease in the parameter k' in Eqs.
(2.11) and (2.12) and in drain current I_D for fixed applied voltages. In turn,
the current consumption of a complete circuit may decrease considerably at high
temperature. Maximum speed of operation will usually decrease in proportion.

   Threshold voltage of both NMOS and PMOS enhancement-mode transistors
decreases in magnitude by 1.5 to 2 mV/°C with increasing temperature, due to
changes in ϕ_{GC} and $2\phi_F$. Usually the mobility variations are more significant for
digital circuit performance, because even a 200-mV change in V_T does not cause
a large percentage change in $V_{GS} - V_T$.

   Temperature effects are not included in the SPICE model for circuit
simulation. Thus there is no point in performing SPICE simulations for different
temperatures unless device parameters at each temperature are entered as new
model data. Some other circuit simulators have built-in models for temperature
effects in MOS devices.

2.8   LAYOUT DESIGN OF MOS TRANSISTOR

The cross section and plan view of an NMOS transistor are shown in Fig. 2.12.
The four masks required to fabricate this device are:

1. Transistor area mask
2. Poly gate mask
3. Contact mask
4. Metal mask

   In preparing these masks there are many constraints, but the goal is to obtain a circuit with the greatest yield in as small an area as possible without compromising the reliability of the circuit.
   For the layout, of course short circuits cannot be tolerated. Neither can voltage breakdowns be permitted between components or successive diffusions. The minimum feature size will reflect the tolerances imposed by a given technology. Among others, these tolerances will be due to registration error in mask alignment going from one pattern to another, process control due to variation in exposure and etching, and sufficient overlap to ensure low ohmic contact where necessary. While the layout rules provide for the minimum dimensions, it is always a wise decision to use larger dimensions where possible.

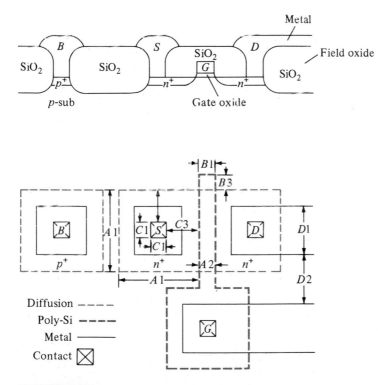

FIGURE 2.12
Layout example of NMOS transistor.

2.8.1   Simplified Design Rules for MOS Transistor

The minimum feature size for the layout of this device is 2 μm.

A. Transistor area mask

1. Minimum area of diffused region	2 $\times$ 2 μm
2. Minimum spacing between diffused regions	2 μm

B. Poly gate mask

1. Minimum area	2 $\times$ 2 μm
2. Minimum spacing between multiple gate stripes	2 μm
3. Minimum overlap of poly gate into field region	2 μm

C. Contact window mask

1. Minimum area	2 $\times$ 2 μm
2. For contact to diffusion, minimum spacing from edge of contact to edge of diffusion	1 μm
3. For contact to diffusion, minimum spacing from edge of contact to edge of poly gate	1 μm
4. For contact to poly, minimum spacing from edge of contact to edge of diffusion	1 μm

D. Metallization mask

1. Minimum width	2 μm
2. Minimum metal-to-metal spacing	2 μm
3. Minimum contact window overlap	1 μm

The example of an NMOS layout in Fig. 2.12 illustrates these design rules.

2.8.2   Simplified Design Rules for CMOS Transistor

With the addition of a well diffusion mask, the fabrication of a CMOS transistor is similar to that for an MOS device. Actually, the well diffusion is the first mask used in the process. Some simplified design rules for this initial mask are given below, and the layout of an n-well CMOS transistor is illustrated in Fig. 2.13.

E. Well Diffusion Mask

1. Minimum spacing from well edge to p^+ within well	2 μm
2. Minimum spacing from well edge to p^+ outside well	4 μm

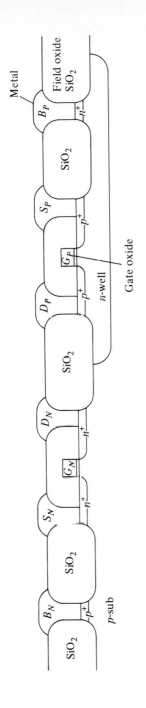

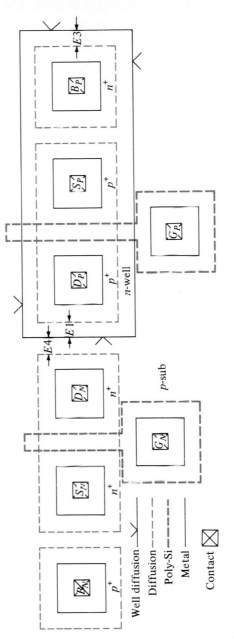

FIGURE 2.13
Layout example of CMOS transistor.

51

3. Minimum spacing from well edge to n^+ within well   2 μm
4. Minimum spacing from well edge to n^+ outside well   4 μm
5. Minimum spacing from well edge to well edge   6 μm

2.9 SUMMARY

This chapter describes the structure and operation of the MOS transistor in simple form based on ideas of electrostatics. Terms which are commonly used in MOS device and circuit work are defined.

The threshold voltage is analyzed. Five factors contribute to the threshold voltage of a MOS device:

- Fixed bulk charge Q_{B0}
- The necessary change in surface potential $2\phi_F$
- The gate-channel work function difference ϕ_{GC}
- The undesired surface charge Q_{ox} or Q_{ss}
- The ion-implanted threshold shifting charge Q_c

The *body effect* in MOS circuits, by which threshold voltage is modified when V_{SB} increases, is very important in operation of NMOS circuits in particular.

A simple analysis of conduction in the MOS transistor shows three regions of operation:

- Cutoff, achieved for $V_{GS} \leq V_T$
- Saturation, for which $V_{GS} - V_T \leq V_{DS}$
- Linear or resistive, for which $V_{GS} - V_T \geq V_{DS}$

The origins and effects of capacitances in MOS transistors are described briefly, and the SPICE model for the device is introduced. Limitations of MOS transistors are described; voltage limitations and latch-up effects due to parasitic bipolar transistors in CMOS are the most troublesome problems.

The basic design rules for the layout of both MOS and CMOS transistors have been presented.

REFERENCES

1. A. S. Grove, *Physics and Technology of Semiconductor Devices,* Wiley, New York, 1967.
2. P. Richman, *MOS Field-Effect Transistors and Integrated Circuits,* Wiley, New York, 1973.
3. R. S. Muller and T. Kamins, *Device Electronics for Integrated Circuits,* Second Edition, Wiley, New York, 1986.
4. Y. P. Tsividis, *Operation and Modeling of the MOS Transistor,* McGraw-Hill, New York, 1987.

DEMONSTRATIONS

D2.1 MOSFET: Electrical Parameters

The objective of this experiment is to determine the major electrical parameters of the NMOS and PMOS transistors in a standard metal-gate CMOS process.

The 4007 consists of three n-channel and three p-channel enhancement-type MOS-FETs in one package. The transistor elements are accessible through the package terminals to provide a convenient means of measuring the MOSFET parameters and for constructing some simple circuits.

1. (a) Use the 4007 and connect the circuit shown in Fig. D2.1a. Record the drain current I_D as a function of $V_{DS} = V_{GS}$ for $V_{DS} = 3, 5,$ and 10 V for the NMOS device. Make these measurements at $V_{SB} = 0, 2,$ and 5 V, making sure the body is *negative* with respect to the source. Also, note that the maximum device dissipation $I_D V_{DS} = 100$ mW.

   (b) Repeat part (a) for a PMOS device. Use the circuit shown in Fig. D2.1b. Note that all the polarities are reversed and the body is now *positive* with respect to the source.

   (c) Connect the circuits shown in Fig. D2.2; then with $V_{SB} = 0$ V and $V_{GS} = 2.5$ V, measure the drain current at $V_{DS} = 2, 4,$ and 6 V for both the NMOS and PMOS devices.

2. (a) From the results of 1(a), plot $\sqrt{I_D}$ versus V_{GS} for the three values of body bias V_{SB}. Draw a straight line through the data points and determine the threshold voltage V_T for each value of body bias.

   Make use of Eq. (2.12), and from the slope of the $V_{SB} = 0$ V plot, determine the conduction coefficient k. [$k = k'(W/L)$.] Graph V_T as a function of $\sqrt{2\phi_F + V_{SB}} - \sqrt{2\phi_F}$. Make use of Eq. (2.6), and from the slope of the plot determine the body-effect coefficient γ. Assume $2\phi_F = 0.6$ V.

   (b) Repeat part (a) using the results from 1(b) and determine V_T, k, and γ for the PMOS device.

   (c) From the results of 1(c) plot I_D versus V_{DS} for both devices, and making use of Eq. (2.13), from the slope determine the channel-length modulation factor λ for each device.

D2.2   MOSFET: Temperature Effects

The objective of this experiment is to determine the effect of temperature on the electrical parameters of the NMOS and PMOS transistors that were measured in D2.1.

1. Repeat the measurements of D2.1 at temperatures of 0 and 100°C. This can be done by immersing the package in an ice bath and in boiling water. Wait 2 or 3 min for the temperature of the package to stabilize before taking any readings.

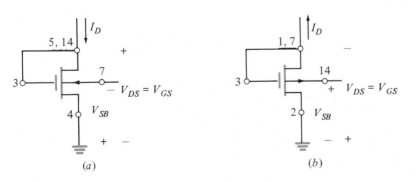

(a)   (b)

FIGURE D2.1

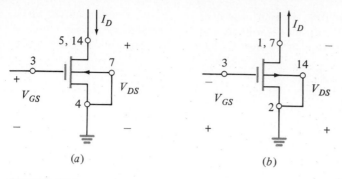

(a)   (b)

FIGURE D2.2

2. From the results determine V_T, k, and γ at 0 and 100°C for each device. What is your estimate of the temperature coefficient for these device parameters?

PROBLEMS

Use the following standard process for silicon-gate MOS devices unless otherwise stated.

$$t_{ox} = 0.10 \ \mu\text{m}$$
$$\text{Substrate doping} = N_A = 5 \times 10^{15} \ \text{cm}^{-3}$$
$$\text{Source, drain doping} = N_D = 5 \times 10^{18} \ \text{cm}^{-3}$$
$$\text{Silicon-gate doping} = N_D = 10^{20} \ \text{cm}^{-3}$$
$$\phi_{GC} = -0.8 \ \text{V} \qquad 2\phi_F = -0.6 \ \text{V}$$
$$N_{ss} = \frac{Q_{ox}}{q} = 5 \times 10^{10} \ \text{cm}^{-2}$$
$$\epsilon_{si} = 11.7 \times 8.85 \times 10^{-14} \ \text{F/cm} \qquad \epsilon_{ox} = 3.9 \times 8.85 \times 10^{-14} \ \text{F/cm}$$
$$\mu_n(300 \ \text{K}) = 580 \ \text{cm}^2/\text{V} \cdot \text{s} \qquad \mu_p(300 \ \text{K}) = 232 \ \text{cm}^2/\text{V} \cdot \text{s}$$

P2.1. For the standard process parameters with $V_{SB} = 0$ V:
  (a) Calculate the unimplanted threshold voltage.
  (b) How many impurities and of what type must be implanted to give a threshold of 1 V? −3 V?
  (c) How much is V_{T0} changed by a 20% change in N_A? In t_{ox}? In Q_{ox}?

P2.2. If worst-case tolerances on t_{ox} and N_A are 50%, what is the range of threshold voltages possible for this process?

P2.3. Calculate body-effect coefficient γ for the standard process. Repeat for $N_A = 2 \times 10^{16}$ cm^{-3}.

P2.4. A new dielectric that has a permittivity 1.5 times that of silicon dioxide is used under the gate.
  (a) What are the new V_{T0} and γ?
  (b) If the new dielectric can also be made 40% thinner (that is, $t_{ox(new)} = 0.6 t_{ox(old)}$), repeat (a).

P2.5. For the new process described in P2.4b:

(a) Find the values of C_{ox} per unit area and C_{sb} per unit area.

(b) Repeat (a) for source-drain doping $N_D = 10^{19}$ cm^{-3}, substrate doping $N_A = 10^{16}$ cm^{-3}.

P2.6. Measured data for a MOS transistor are given below. Determine whether it is an enhancement or depletion device, NMOS or PMOS. Calculate k, V_{T0}, and γ.

MOS measured data

V_{GS}, V	V_{DS}, V	V_{BS}, V	I_D, μA
3	3	0	77
5	5	0	389
3	3	−3	33
5	5	−5	237

P2.7. For the data below calculate k, V_{T0}, γ, and λ.

NMOS measured data

V_{GS}, V	V_{DS}, V	V_{BS}, V	I_D, μA
2	5	0	10
5	5	0	400
5	5	−5	250
5	8	0	480
5	5	−3	300

P2.8. For the data in P2.7, assuming the source and drain diffusions can be modeled as abrupt one-sided junctions, calculate approximate W and L values at $V_{DS} = 0$ V.

P2.9. For a MOS transistor with drain current I_D at 25°C, what range of temperatures can it operate over and have I_D remain within 10% of nominal? (Mobility exponent $a = 1.5$.)

P2.10. A PMOS process with an n^+ silicon gate has gate doping $N_D = 10^{18}$ cm^{-3}, substrate doping $N_D = 2 \times 10^{15}$ cm^{-3}, and $t_{ox} = 0.1$ μm.

(a) Calculate V_{T0}, γ, k', and C_{ox} (oxide capacitance per unit area).

(b) Repeat (a) if $N_D = 5 \times 10^{15}$ cm^{-3} and $t_{ox} = 0.06$ μm.

P2.11. Prove that two identical MOS transistors in series with their gates tied together have the same I-V characteristics as a single device with twice the length. Neglect body effect.

P2.12. (a) Sketch a minimum sized layout of the NMOS transistor described in P2.7 and P2.8. Use the simplified design rules as described in Sec. 2.8.1. and a minimum feature size of 3 μm. Your layout should be to scale on grid paper, with a level of detail like that of Fig. 2.12. Neglect the substrate connection.

(b) Calculate approximate values for C_g, C_{sb}, and C_{db}.

CHAPTER
3

MOS INVERTERS AND GATE CIRCUITS

3.0 INTRODUCTION

The inverter is the basic circuit from which most MOS logic circuits are developed. The MOS inverter exhibits all of the essential features of MOS logic gates except for logic function. DC and transient analysis and design techniques can be developed using the inverter as a vehicle. Subsequent extension of MOS inverter concepts to NOR and NAND gates is very simple. In this chapter we will spend a good deal of time establishing analysis techniques for dc voltage transfer characteristics, noise margins, and propagation delay. Both NMOS and CMOS circuits are considered. Alternative load elements are compared, considering power consumption, circuit density, transfer characteristics, and transient performance. The design of NMOS and CMOS logic gates based upon prototype inverters is described.

Figure 3.1 shows a single NMOS transistor connected with a resistor load to form an inverter, together with drain current-voltage (I-V) characteristics, load line construction, and voltage transfer characteristic. The quantitative design of such an inverter is guided by several considerations in addition to specifications for voltage levels and noise margins.

For any given process, the cost of an integrated circuit is proportional to the chip area, so designers usually attempt to minimize the area occupied by each circuit element. MOS transistors achieve minimum size when channel dimensions W and L are as small as can be achieved within the particular technology. As a practical matter, the smallest possible values of W and L are similar because they are determined by similar pattern-definition processes. This implies that the ratio W/L should not depart greatly from unity, unless there is a need to drive high-capacitance or low-resistance loads.

Power consumption should be minimized in the design of most integrated circuits. Depending on their pin count and construction, dual in-line integrated circuit packages can dissipate 0.5 to 2 W without excessive temperature rise above normal ambient room temperature. Since 10,000 or more gate circuits can be accommodated on one large-scale integrated (LSI) circuit chip, average power consumption per gate in LSI components should be in the range of 100 μW or less. Of course, circuits for battery-powered operation should consume much less power.

Power consumption of most circuits can be reduced by reducing the operating voltage. Circuits for use in wristwatches and similar applications operate at 1 to 1.5 V. However, circuits for commercial and industrial applications preferably operate from a single 5-V power supply. A 5-V supply is assumed in this chapter unless otherwise stated.

Digital MOS circuits can be classified into two categories, depending on whether periodic *clock signals* are necessary to achieve combinational logic functions. *Static circuits* require no clock or other periodic signals for operation in combinational logic networks. Clocks are required for static circuits in sequential logic applications, but in these cases, clock signals are usually applied only to normal logic gate inputs. *Dynamic circuits* require periodic clock signals, synchronized with data signals, for proper operation even in combinational logic applications. In dynamic circuits, clock signals are applied to load elements and so-called *transmission*, or *transfer*, *gates* and not to normal logic gate inputs.

In the remainder of this chapter, Secs. 3.1 to 3.4 explain dc and transient analysis methods for NMOS digital circuits. Complementary MOS circuits are covered in Sec. 3.5. NMOS and CMOS logic gate design is presented in Secs. 3.6 and 3.7. Description of dynamic circuits is presented in Sec. 3.8. Scaling of MOS circuits to smaller internal dimensions for increased density and improved performance is described in Sec. 3.9.

3.1   STATIC NMOS INVERTER ANALYSIS

Consider the nominal characteristics of the NMOS inverter shown in Fig. 3.1a. We know from Chap. 1 that a voltage transfer characteristic (VTC) in the form shown by Fig. 3.1c is sought. Nominal logic 0 and logic 1 voltage levels should fall in ranges such that small variations in logic input voltages have little or no effect on output voltage. The desired result can be achieved at the low input level if the input voltage remains below the transistor threshold voltage V_T. Then

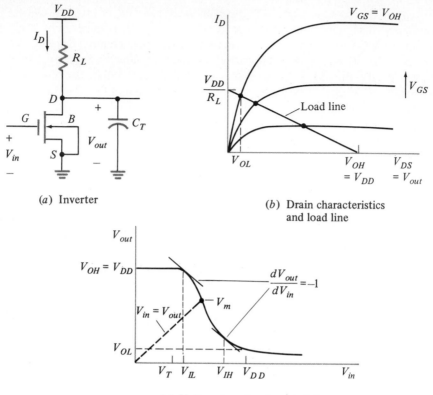

(a) Inverter

(b) Drain characteristics and load line

(c) Voltage transfer characteristic

FIGURE 3.1
NMOS inverter, resistor load.

no current flows and inverter output voltage V_{out} remains at V_{DD}. The nominal voltage representing a logic high level is $V_{OH} = V_{DD}$.

The VTC can be found from the load line construction by noting that each value of $V_{GS} = V_{in}$ for the inverter transistor gives a different drain current curve in Fig. 3.1b. Each intersection of an inverter transistor drain characteristic curve with the load line gives a value of $V_{DS} = V_{out}$ for one value of input voltage. The plot of V_{out} versus V_{in} is the desired VTC as seen in Fig. 3.1c.

When a logic 1, represented by V_{OH}, appears at the input of this inverter, the transistor is driven into conduction along the upper line on the drain I-V characteristic of Fig. 3.1b. With proper design, the resulting logic low level V_{OL} falls safely below transistor threshold voltage, and a following identical inverter stage will be nonconducting.

DIRECT CALCULATION OF VTC CRITICAL POINTS. We will not continue with this graphical approach to determination of voltage transfer characteristics. Instead, we will formulate an explicit solution for each of the five critical points

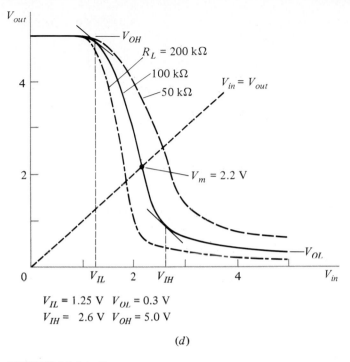

$V_{IL} = 1.25$ V $\quad V_{OL} = 0.3$ V
$V_{IH} = 2.6$ V $\quad V_{OH} = 5.0$ V

(d)

FIGURE 3.1 (*cont.*)
NMOS inverter, resistor load.

on the VTC of Fig. 3.1c. This much information is all that is required to define voltage levels and noise margins for an NMOS inverter or gate circuit. In the following sections, we assume that channel-length modulation factor λ is too small to have a significant effect on voltage transfer characteristics. This is a good assumption for most practical MOS digital circuits.

The output high level V_{OH} is already known to be V_{DD}. The output low level is found by equating the currents in the transistor and load, assuming the input of this inverter is driven by the output high level V_{OH} of another identical inverter. The transistor is assumed to be in the linear range of its *I-V* characteristics under these conditions; the assumption must be verified when the calculation is complete.

$$I_{D(lin)} = I_{RL} \tag{3.1a}$$

$$\frac{k}{2}[2(V_{OH} - V_T)V_{OL} - V_{OL}^2] = \frac{V_{DD} - V_{OL}}{R_L} \tag{3.1b}$$

Restating this in terms of V_{OL},

$$V_{OL}^2 - 2\left(\frac{1}{kR_L} + V_{DD} - V_T\right)V_{OL} + \frac{2V_{DD}}{kR_L} = 0 \tag{3.1c}$$

This quadratic equation may be solved for V_{OL}. Only the positive root has physical significance. If V_{OL} is sufficiently small (< 0.4 V), the squared term may be neglected to simplify the solution. Using this approximation,

$$V_{OL} \approx \frac{V_{DD}}{1 + kR_L(V_{DD} - V_T)} \tag{3.1d}$$

Critical input voltages V_{IL} and V_{IH} are defined as the points at which

$$\frac{dV_{out}}{dV_{in}} = -\frac{dI_D}{dV_{in}}(R_L \parallel r_{ds}) = -1.0 \tag{3.2a}$$

The minus sign in this equation arises because V_{out} decreases as V_{in} increases.

At $V_{in} = V_{IL}$ the output voltage is near V_{DD} and the transistor is operating in the saturation region. Thus the differential transistor output resistance r_{ds} is very high and may be neglected compared to R_L in Eq. (3.2a). By substituting the saturated drain current equation [Eq. (2.12)] in the above and differentiating, we obtain

$$k(V_{IL} - V_T)R_L = 1 \tag{3.2b}$$

This is easily solved for V_{IL}.

At $V_{in} = V_{IH}$ the output voltage is now near 0 V and the transistor is operating in the linear region. Now, the effective load resistance is R_L in parallel with the (typically) much smaller differential output resistance of the transistor. Neglecting the effect of R_L (another assumption which should be verified for the specific case), we now use the linear drain current equation [Eq. (2.11)] in Eq. (3.2a) and write

$$\frac{dV_{out}}{dV_{in}} = -\frac{dI_D}{dV_{in}}\frac{dV_{out}}{dI_D} = -1.0$$

$$= -\frac{dI_D}{dV_{GS}}\frac{dV_{DS}}{dI_D} = -1.0$$

$$= -\frac{k_I V_{DS}}{k_I(V_{GS} - V_T - V_{DS})} = -1.0 \tag{3.2c}$$

Since $V_{in} = V_{GS}$ and $V_{out} = V_{DS}$, we obtain

$$V_{out} = \frac{V_{in} - V_T}{2} \tag{3.2d}$$

The explicit value of $V_{in} = V_{IH}$ and the corresponding value of V_{out} are found by substituting Eq. (3.2d) in the linear region drain current equation, namely,

$$\frac{k}{2}[2(V_{IH} - V_T)V_{out} - V_{out}^2] = \frac{V_{DD} - V_{out}}{R_L} \tag{3.2e}$$

An additional point on the VTC that is often useful is the midpoint voltage V_M at which $V_{in} = V_{out} = V_{GS} = V_{DS}$ for which the transistor is in saturation. This is found by equating currents:

$$\frac{k}{2}(V_M - V_T)^2 = \frac{V_{DD} - V_M}{R_L} \tag{3.3a}$$

Rearranging to write this in terms of V_M, we obtain

$$V_M^2 - 2\left(V_T - \frac{1}{kR_L}V_M\right) + \left(V_T^2 - \frac{2V_{DD}}{kR_L}\right) = 0 \tag{3.3b}$$

Accuracy expected in analysis of MOS circuits is on the order of 0.1 V for all voltages and $\pm 20\%$ for currents. Higher accuracy is not necessary because typical variations from circuit to circuit are usually greater than this.

Example 3.1. Given the following data for an inverter like the one shown in Fig. 3.1a, find the critical voltages on the VTC and the noise margins.

$$k' = 20 \ \mu A/V^2 \qquad\qquad V_T = 1.0 \ V$$
$$W/L = 2.0 \qquad\qquad V_{DD} = 5 \ V$$
$$R_L = 100 \ k\Omega$$

Solution
(*a*) $V_{OH} = V_{DD} = 5$ V. (Check later to be sure that $V_{OL} < V_T$.)
(*b*) Find V_{OL} by solving Eq. (3.1c) with the given data.

$$V_{OL}^2 - 8.5V_{OL} + 2.5 = 0 \qquad V_{OL} = 0.31 \ V \text{ or } 8.2 \ V$$

Only the first root is physically meaningful. Note that V_{OL} is safely below V_T. Also, the assumption that the transistor is in the linear region is valid because $V_{DS} < V_{GS} - V_T$.
(*c*) Find V_{IL} by using Eq. (3.2b).

$$V_{IL} = \frac{1}{kR_L} + V_T = 1.25 \ V$$

A check confirms that the transistor is in the saturation region, as assumed.
(*d*) Find V_{IH} by substituting Eq. (3.2d) in Eq. (3.2e):

$$(V_{IH} - V_T)^2 - \frac{V_{IH} - V_T}{kR_L} - \frac{2V_{DD}}{kR_L} = 0 \qquad V_{IH} = 2.5 \ V \text{ or } -0.7 \ V$$

Only the positive root is meaningful. V_{out} at this point is 0.75 V, so the transistor is in the linear region, as assumed.
(*e*) Finally, find V_M by using Eq. (3.3b):

$$V_M^2 - 1.5V_M - 1.5 = 0 \qquad V_M = 2.2 \ V \text{ or } -0.7 \ V$$

Only the positive root is physically meaningful.
(*f*) The noise margins are then

$$NM_L = V_{IL} - V_{OL} = 0.94 \ V \qquad NM_H = V_{OH} - V_{IH} = 2.5 \ V$$

Computer circuit simulators such as SPICE are very helpful for MOS circuit analysis and design work. Indeed, they are a virtual necessity for more complex circuits. Voltage transfer characteristics were computed using SPICE for Example 3.1, using R_L of 50 and 200 kΩ as well as 100 kΩ. The results are shown in Fig. 3.1d. For the case of $R_L = 100$ kΩ, the critical voltages as read from this plotted VTC agree with the results of hand analysis to within 0.1 V at each point.

The 100-kΩ resistor of the above example is needed to limit power consumption, but it would require a large amount of chip area if realized in a standard MOS process. Sheet resistances available in standard processes are in the range 20 to 100 Ω per square. Assuming 100 Ω per square and a resistor width of 5 μm, R_L would be 5,000 μm long and the area it would occupy would be more than 100 times that of the transistor. Of course, the required resistor value is inversely proportional to transistor W/L. But if transistor W/L is increased to reduce the size of R_L, transistor size as well as operating current will increase in proportion. For these reasons, conventional resistors are rarely used as loads in MOS digital circuits. A possible solution to this problem is use of small-area, high-valued resistors formed by additional special processing methods. While this approach is employed in certain static LSI memory components, it is not common for logic circuits because of relatively poor control of resistance value. The following section describes several alternative ways of using small transistors to perform the function of a load resistor.

3.2   TRANSISTORS AS LOAD DEVICES

Enhancement-mode NMOS transistors can be used as load elements operating in either saturation or linear regions. These were the only forms of transistor load in single-polarity MOS circuits before depletion-mode transistors became feasible through ion implantation in the early 1970s. Better circuit performance and smaller circuit area are obtained using depletion-mode NMOS transistors as load elements. In CMOS technology, active or current-source loads are obtained using a device of polarity opposite to that of the inverting amplifier. Pulsed (or clocked or dynamic) load devices are used in NMOS dynamic digital circuits.

3.2.1   Saturated Enhancement Load

A single NMOS transistor with gate connected to drain can be used as a load device, as shown in Fig. 3.2a. Note that the body is grounded because it is common to all transistors in a single chip. Because $V_{GS} = V_{DS}$, this load transistor can operate only in saturation or cutoff. Let us calculate the approximate value of k needed for an enhancement load transistor which will provide the same current for $V_{DD} = 5$ V and $V_{OL} = 0.3$ V as the 100-kΩ load resistor described above. For simplicity, we continue to ignore the small effect of λ and use Eq. (2.12) to represent the drain current in saturation. That is,

$$\frac{k_L}{2}(V_{GS} - V_T)^2 = \frac{V_{DD} - V_{OL}}{R_L} = 47 \ \mu\text{A} \qquad (3.4a)$$

Now substituting the known values (ignoring for simplicity the small body effect due to $V_{SB} = 0.3$ V), we obtain

$$k_L = \frac{2(47)}{(4.7 - 1.0)^2} = 6.9 \ \mu A/V^2 \tag{3.4b}$$

The load line construction and the VTC for the inverter with this enhancement load are shown in Fig. 3.2b and c.

This load transistor requires $W/L = 6.9/k' = 0.34$ and is larger in area than the inverter device. If the minimum allowed dimension is 5 μm, the inverting device assumed above will have $W/L = \frac{10}{5}$, while the load device will have $W = 5 \ \mu$m and $L = 5/0.34 = 15 \ \mu$m.

Note from Eq. (3.4) that the values of k and W/L scale linearly with I_D, as they do for the inverting transistor to which this load device is connected. Because of this fact, W/L (sometimes known as the *aspect ratio*) for each transistor may be multiplied by the same factor without affecting voltage levels. Of course, the operating current will be multiplied by the same factor. The geometry ratio K_R for a MOS inverter is defined as

$$K_R = \frac{(W/L)_{inverter}}{(W/L)_{load}} = \frac{k_I}{k_L} = \frac{k'(W/L)_I}{k'(W/L)_L} \tag{3.5}$$

Some books use the term *beta ratio* (β_R) in place of the K_R just defined. We avoid this terminology to avoid confusion with the current gain *beta* of a bipolar transistor.

If the designer is free to adjust the operating current of inverters and gates, a minimum-area layout will usually be achieved with device sizes chosen so that the geometric mean of inverter and load W/L's is unity. It should be clear also that reducing the value of K_R reduces circuit area.

One serious deficiency of the enhancement load has been overlooked so far. The output high level V_{OH} is not equal to V_{DD} as it was for the resistor load. The load transistor ceases to conduct after its gate-source voltage decreases to the threshold voltage. In this case, the result is that the output node does not rise above $V_{DD} - V_{TL}$. The threshold voltage of the load device is no longer V_{T0}, as it is for the inverter device, because the full output voltage appears as a body bias between source and body of the load device. The threshold voltage is now given by Eq. (2.6), and $V_{SB} = V_{OH}$ is on the order of 3.5 V based on the parameters given above. A recomputation shows that the value of K_R needed to achieve $V_{OL} = 0.3$ V is considerably increased if V_{in} is reduced from 5 to 3.5 V. These circumstances make it difficult to design simple enhancement load static inverters and gates which will operate with safe noise margins on a 5-V supply. This is one important reason why early MOS circuits required higher voltage supplies.

Example 3.2. An NMOS inverter with a saturated enhancement load, as shown in Fig. 3.2, operating from $V_{DD} = 12$ V, is to be analyzed to determine its nominal voltage transfer characteristics and noise margins. $K_R = 15$, $V_{T0} = 1.5$ V, $2\phi_F = 0.6$ V, and $\gamma = 0.37$ V$^{1/2}$.

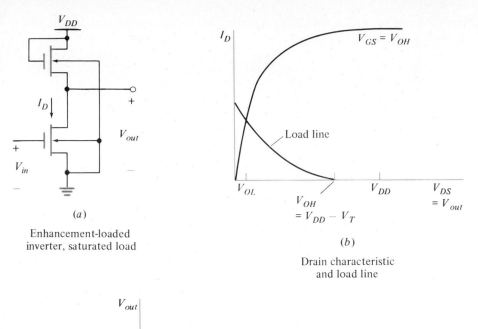

(a)

Enhancement-loaded
inverter, saturated load

(b)

Drain characteristic
and load line

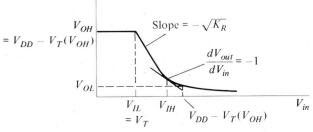

(c) Voltage transfer characteristic

FIGURE 3.2
NMOS inverter, saturated enhancement load.

(a) First find V_{OH}. Remember that the output node can only rise to a threshold drop below V_{DD} before the load device ceases to conduct.

$$V_{OH} = V_{DD} - V_T(V_{OH})$$

$$= V_{DD} - [V_{T0} + \gamma(\sqrt{V_{OH} + 2\phi_F} - \sqrt{2\phi_F})]$$

Use a substitute variable, $x = \sqrt{V_{OH} + 2\phi_F}$, and solve the quadratic equation in x. Only the positive root is meaningful from a physical standpoint. The result is $V_{OH} = 9.6$ V. The output voltage cannot rise above this value because the load transistor ceases to conduct.

(b) Find V_{OL}. Note that the currents in the two transistors are equal in the steady state. V_{OL} is the output voltage from this inverter when its input voltage is V_{OH}, the output high level of another (identical) inverter. For this point on the VTC and any additional ones, it is necessary to solve an expression formed by

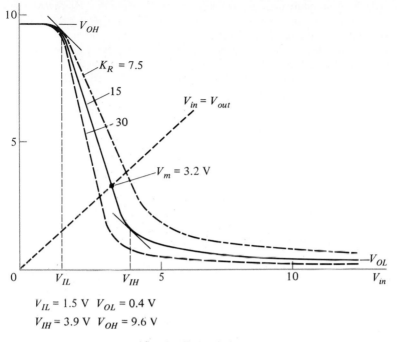

$V_{IL} = 1.5$ V $V_{OL} = 0.4$ V
$V_{IH} = 3.9$ V $V_{OH} = 9.6$ V

(*d*) SPICE simulation

FIGURE 3.2 (*cont.*)
NMOS inverter, saturated enhancement load.

equating the drain currents of the lower and upper (inverter and load) transistors. Subscripts I and L are added to current and voltage variables to distinguish between inverter and load devices. For correct results, we need to know which form of the drain current equation (e.g., linear or saturated) to use for each transistor.

When conducting, the load transistor in this circuit is always saturated, since $V_{DS} = V_{GS}$. As for the inverter transistor, with $V_{GS} = 9.6$ V the output voltage should be lower than V_{T0} if there is to be any noise margin. Therefore, our initial assumption (to be confirmed later) is that the lower transistor is in the linear region with $V_{DS} < V_{GS} - V_T$. Equate the drain currents, using the appropriate equations:

$$I_{DI(lin)} = I_{DL(sat)}$$

Since only K_R is given, we may write

$$K_R[2(V_{in} - V_{TI})V_{out} - V_{out}^2] = [V_{DD} - V_{out} - V_{TL}(V_{out})]^2$$

This expression is correct whenever the load is saturated and the inverter is in the linear region. But note that the load transistor threshold voltage V_{TL} is modified by the body effect, due to the voltage V_{out} which appears between its source and the (grounded) common substrate. When Eq. (2.6) is used to evaluate $V_{TL}(V_{out})$ in the above, the resulting expression is a fourth-order polynomial. Direct solution by hand to find V_{OL} is tedious and susceptible to error. Simulation using SPICE or an equivalent computer program is a desirable alternative. If hand analysis is necessary, choose several values of V_{out}, substitute in the above equation, and solve the resulting linear first-order equation in V_{in}.

In a well-designed inverter V_{OL} will be about 5% of V_{DD}. For output voltages of 0.8, 0.6, 0.5, 0.4, and 0.3 V, the corresponding values of input voltage for this inverter are 5.7, 7.1, 8.3, 10.1, and 13.1 V, respectively. After one or two iterations, we have $V_{out} = V_{OL} = 0.42$ V when $V_{in} = V_{OH} = 9.6$ V.

(c) Find V_{IL}. The values of V_{in} at which $dV_{out}/dV_{in} = -1.0$ are needed for definition of the noise margins. From Eq. (3.2a),

$$\frac{dV_{out}}{dV_{in}} = -\frac{dI_D}{dV_{in}}(r_{sdL} \parallel r_{dsI}) = -1.0$$

Since V_{IL} is usually only slightly greater than V_{TI}, we initially assume (and later confirm) that the inverter device is saturated. The load device is also saturated. For this case, a good assumption is that $r_{dsI} \gg r_{sdL}$. (Again, this should be checked later.) Hence,

$$\frac{dV_{out}}{dV_{in}} = -\frac{dI_{DI}}{dV_{GSI}}\frac{dV_{DSL}}{dI_{DL}} = -1.0$$

but $V_{DSL} = V_{GSL}$; therefore, with both devices saturated,

$$\frac{dV_{out}}{dV_{in}} = -\frac{k_I(V_{GSI} - V_{TI})}{k_L(V_{GSL} - V_{TL})} = -\frac{\sqrt{2k_II_{DI}}}{\sqrt{2k_LI_{DL}}} = -\sqrt{K_R}$$

The conclusion is that the slope of the VTC is a constant, equal to the negative of the square root of K_R, whenever both transistors are saturated. In this example, $\sqrt{K_R} = 3.9$. Consequently, V_{IL} is reached immediately when the lower transistor begins to conduct, or at $V_{TI} = 1.5$ V.

(d) Find V_{IH}. At V_{IH}, V_{out} is approaching V_{OL}, so we assume the inverter transistor is now in the linear region. The load transistor is still saturated. We now assume $r_{dsI} \ll r_{sdL}$. Consequently, the earlier analysis leading to Eq. (3.2d) is still applicable. In this case, simultaneous solution of two equations is needed.

$$I_{DI(lin)} = I_{DL(sat)}$$

$$V_{out} = \frac{V_{in} - V_{TI}}{2}$$

Substituting for V_{out} in both $I_{DI(lin)}$ and $I_{DI(sat)}$, we finally obtain

$$V_{IH} = \frac{2(V_{DD} - V_{TL})}{\sqrt{3K_R} + 1} + V_{TI}$$

The above analysis neglects the shift in V_{TL} due to body effect. However, the result is only a weak function of V_{TL}. Hence if we "guess" that $V_{TL} = 2.0$ V, we find $V_{IH} = 4.1$ V. The corresponding output voltage $V_{out} = 1.8$ V. An iteration to correct V_{TL} changes V_{IH} by less than one-tenth of a volt.

(e) The midpoint voltage V_M is found by noting that in this condition both transistors are saturated, with $V_{GS} = V_{DS}$ for each. Equating drain currents for saturated devices,

$$I_{DI(sat)} = I_{DL(sat)}$$

$$K_R(V_{out} - V_{TI})^2 = [V_{DD} - V_{out} - V_{TL}(V_{out})]^2$$

Solving for V_{out} in terms of known parameters,

$$V_{out} = \frac{V_{DD} - V_{TL}(V_{out}) + V_{TI}\sqrt{K_R}}{1 + \sqrt{K_R}}$$

Again we have the problem that V_{TL} is a function of V_{out}. Exact solution would involve a higher-order polynomial. If we guess that at this point $V_{out} = (V_{OH} + V_{OL})/2 = 5$ V, we find that $V_{TL}(V_{out}) = 2.1$ V, and

$$V_{out} = V_M = 3.2 \text{ V}$$

Another iteration to correct the value of V_{TL} changes this result by less than one-tenth of a volt.

Figure 3.2*d* shows the output from a SPICE analysis for the VTC of this inverter, as well as variations for which $K_R = 7.5$ and 30. The SPICE results for $K_R = 15$ are tabulated in parentheses below. They are seen to be very close to the results from the hand solution.

Summarizing the results of this analysis, we have

$$V_{OH} = 9.6 \text{ V } (9.6) \qquad V_{OL} = 0.4 \text{ V } (0.42)$$

$$V_{IH} = 4.1 \text{ V } (3.9) \qquad V_{IL} = 1.5 \text{ V } (1.5)$$

$$NM_H - 5.5 \text{ V } (5.7) \qquad NM_L - 1.1 \text{ V } (1.1)$$

$$V_M = 3.2 \text{ V } (3.2)$$

Exercise 3.1. Calculate $V_{OH}, V_{OL}, V_{IH}, V_{IL}$, and noise margins for an inverter with a saturated enhancement load as shown in Fig. 3.2*a*, operating from $V_{DD} = 5$ V. Use $V_{T0} = 1.0$ V, $2\phi_F = 0.6$ V, $\gamma = 0.37$ V$^{1/2}$, and $K_R = 10$.

Exercise 3.2. Calculate the nominal value of K_R needed to obtain $V_{OL} = 0.2$ V in the inverter of Fig. 3.2*a*, assuming $V_{DD} - 5$ V, $V_{I0} = 1$ V, $V_{IH} - 3$ V, and the same values of γ and $2\phi_F$ as used above. Remember that V_{OL} is present for $V_{in} = V_{OH}$.

3.2.2 Linear Enhancement Load

The output high level V_{OH} of the previous example can be increased simply by connecting the gate of the load transistor to a dc voltage V_{GG} greater than V_{DD}. This is shown in Fig. 3.3. The desirable value for V_{GG} is given by

$$V_{GG} > V_{DD} + V_T(V_{DD})$$

where the last term is simply the value of V_T with a body bias of V_{DD}. When this condition is met, the load device operates in the linear region over the entire range of V_{out}, since throughout this range the load device is operating with $V_{DS} < V_{GS} - V_T$. Figure 3.3*b* shows the load line construction for this type of inverter. The names *linear load, nonsaturated load,* or *triode load* are applied to this mode of operation. Despite the first name, there remains a noticeable curvature to the load line, as seen in Fig. 3.3*b*.

The linear enhancement load has several disadvantages when used in static inverters and gates. More chip area is required, since an extra voltage source V_{GG}

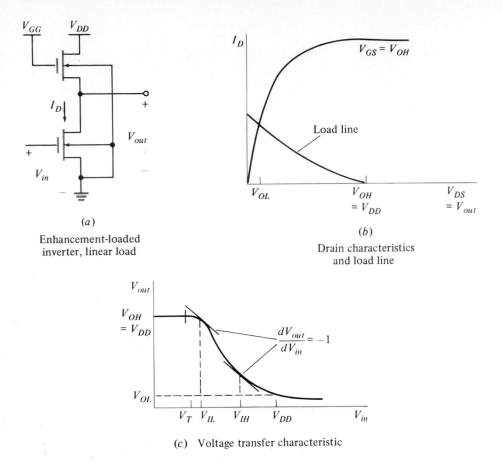

(a)

Enhancement-loaded
inverter, linear load

(b)

Drain characteristics
and load line

(c) Voltage transfer characteristic

FIGURE 3.3
NMOS inverter, linear enhancement load.

with associated additional interconnections on the chip is needed. The required value of K_R is even larger than for a saturated enhancement load. Some modern dynamic memory circuits use linear enhancement loads on a pulsed basis. A dynamic design technique known as *bootstrapping* eliminates the need for an additional supply.

3.2.3 Depletion Load

Depletion transistors as load elements in static NMOS circuits overcome the disadvantages of the alternatives described above at the relatively minor expense of a special masked ion implantation processing step to create the depletion devices. So-called *enhancement-depletion (E-D)* NMOS technology is the basis for most modern microprocessors, microprocessor peripheral devices, and static NMOS memories.

The E-D inverter is illustrated schematically in Fig. 3.4a. The load line construction of Fig. 3.4*b* shows the *I-V* characteristic of ideal and practical depletion load devices. The differences between these two deserve comment.

A constant-current source would be a truly ideal load device because the full static load current (47 μA in Example 3.2) would be available to charge the load capacitance from V_{OL} to V_{OH} when the inverter input changes from high to low. An ideal depletion device with gate connected to source ($V_{GS} = 0$) approaches current-source performance. Suppose we chose a load device with $W/L = 1$ to minimize area and set the depletion threshold voltage V_{TD} to provide $I_D = 47$ μA at $V_{OL} = 0.3$ V. The required value is $V_{TD} = -2.2$ V. In the absence of body effect ($\gamma = 0$) and channel-length modulation ($\lambda = 0$), this depletion device will provide the upper load line shown in Fig. 3.4*b*. The load device remains in saturation until its V_{DS} falls to $V_{GS} - V_{TD} = 2.2$ V. This occurs at $V_{out} = 2.8$ V, assuming a 5-V supply. Above this output voltage, I_D of the load device decreases because it enters the linear region. However, note that the load device remains conducting all the way to $V_{out} = V_{DD}$. Therefore $V_{OH} = V_{DD}$ as desired.

Now consider a practical depletion load transistor with normal body effect. With a typical body-effect coefficient $\gamma = 0.37$, there is a substantial degradation in load device characteristics. First, the body effect makes the depletion transistor threshold voltage more positive. For the parameters in this example, $V_{TD}(5)$ is shifted by 0.6 V under nominal conditions compared to its value with zero body bias. In order to provide a load current that is not excessively sensitive to normal production variations in V_{TD} and γ, the design center value for $V_{TD}(0)$ is set at about -3.0 V during fabrication. We assume this value for further calculations.

At $V_{OL} = 0.3$ V, this depletion load with a 3.0-V threshold suffers negligible threshold shift from body effect. Calculation of the saturated drain current shows that this load transistor provides considerably more current per unit W/L than the ideal device with $V_T = -2.2$ V. The result is that W/L should be decreased to less than 1 for the load of a minimum-area inverter. A typical value for practical design is $K_R = 4$. Figure 3.4*d* shows the results of SPICE simulations for depletion load inverter VTC with $K_R = 2$, 4, and 8.

The *I-V* characteristics for the four types of load device for NMOS discussed above are compared by means of SPICE simulation results in Fig. 3.5*a*. The four load devices were adjusted so that each supported a current of 50 μA when connected between 5 V and ground. Note that the load line for the depletion device falls far short of an idealized constant-current source.

The consequences of body effect in depletion load devices are quite significant. While there is negligible body effect at low output voltages, as the output voltage rises the load transistor threshold voltage becomes less negative. Therefore, the drain current for the depletion load decreases as V_{out} rises, even after this load device enters the saturation region. The *I-V* characteristics of Fig. 3.5*a* show that a linear resistor more closely approximates the depletion load than does an ideal current source.

The voltage transfer characteristics for NMOS inverters employing the four load devices characterized in Fig. 3.5*a* were simulated using SPICE. The results are shown in Fig. 3.5*b*. Note that the depletion load produces an inverter with the most nearly ideal VTC.

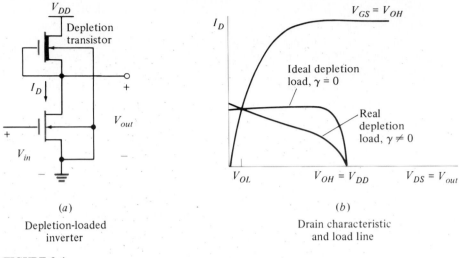

(a)

Depletion-loaded
inverter

(b)

Drain characteristic
and load line

FIGURE 3.4
NMOS inverter, depletion load.

Example 3.3. Consider a small-area E-D inverter, shown schematically in Fig. 3.4, with $(W/L)_I = 2$ and $(W/L)_L = 0.5$ and operating from a 5-V supply. Assume the following device parameters:

$$V_{TE} = 1.0 \text{ V} \qquad V_{TD} = -3.0 \text{ V}$$

$$k' = 20 \text{ } \mu\text{A/V}^2 \qquad \gamma = 0.37 \text{ V}^{1/2}$$

$$2\phi_F = 0.6 \text{ V}$$

(a) Confirm that V_{OH} is 5 V by calculating the threshold voltage for the depletion load device when source-to-body voltage is 5 V. Use Eq. (2.6); the result is $V_{TD}(5) = -2.4$ V. This means that the load device with gate connected to source has a conducting channel present for any output voltage between 0 and 5 V.

(b) Find V_{OL} by equating drain currents of inverter and load transistors. V_{GS} for the load device is zero. For the inverter device, $V_{GS} = V_{in} = V_{OH} = 5$ V. With the output voltage low, the inverter is in the linear region; the load device, as long as $V_{DS} > 0 - V_{TD}$, is saturated.

$$\frac{k_I}{2}[2(V_{OH} - V_{TE})V_{OL} - V_{OL}^2] = \frac{k_L}{2}[0 - V_{TD}(V_{OL})]^2$$

Instead of calculating several values of V_{in} from several values of V_{out} (as shown in Example 3.2), it is often quicker to assume a nominal value of V_{OL} which is used to evaluate $V_{TD}(V_{out})$. Body effect is small for small output voltages; at worst, one iteration is all that is required to obtain V_{OL} accurate to 0.1 V. Assuming $V_{OL} = 0.4$ V, $V_{TD} = -3.0 + 0.1 = -2.9$ V. Now the above equation may be solved; the result is $V_{OL} = 0.3$ V. Iteration does not significantly change this result. Since k_L is known, the drain current may be calculated from Eq. (2.12):

$$I_{DL} = I_{DI} = \frac{20}{2}(0.5)(0 + 2.9)^2 = 42 \text{ } \mu\text{A}$$

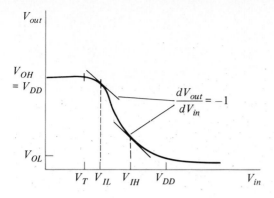

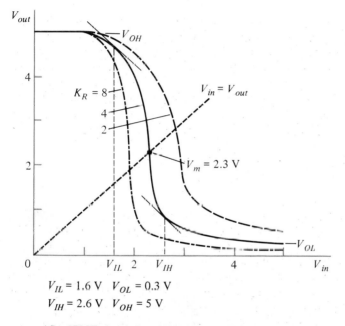

(c) Voltage transfer characteristic

$V_{IL} = 1.6$ V $V_{OL} = 0.3$ V
$V_{IH} = 2.6$ V $V_{OH} = 5$ V

(d) SPICE simulation NMOS inverter, depletion load

FIGURE 3.4 (*cont.*)
NMOS inverter, depletion load.

(c) Find V_{IL}. We again make use of Eq. (3.2*a*), but with the load transistor in the linear region and the inverting transistor in the saturation region:

$$\frac{dV_{out}}{dV_{in}} = -\frac{I_{DI}}{dV_{GSI}}\frac{dV_{DSL}}{dI_{DL}} = -1.0$$

$$= -\frac{k_I(V_{GSI} - V_{TI})}{k_L(V_{GSL} - V_{TL} - V_{DSL})} = -1.0$$

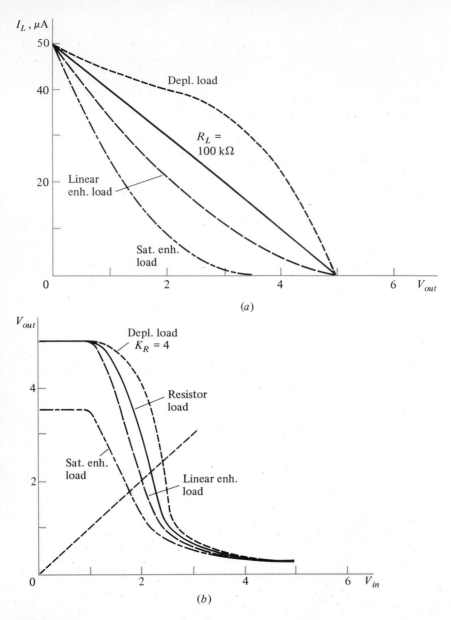

FIGURE 3.5
(*a*) Comparison of *I-V* characteristics for load devices. (*b*) Comparison of NMOS VTC for four load devices.

If we recognize that $V_{DSL} = V_{DD} - V_{out}$, we may solve for V_{out}:

$$V_{out} = K_R(V_{GSI} - V_{TI}) - V_{GSL} + V_{TL} + V_{DD}$$

Note that $V_{GSL} = 0$ V; but V_{TL} is a function of V_{out}. Hence some iteration may be required, but solving this equation simultaneously with the drain current equations,

$$I_{DL(lin)} = I_{DI(sat)}$$

the results are

$$V_{IL} = 1.5 \text{ V} \qquad \text{and} \qquad V_{out} = 4.7 \text{ V}$$

From a SPICE simulation of this inverter (Fig. 3.4d), we obtain

$$V_{IL} = 1.6 \text{ V} \qquad \text{at } V_{out} = 4.7 \text{ V}$$

(d) Find V_{IH}. As above in Eqs. (3.2c and d), the VTC for a saturated load device and linear inverter device has a slope of -1 when $V_{out} = (V_{in} - V_T)/2$ is satisfied simultaneously with the linear region drain current equation for the inverter transistor. We assume that V_{out} will still be low so that the drain current will not have changed significantly.

$$I_{DI} = \frac{20}{2}(2)[2(V_{IH} - V_{TI})V_{DS} - V_{DS}^2]$$

$$= 20[2(2V_{DS})V_{DS} - V_{DS}^2] = 42 \ \mu A$$

Solving,

$$V_{DS} = V_{out} = 0.84 \text{ V} \qquad V_{IH} = 2.7 \text{ V}$$

Using SPICE, $V_{IH} = 2.6$ V at $V_{out} = 0.86$ V. Finally, the noise margins are

$$NM_L = 1.5 - 0.3 = 1.2 \text{ V} \qquad NM_H = 5.0 - 2.7 = 2.3 \text{ V}$$

Exercise 3.3. Calculate the drain current at $V_{out} = 1, 2, 3,$ and 4 V for a depletion load device which has $I_D - 42 \ \mu A$ at $V_{OL} = 0.2$ V. Assume $2\phi_F = 0.6$, $\gamma = 0.37$, and $V_{TD} = -3.0$ V at zero body bias.

3.3 CIRCUIT LAYOUT AND CAPACITANCES

Before we can analyze the transient characteristics of MOS inverters, we must know the actual physical dimensions of the transistors and their interconnections so that the capacitances which limit switching speed can be calculated. For manual analysis, device dimensions together with the capacitance per unit area for junctions (see Sec. 4.3) and oxides [see Eq. (2.8)] permit calculation of all capacitances. Circuit simulators such as SPICE usually require entry of dimensions so that the program may calculate capacitances on circuit nodes.

3.3.1 Layout Example and SPICE Input Data

The essential features of a practical NMOS silicon-gate inverter circuit layout are shown in Fig. 3.6. As is usual in integrated circuits work, vertical dimensions are exaggerated compared to horizontal dimensions so that small details of vertical structure may be seen. Horizontal dimensions are approximately to the scale shown in Fig. 3.6. For clarity, the final layer of metal interconnections is not drawn in this figure.

Several features of Fig. 3.6 deserve comment. The drain of the inverter shares a common n^+ diffused region with the source of the load device. This saves area over using separated diffusions, each with its own contact. For the depletion load circuit shown here, the gate of the load device must be connected to its source by metal. The output of this inverter may be connected to the following circuitry using diffusion or polysilicon or metal, since all of these are in contact with the output node. Note that diffused and polysilicon conductors cannot cross, because the intersection of these layers forms a transistor. Often ground and/or

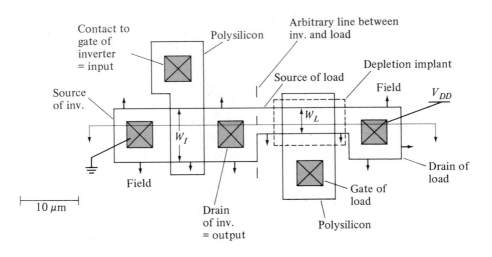

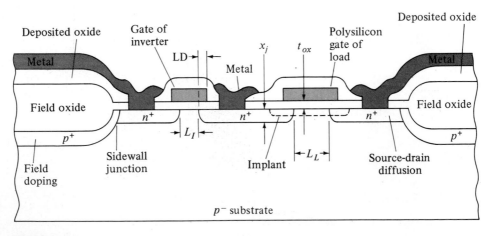

FIGURE 3.6
Layout and section view: NMOS inverter.

V_{DD} nodes are carried a short distance through a diffused path before making contact with metal.

Figure 3.6 shows gate widths of 10 and 5 μm for inverter and load devices, respectively. These are the distances between the walls of the field diffusion and are entered on SPICE device cards as channel width W. Also seen in Fig. 3.6 are gate lengths of 7 and 12 μm for inverter and load devices, respectively. These are assumed to be actual dimensions of the polysilicon gate electrodes after fabrication, and are entered on SPICE device cards as channel length L. During fabrication, the n^+ source and drain diffuse toward each other *(lateral diffusion)* under the gate electrode, resulting in an overlap of the nominally self-aligned gate above source and drain, and an electrical channel length shorter than the gate length. For this example, we assume 1 μm of lateral diffusion at each side (entered as LD on SPICE.MODEL card), so final electrical channel lengths are 5 and 10 μm. Gate-source and gate-drain overlap are 1 μm each. Gate length and channel length are frequently spoken about interchangeably, even though lateral diffusion is usually significant and should not be neglected.

The SPICE program calculates all device capacitances provided the necessary data is entered. On the .MODEL card must be entered values for CJ and CJSW, the zero-bias capacitances for the bottom and sidewalls of the source-body and drain-body junctions. The capacitance C_{j0} is the zero-bias depletion-layer capacitance derived in Sec. 4.3.4. From Eq. (4.3.8), the capacitance per unit area of an abrupt n^+p junction is

$$C_{j0} = \sqrt{\frac{q\epsilon_{si}N_A}{2\phi_0}} = CJ \tag{3.6}$$

The sidewall capacitance per unit area is higher than CJ because the n^+ source and drain abut the p^+ field diffusion. Unless other data are available, we assume that field doping is 10 times higher than body doping, so sidewall capacitance per unit area is higher by the square root of 10 (abrupt junction assumed). Finally, SPICE requires that CJSW be given per unit of diffusion perimeter. Therefore, sidewall capacitance per unit area is multiplied by junction depth XJ (X_j) to obtain the value for CJSW to be entered on the .MODEL card.

The built-in junction potential PB (ϕ_0) is derived from physical parameters in Sec. 4.2. From Eq. (4.2.12),

$$\phi_0 = V_T \ln \frac{N_A N_D}{n_i^2} = PB \tag{3.7}$$

SPICE takes the same value of PB for bottom and sidewall junctions, causing only a very minor error. It is required that TOX (t_{ox}) be entered on the .MODEL card if SPICE is to calculate the gate capacitances.

To summarize, the minimum set of data to enter on the .MODEL card for MOS devices comprises the dc parameters VTO, KP, and GAMMA (from measured data), gate oxide thickness TOX, and the capacitance parameters CJ, CJSW, and PB. Other parameters may take on default values without serious errors.

To complete the required input for SPICE, device cards must be prepared with data on source and drain areas and perimeters. Where two devices share a common node (e.g., drain of inverter, source of load), it does not matter how node area and perimeter are allocated between the devices. The dashed line across the common diffusion in Fig. 3.6 shows a reasonable dividing line between the two transistors.

Example 3.4. Prepare the .MODEL cards and device cards for the inverter shown in Fig. 3.6. The following data are available:

Name	Text symbol	Value	SPICE name
Threshold voltages	V_{TE}	$= +1.0$ V	VTO
	V_{TD}	$= -3.0$ V	VTO
Transconductance	k'	$= 20 \ \mu\text{A/V}^2$	KP
Body factor	γ	$= 0.37$	GAMMA
Body doping	N_A	$= 5 \times 10^{14} \ \text{cm}^3$	NSUB
Source, drain doping	N_D	$= 1 \times 10^{20} \ \text{cm}^3$	
Gate oxide thickness	t_{ox}	$= 0.1 \ \mu\text{m}$	TOX
Junction depth	X_j	$= 1.0 \ \mu\text{m}$	XJ
Lateral diffusion		$= 1.0 \ \mu\text{m}$	LD

Capacitances per unit area may now be calculated in the form required for input to SPICE. Field doping along junction sidewalls is assumed to be 10 times greater than substrate doping, so capacitance per unit area is larger by the square root of 10. The sidewall capacitance C is stated *per meter of perimeter*. It is calculated using the given junction depth of 1 μm. The gate and drain overlap capacitances *per meter of gate width* are calculated from the given gate oxide capacitance and an assumed 1-μm lateral diffusion.

Symbol	Value	Where from?	SPICE name
ϕ_0	$= 0.86$ V	Eq. (3.7)	PB
C_{j0}	$= 70 \ \mu\text{F/m}^2$	Eq. (3.6)	CJ
C_{jsw}	$= 220$ pF/m	$X_j \sqrt{10} \ C_{j0}$	CJSW
C_{ox}	$= 345 \ \mu\text{F/m}^2$	Eq. (2.8)	
$C_{ox}LD$	$= 345$ pF/m		CGSO = CGDO

Areas and perimeters are calculated from Fig. 3.6.

Inverter:

$$\text{Source, drain areas} \ = \ 10 \ \mu\text{m} \times 10 \ \mu\text{m} \ = \ 100 \ \text{pm}^2$$

$$\text{Source perimeter} \ = \ 4 \times 10 = 40 \ \mu\text{m}$$

$$\text{Drain perimeter} \ = \ 10 + 10 + 10 + 5 = 35 \ \mu\text{m}$$

Load:

$$\text{Source area } = 5 \times 5 = 25 \text{ pm}^2$$

$$\text{Drain area } = 10 \times 10 = 100 \text{ pm}^2$$

$$\text{Source perimeter } = 5 + 5 + 5 = 15 \ \mu m$$

$$\text{Drain perimeter } = 4 \times 10 = 40 \ \mu m$$

Several comments are needed to explain the above. The source of the inverter and the drain of the load do not change in voltage, so capacitances from body to these elements do not limit performance. Nevertheless, data concerning these elements are entered so no error will be made if the circuit is later changed. The way in which junction area is assigned between inverter and load transistors for the shared output node does not affect the results. The dashed line marked at the top of Fig. 3.6 shows how the division was made in this case. The high value of sidewall capacitance is not present along the edge of diffusion bordering on active channels, but for simplicity the entire perimeter is multiplied by this capacitance value.

SPICE input data:

.MODEL NE NMOS (VTO=1.0 KP−20E-6 GAMMA=0.37
| NSUB=5E14 TOX=0.1U XJ=1.0U LD−1.0U
+CJ=70U CJSW=220P CGSO=345P CGDO=345P)
.MODEL ND NMOS (VTO=-3.0 KP=20E-6 GAMMA=0.37
+ NSUB=5E14 TOX=0.1U XJ=1.0U LD=1.0U
+ CJ=70U CJSW=220P CGSO=345P CGDO=345P)

M1 2 1 0 0 NE W=10U L=7U AD=100P PD−35U AS=100P PS=40U
M2 3 2 2 0 ND W−5U L=12U AS=25P PS=15U AD=100U PD=40U

3.3.2 Capacitance Calculations for Hand Analysis

A number of simplifications are necessary to facilitate hand analysis of MOS digital circuits because of the many nonlinear dc parameters and nonlinear capacitances in even a simple inverter. A useful approach to hand analysis will now be described.

As explained in Sec. 2.5, each MOS transistor has five separate voltage-dependent capacitances coupling its four electrodes. Manual analysis of MOS transistor circuits in which each capacitor is considered individually is virtually impossible. However, approximate calculations of switching times becomes feasible if all capacitance effects are lumped into a single total capacitor C_T which is connected to the output node of each inverter or gate.

Voltage-dependent effects of junction capacitances are removed by defining equivalent linear capacitances C_{eq} which require the same change in charge as the nonlinear capacitors for a transition between two voltage levels V_1 and V_2. With $V_2 > V_1$,

$$C_{eq} = \frac{\Delta Q}{\Delta V} = \frac{Q(V_2) - Q(V_1)}{V_2 - V_1} \equiv K_{eq} C_{j0} \tag{3.8}$$

The depletion-layer capacitance per unit area C_{j0} is calculated according to Eq. (3.6). Sidewall capacitances cannot be ignored for modern MOS processes; sidewall capacitance per unit area is calculated as described above. The dimensionless parameter K_{eq} is derived in Sec. 4.3.6. From Eq. (4.3.14), for an abrupt junction,

$$K_{eq} = \frac{-2\phi_0^{1/2}}{V_2 - V_1} \left[(\phi_0 - V_2)^{1/2} - (\phi_0 - V_1)^{1/2} \right] \tag{3.9}$$

Note that the voltage applied to the junction is $V_2 = -(V_{OL} - V_{BB})$ in the low state and $V_1 = -(V_{OH} - V_{BB})$ in the high state. V_{BB} is the (zero or negative) *body bias* voltage applied to the body with respect to the sources of inverter transistors. By convention, voltages applied to junctions are defined as positive for forward bias and negative for reverse bias.

Figure 3.7 shows calculated values of K_{eq} as a function of supply voltage V_{DD} and body voltage V_{BB}, on the assumption that $V_{OL} = 0$, $V_{OH} = V_{DD}$, and $\phi_0 = 1$ V. These data are used in calculating C_T. The result of the linearization of junction capacitances using K_{eq} is a minor distortion of the shape of transient voltage waveforms at circuit nodes. Typically, logic propagation delays are not significantly changed by this approximation.

Figure 3.8 shows how the device capacitances in a circuit comprising two cascaded inverters can be lumped at the inverter output nodes. First, all capacitances across which there is no voltage change are ignored, since they have no effect on circuit performance. The device capacitances which must be considered are shown in Fig. 3.8a. In this figure, the capacitances C_L represent the capacitance of interconnecting wiring connected to inverter outputs.

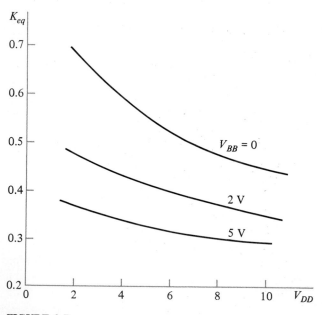

FIGURE 3.7
Equivalent capacitance calculation. Lumped linear equivalent capacitance. $C_{eq} + K_{eq}C_{j0}$.

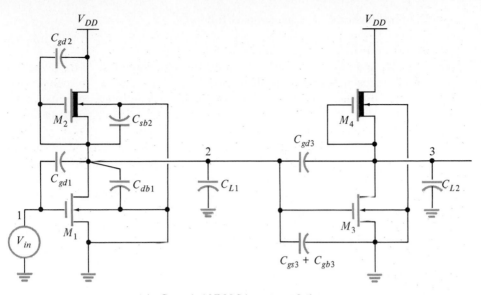

(a) Cascaded NMOS inverters. Only the capacitances to be included in the hand analysis are drawn

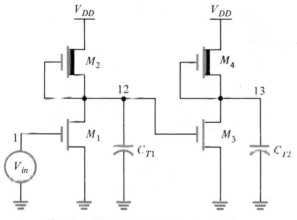

(b) Simplified circuit for hand analysis

FIGURE 3.8
Method of lumping capacitances to inverter output nodes. Numbered nodes (2, 12, etc.) are keyed to Fig. 3.9.

The capacitance C_{T1} is made up as follows:

$$C_{T1} = K_{eq}(C_{db1} + C_{sb2}) + C_{gd1} + C_{gd2} + C_{g3} + C_L \qquad (3.10)$$

Transistors M_1 and M_2 are saturated or cut off during a large part of the switching cycle. Hence capacitances C_{gd} for these two devices are defined to include only the gate overlap capacitance to the drain of each. The total gate capacitance C_{g3} loads the output of the first inverter, so it is included without breaking it into separate components.

The major consequence of lumping all capacitance to ground is that the effects of capacitive coupling between input and output are ignored. This simplification is necessary for hand calculations. The effects of direct input-output coupling, which are serious particularly for dynamic circuits, can be best studied by computer simulation.

Specific values of all capacitances are found from calculated values of capacitance per unit area and areas of nodes determined from the circuit layout at hand. In addition to the planar areas which are obvious in Fig. 3.6b, sidewall areas where source and drain diffusions meet field doping (as shown in Fig. 3.6a) are very significant in modern circuits. Sidewall capacitance per unit area is typically 3 to 5 times greater than capacitance along the bottom of the source-drain diffusion because the doping in the field regions is typically 10 to 25 times greater. Adequate accuracy is achieved by taking the sidewall area as the product of diffusion perimeter and junction depth, neglecting the curvature of the sidewall

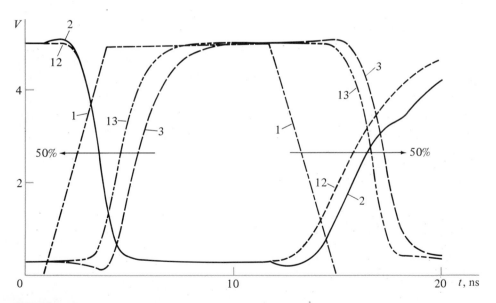

FIGURE 3.9
SPICE simulation of transient response for two cascaded NMOS inverters of Fig. 3.8. Two cases:
(a) full model—all nonlinear capacitances included (2 and 3); (b) lumped model—one capacitance at output (12 and 13).

and the gradient in field doping. The capacitance for small areas of metal or polysilicon (e.g., as in Fig. 3.6) outside transistor regions, including the gate-body overlap capacitance CGBO, may be neglected.

Example 3.5. Calculate the lumped equivalent capacitances C_{T1} and C_{T2} for the cascade connection of the two NMOS inverters shown in Fig. 3.8b. Use SPICE to compare the transient response of the lumped-capacitance circuit with the circuit with completely modeled capacitances. Assume each inverter has a layout as shown in Fig. 3.6.

The layout and data from Example 3.4 may now be used to calculate C_{T1} and C_{T2} for the circuit of Fig. 3.8b.

$$C_{T1} = K_{eq}(C_{db1} + C_{sb2}) + C_{gd1} + C_{gd2} + C_{g3} + C_{L1}$$
$$= 0.54(14.7 + 5.05) + 3.45 + 1.73 + 24.2 + 0 = 40.05 \text{ fF}$$

$$C_{T2} = K_{eq}(C_{db3} + C_{sb4}) + C_{gd3} + C_{gd4} + C_{L2}$$
$$= 0.54(14.7 + 5.05) + 3.45 + 1.73 + 0 = 15.85 \text{ fF}$$

Units of capacitance are femtofarads (fF), equal to 10^{-15} F. Note that interconnect capacitances C_L are set equal to zero in this example and that the second inverter is loaded only by its own parasitic capacitances. In practice, C_L is usually the largest part of C_T, and it is less voltage-dependent and more easily approximated than the device capacitances. Because only device capacitances are included in this example, it is a severe test of model accuracy.

The accuracy of transient analysis using this lumped-capacitance model was checked by using program SPICE to simulate the propagation delays of this two-inverter circuit. Two separate circuits were simulated simultaneously using identical dc device equations and dc parameters for both cases. The first simulation allowed SPICE to compute and use all nonlinear capacitances for every transistor (full model). In the second case, capacitance calculations by the program were deliberately forced to small values or zero. Instead, the lumped linear capacitances C_{T1} and C_{T2}, calculated above, were added to the second circuit as separate elements (lumped model).

Figure 3.9 shows the computed transient response for the two cases. Note the very close alignment of results; comparative propagation delays are listed in Table 3.1. The difference in propagation delays, measured at the 50% points on voltage transitions, is 20% or less in most cases. The full model and lumped model were compared also for the case of one inverter driving inputs of 10 identical inverters in parallel; again there was close agreement. These results are good considering the extreme simplifications in the lumped-capacitance model.

TABLE 3.1
Transient performance

Propagation delay	Full model, ns	Lumped model, ns
t_{PHL}	1.1	1.1
t_{PLH}	3.1	2.4

3.4   SWITCHING TIME ANALYSIS AND POWER-DELAY PRODUCT

The currents available to charge and discharge C_T are the drain currents of the inverter and load transistors. The drain current of the load transistor is a function of V_{out}. The drain current of the inverter transistor is a function of both V_{in} and V_{out}. The true situation, in which V_{in} and V_{out} are both changing with time while satisfying the nonlinear dc device equations, is simulated point by point in the time domain using SPICE or a similar program. With suitable simplifying approximations, a first-order hand analysis is possible. Simplified analyses of this sort are helpful in developing insight into circuit performance and in making the most effective use of subsequent computer simulations.

For hand analysis of transient circuit response, it is very helpful to approximate the input time waveform by step functions, as shown in Sec. 1.7. In this approximation, the input step is assumed to occur at the 50% point of the actual input waveform, as shown in Fig. 1.20. Below we will see that usually only small errors are caused by this important simplifying assumption.

3.4.1   Propagation Delay Times

Consider now the NMOS inverters of Figs. 3.1 through 3.4. When the input makes an (idealized) instantaneous change from V_{OH} to V_{OL}, the inverter transistor turns off while the load current continues to flow. The situation is illustrated by the (b) portions of Figs. 3.1 to 3.4. The time required for V_{out} to charge from V_{OL} to the 50% point can be calculated by assuming the lumped load capacitance is charged by a constant current equal to the average current through the load device. Calculation of an exact average load current is rather complex and different for every choice of load device. However, an approximation which is adequate for hand analysis is obtained by considering only the two currents at the endpoints of the voltage transition.

The approximate average capacitor charging current $I_{LH(av)}$ is found simply by averaging the load device current at $V_{out} = V_{OL}$ and at $V_{out} = (V_{OH} + V_{OL})/2$. The propagation delay time is then calculated simply as

$$\Delta t = \frac{C\Delta V}{I_{av}}$$

In this specific case, the propagation delay time is given by

$$t_{PLH} = \frac{C_T(V_{OH} - V_{OL})/2}{I_{LH(av)}} \tag{3.11a}$$

Just after the input makes an (idealized) instantaneous change from V_{OL} to V_{OH}, both inverter and load devices are conducting. The current available to discharge the load capacitance from V_{OH} to the 50% point is the difference between inverter and load device currents. These currents are calculated separately for each device at V_{OH} and at the 50% point. Then I_{DL} is subtracted from I_{DI} at each voltage endpoint to obtain the net current available to discharge the load capacitance. The two net currents at the endpoints are used to obtain an approximate average value $I_{HL(av)}$. Then the time for the transition from V_{OH} to the 50% point is calculated in the same manner as above.

$$t_{PLH} = \frac{C_T(V_{OH} - V_{OL})/2}{I_{HL(av)}} \qquad (3.11b)$$

The average propagation delay is defined as

$$t_P = \frac{t_{PLH} + t_{PHL}}{2} \qquad (3.11c)$$

Example 3.6. Calculate the approximate propagation delays for the first stage of the circuit shown in Fig. 3.8, using the lumped load capacitances found in the previous section. Compare the results with SPICE simulation results. Use the device data from Example 3.3.

Operating conditions

	V_{out}	I_{DL}	I_{DI}	I_{av}
For t_{PLH}	0.3	42.	0.0	
	2.6	34.	0.0	38.
For t_{PHL}	5.0	0.0	320.	
	2.6	34.	281.	283.

From Example 3.5, we have $C_{T1} = 40.05$ fF. Now the propagation delay can be calculated, using Eq. (3.11a).

$$t_{PLH} = \frac{40.05(5 - 0.3)}{2(38)} = 2.48 \text{ ns}$$

From the SPICE simulation run with lumped capacitance, we obtain $t_{PLH} \approx 2.4$ ns, whereas with the full model, the SPICE result is $t_{PLH} \approx 3.1$ ns. As the fall time of V_{in} is varied between 0 and 5 ns, t_{PLH} does not change significantly. The inverter transistor turns off quickly, leaving only the predictable drain current of the load device flowing. The hand analysis is in good agreement with simulation results, considering the simplifications used.

The agreement is not so good in the case of t_{PHL}, because the inverter drain current is changing drastically during the finite rise time of V_{in}. Remember that we assume an instantaneous step input voltage for hand calculation. From Eq. (3.11b), the hand calculation gives

$$t_{PHL} = \frac{40.05(5 - 0.3)}{2(283)} = 0.34 \text{ ns}$$

From the SPICE simulation with rise time $t_r = 3$ ns for V_{in}, $t_{PHL} = 1.1$ ns. The hand analysis gives a fall time which is considerably shorter than the 3-ns rise time at the input. Under these circumstances a correction should be made to take into account the influence of the rise time t_r of V_{in}. Analytical studies suggest a root-mean-square dependence of the form

$$t_{PHL(actual)} = \sqrt{t_{PHL(step\,input)}^2 + \left(\frac{t_r}{2}\right)^2} \tag{3.12}$$

A series of SPICE simulations on several MOS circuits shows that this expression achieves a good fit between results of lumped-capacitance hand analysis and of SPICE simulations over a range of typical conditions. For the particular case of this example,

$$t_{PHL(actual)} = \sqrt{0.34^2 + 1.5^2} = 1.54 \text{ ns}$$

The SPICE result, as cited above, is 1.1 ns. This sort of agreement between analysis and simulation is typical.

3.4.2   Power-Delay Product

A useful and often-quoted figure of merit for digital circuits is the *power-delay product (PDP)*. The confusing term *speed-power product* is often used synonymously. Measured values of average power consumption and average delay (t_P) may be used to calculate the PDP for any gate circuit.

$$PDP = P_{D(av)}t_P \tag{3.13}$$

The units of PDP are (watts)(seconds) = joules. Thus PDP may be interpreted as energy consumed per logic decision. Modern digital gates show PDP in the range 1 to 10 pJ. Obviously a minimum value of power-delay product is desired.

Insight on means of reducing PDP may be obtained by stating it in terms of design parameters. Average power is only the product of supply voltage V_{DD} and average supply current $I_{D(max)}/2$. (For the NMOS circuits considered so far, average power is independent of data rate. This is not true for CMOS nor for certain NMOS dynamic and buffer circuits.) For a fixed inverter geometry ratio K_R and fixed supply voltage, the average load charging and discharging currents $I_{LH(av)}$ and $I_{HL(av)}$ are proportional to the maximum supply current $I_{D(max)}$. As shown in Eq. (3.11), delay times are thus inversely proportional to $I_{D(max)}$ and proportional to $C_T(V_{OH} - V_{OL})$. By bringing all this together, we may state PDP as a proportionality:

$$PDP \propto \frac{C_T(V_{OH} - V_{OL})}{I_{D(max)}} \frac{V_{DD}I_{D(max)}}{2} = \frac{C_T(V_{OH} - V_{OL})V_{DD}}{2} \tag{3.14}$$

Thus PDP may be reduced by reducing capacitance, logic swing, and/or supply voltage. Later we will return to consider these matters further.

This chapter describes the operating characteristics of MOS digital circuits based upon nominal or average device characteristics. In reality, of course, device parameters have statistical variations which can be very significant. Most notable are the variations in threshold voltage and k'. Threshold voltage frequently varies as much as $\pm 40\%$ around design center, while k' may vary by $\pm 20\%$. Production variations and temperature variations can both contribute to these parameter changes. Percentage variations in capacitances are typically smaller than those for dc parameters.

When designing systems based upon standard off-the-shelf components, the manufacturer's published information will give *worst-case* limits on dc and transient characteristics. This information should be used to achieve a reliable system design. Statistical design is not acceptable if it is expected that every completed system operates properly.

When designing integrated circuit chips, good practice requires that worst-case or statistical information on process and device characteristics be used for design calculations and simulations in order to achieve satisfactory production yield of circuits meeting a desired set of dc and transient performance parameters.

3.5 COMPLEMENTARY MOS (CMOS) INVERTER ANALYSIS

All static parameters of CMOS inverters are superior to those of NMOS inverters. The price paid for these substantial improvements is increased process complexity (to provide isolated transistors of both polarity types) and increased area per circuit.

3.5.1 DC Analysis of CMOS Inverter

A CMOS inverter is shown in Fig. 3.10 along with its load line construction and VTC. Note in Fig. 3.10*a* that the gate of the PMOS upper device is connected to the gate of the NMOS lower device. When the input is at V_{DD}, the NMOS device is conducting while the PMOS device (which has $V_{GS} = 0$ V) is cut off. Hence the drain current of the NMOS is limited to the very small leakage current of the PMOS, even though a highly conductive channel is present in the NMOS device. The result of the very small leakage current flowing through the highly conductive channel is that V_{out} is approximately 0 V. When the input is at ground, the NMOS device is cut off while the PMOS device is conducting. Only the small leakage current of the NMOS can flow, so V_{out} is very close to V_{DD}. Therefore $V_{OH} = V_{DD}$ and $V_{OL} = 0$ V.

We see from the above that in either logic state, one transistor in the series path from V_{DD} to ground is nonconducting. The only current which flows in either steady state is the infinitesimal leakage current of reverse-biased *pn* junctions. Quiescent power dissipation is in the nanowatt range. Yet, although no steady-state current flows, the on transistor supplies current to drive any load resistance or capacitance whenever the output voltage differs from 0 V or V_{DD}.

The tiny steady-state power consumption of CMOS is its most attractive feature. In addition, noise margins are relatively large because $V_{OL} = 0$ V and $V_{OH} = V_{DD}$. A completely symmetrical VTC is obtained if $V_{TP} = -V_{TN}$ and $k_P = k_N$.

The load line construction for the CMOS inverter requires some explanation. Shown in Fig. 3.10*b* are two I_D versus V_{DS} curves for each transistor. When V_{GS}

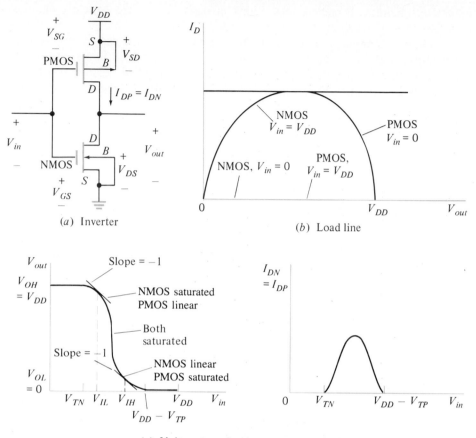

(a) Inverter

(b) Load line

(c) Voltage transfer characteristic and current

FIGURE 3.10
CMOS inverter.

is equal in magnitude to V_{DD} for either device, drain current curves are the upper curves on Fig. 3.10b. When gate-source voltage is zero for either device, drain current is zero. Thus these drain current curves lie along the horizontal axis in Fig. 3.10b. The two logic states lie at $V_{DS} = 0$ and V_{DD} along the horizontal V_{DS} axis, both at $I_D = 0$.

The drain current equations (2.11) and (2.12) apply to p-channel devices without change provided the polarity definitions shown in Fig. 3.10a are observed. In normal operation of p-channel devices, V_{GS}, V_{DS}, and V_{TP} all have negative values. Drain current I_D, defined as flowing *out* of the drain for PMOS, has a positive value. The parameters k_P, γ, and λ retain positive values.

Another way to handle the change of signs for PMOS analysis is simply to use absolute values of all voltages in the drain current equations. If you do this, check to be sure that the PMOS devices are properly biased for normal operation. The gate must be negative with respect to source by more than a threshold voltage if current is to flow.

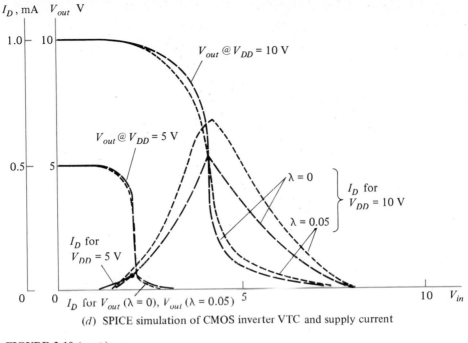

(d) SPICE simulation of CMOS inverter VTC and supply current

FIGURE 3.10 (cont.)
CMOS inverter.

Example 3.7. As an example, we will calculate and plot the VTC and the current that flows during a transition between states for a CMOS inverter like the one shown in Fig. 3.10, using the parameters given below. Although the two threshold voltages are usually equal in magnitude and $k_N = k_P$, for generality a different case is shown here.

$$V_{DD} - 5 \text{ V}$$
$$k_N = 100 \ \mu\text{A/V}^2 \qquad V_{TN} - +1.0 \text{ V}$$
$$k_P = 50 \ \mu\text{A/V}^2 \qquad V_{TP} = -1.5 \text{ V}$$

(a) $V_{OH} = 5$ V and $V_{OL} = 0$ V, as explained above.
(b) V_{IL} and V_{IH} are determined from the $dV_{out}/dV_{in} = -1$ condition. At $V_{in} = V_{IH}$, we assume the NMOS device is linear while the PMOS device is saturated. (In fact, there is only one value of input voltage for which both transistors are saturated.) Equations (2.11) and (2.12) may be used to obtain

$$I_{DN(lin)} = I_{DP(sat)}$$

Equations (3.2c and d) may be applied here, since the output conductance of a saturated transistor is much smaller than that of one in the linear region. These define the operating point at which $dV_{out}/dV_{in} = -1$. However, Eq. (3.2c) must be modified to include the derivative of both drain currents with respect to V_{in}.

$$\frac{dV_{out}}{dV_{in}} = \frac{dI_{DN} - dI_{DP}}{dV_{in}} \frac{dV_{out}}{d(-I_{DN})} = -1.0 \qquad (3.15a)$$

When solved, this yields

$$V_{IH} = \frac{2V_{out} + V_{TN} + (k_P/k_N)(V_{DD} - |V_{TP}|)}{1 + (k_P/k_N)} \qquad (3.15b)$$

This expression is substituted in the above drain current equation to obtain explicit solutions for V_{IH} and the corresponding value of V_{out}. The resulting equation has one root between 0 and V_{DD}. It gives the solution $V_{IH} = 2.3$ V, $V_{out} = 0.32$ V. SPICE gives the same result.

At $V_{in} = V_{IL}$, the NMOS device is saturated while the PMOS device is in the linear region.

$$I_{DN(sat)} = I_{DP(lin)}$$

To solve for V_{IL}, Eq. (3.15a) must be rewritten to incorporate the output conductance of the PMOS device, which has its source connected to V_{DD} instead of ground. The result is

$$V_{IL} = \frac{2V_{out} - V_{DD} - |V_{TP}| + (k_N/k_P)(V_{TN})}{1 + (k_N/k_P)} \qquad (3.15c)$$

Again substituting in the drain current equation, the resulting solution is $V_{IL} = 1.7$ V at $V_{out} = 4.7$ V. The result from SPICE is the same.

(c) Find the nearly vertical segment of the VTC within which both transistors are saturated by equating the saturated drain current equations for the two transistors. The voltage V_M at which $V_{in} = V_{out}$ falls within this segment of the VTC.

$$\frac{k_N}{2}(V_{in} - V_{TN})^2 = \frac{k_P}{2}(V_{DD} - V_{in} - |V_{TP}|)^2$$

Substituting the known voltages and rearranging,

$$\sqrt{\frac{k_N}{k_P}} = \frac{3.5 - V_M}{V_M - 1.0}$$

After substituting the given values of k_N and k_P, $V_M = 2.04$ V. SPICE gives the same result.

Now the peak current I_{DP} can be found by substituting $V_{in} = 2.04$ V in either of the equations for saturated I_D. The result is $I_{DP} = 54$ μA. The output voltage range over which both transistors are saturated is found by noting that with $V_{in} = 2.04$ V, the NMOS device remains saturated down to $V_{DS} = 1.04$ V and the PMOS device remains saturated until its V_{DS} becomes smaller in magnitude than $5 - 2.04 - 1.5 = 1.46$ V. Therefore V_{out} traces a nearly vertical path from 1.04 to 3.54 V, while V_{in} is essentially constant at 2.04 V. Only the presence of finite output conductance for the transistors in saturation prevents the VTC from being vertical in this range. Because we assume $\lambda = 0$ for simplicity, we obtain a vertical segment of the VTC in this hand analysis.

Figure 3.10d shows the output from four SPICE runs for this example circuit, with $\lambda = 0$ and with $\lambda = 0.05$ (a typical value) used for each device, for V_{DD} of 5 and 10 V. The peak current during the transition between states is much higher for the 10-V supply, as expected. In each case, peak supply current is about 20% higher with $\lambda = 0.05$. There are relatively small differences between the voltage transfer characteristics as a function of λ.

3.5.2 Transient Analysis of CMOS Inverter

Transient analysis of CMOS inverters is carried out in very much the same way as for NMOS inverters in Sec. 3.3. A single lumped linear load capacitance at each output node is defined. Then the average currents available for charging and discharging are calculated. Propagation delays are found using Eq. (3.11). Accuracy is comparable to that obtained in the NMOS example above.

Capacitances may be calculated from a layout such as that shown in Fig. 3.11, provided that information on junction depths and doping concentrations is available. Either the NMOS or the PMOS device (sometimes both) must be formed in an appropriately doped *well*, since the devices require bodies of opposite conductivity type. Figure 3.11 shows the NMOS in a *p*-type well, with the PMOS formed in the *n*-type substrate. The well will always be more heavily doped

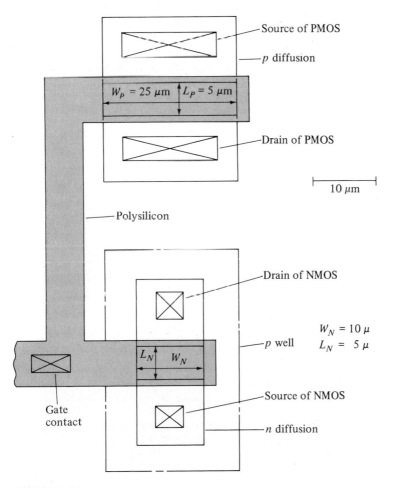

FIGURE 3.11
CMOS inverter layout.

than the substrate, because it must be formed by overcompensating the initial substrate doping concentration. Consequently, junction capacitances per unit area are higher for the devices formed in wells.

Channel widths and lengths W and L shown in Fig. 3.11 are typical for a near-minimum-size CMOS inverter. W_P/L_P must be about 2.5 times W_N/L_N to obtain approximately equal values of k_P and k_N. This equality is often desired in order to obtain equal rise and fall times when driving capacitive loads.

In Fig. 3.12 are shown the results of two SPICE simulations of a cascade of two CMOS inverters of the design shown in the layout of Fig. 3.11. One simulation used the full SPICE capacitance model, including all nonlinearities. For the second simulation, all model capacitances were forced to zero. Capacitive effects were calculated by hand and lumped at the output node of each inverter in the manner described in Sec. 3.3.2 for NMOS circuits. Note that the two simulations give similar results.

3.5.3 Power-Delay Product for CMOS Circuits

The power consumption of a conventional CMOS gate approaches zero under dc conditions. Current is drawn from the supply to charge the load capacitance C_T when the output switches from low to high. Each time there is such a transition, the amount of energy placed on the load capacitance is $C_T V_{DD}^2/2$. This energy may be compared with Eq. (3.14) for NMOS. The expressions are identical, considering that $V_{OH} - V_{OL}$ for CMOS equals V_{DD}. The power consumption for CMOS may be calculated by multiplying this PDP (energy consumed per logic decision) by the data rate f. Thus the power consumption of CMOS varies linearly with operating frequency.

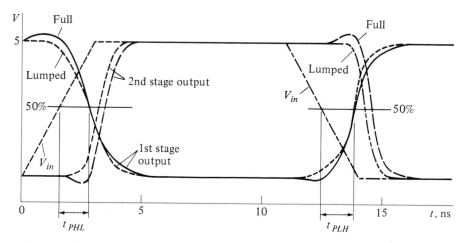

FIGURE 3.12
SPICE simulation of transient response of two cascaded CMOS inverters. Two cases: (a) full model; (b) lumped model.

In most digital systems, only a small fraction of the logic gates change state on each clock cycle. Under these conditions, the power consumption of a CMOS system will be far lower than an NMOS system, even though the PDP for CMOS gates is typically somewhat larger in magnitude than for NMOS.

3.6 NMOS GATE CIRCUITS

NMOS gates are not available as separately packaged individual circuits, but they are used extensively in large-scale integrated microprocessors, memories, etc., as described in Chaps. 9 and 10.

NOR gates in NMOS are formed simply by paralleling additional inverter devices, as illustrated for a case of depletion load circuits in Fig. 3.13a. The static and transient analysis proceeds exactly as for the corresponding inverter circuit. Each inverting device must have the same W/L as in the prototype inverter,

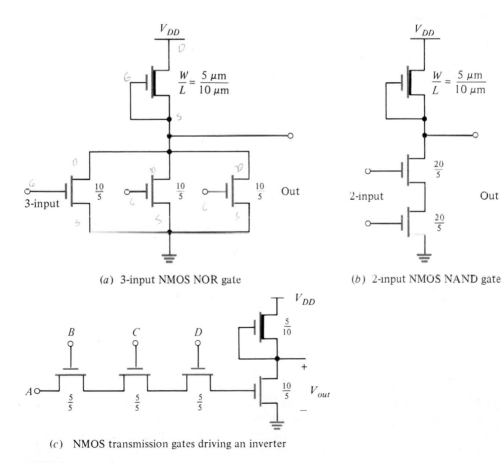

(a) 3-input NMOS NOR gate   (b) 2-input NMOS NAND gate

(c) NMOS transmission gates driving an inverter

FIGURE 3.13
NMOS NOR and NAND gates with typical W/L.

because the output must reach V_{OL} when only one input is high. Of course, this means that V_{OL} is lower when more than one input is high. In transient analysis, the appropriate capacitances of all inverting devices must be included in the lumped total capacitance C_T.

NAND gates in NMOS are formed by connecting additional inverter devices in series, as shown in Fig. 3.13b. Assuming inverter device length L is fixed, the width W must be N times the width of the single device of a simple inverter for an N input gate. This must be so in order that the combined W/L of all N inverter devices is adequate to achieve the desired V_{OL} for the gate when all inputs are high. Because of the need for increased area when adding NAND inputs, NAND logic with more than 2 inputs is not economically attractive in NMOS. NOR logic is preferable.

The analysis for V_{OL} of a NAND gate is very difficult if one attempts to find the node voltages between the transistors in the series path. Each transistor has a different value of V_{SB} and threshold voltage. Fortunately, it is unnecessary to solve for every node voltage. Provided all inverter transistors have the same input voltage and the same width W and length L, the output voltage can be found by assuming that the series string of N inverter transistors is a single transistor of width W and length NL.

When forming C_T for a NAND gate, a conservative result is obtained if the capacitances for all inverting devices are lumped at the output node.

The transmission gate of Fig. 3.13c has the advantage that each additional input requires only a minimum-size transistor ($W/L = 1$) because no steady-state current flows. However, there are several problems with this form of gating. First, the voltage at the input to the inverting transistor will be one threshold voltage lower than the lowest of the transmission device input voltages. When input D goes low, the gate-drain capacitance couples a negative voltage step to the inverter input.

These problems with transmission gates make it difficult to achieve adequate drive to maintain V_{OL} of the inverter. Also troublesome is the fact that the logic function depends upon the time sequencing of input signal A relative to the other inputs. For instance, if B or C or D goes low while A remains high, a high level will be stored on the gate capacitance of the inverter. On the other hand, if A goes low before any of the others, a low level will be placed at the inverter input. Thus design of transmission gates is subject to additional constraints compared to NOR and NAND gates of Fig. 3.13a and b.

3.7 CMOS GATE CIRCUITS

CMOS circuits are finding increasing use in a variety of applications. Small-scale integrated (SSI) gate circuits in the 4000, 4000B, and 74C Series have been in use for many years. Improved versions of these SSI circuits are being developed and introduced. Large-scale integrated (LSI) CMOS circuits are used in almost all of the modern electronic watches and calculators. CMOS technology is finding increasing use in microprocessors and memories.

3.7.1 Basic CMOS Gate Circuits

NOR, NAND, and transmission gates in CMOS are illustrated in Fig. 3.14. Note that two transistors must be added for each additional input. This is one important reason for the lower circuit density of CMOS compared to NMOS. The static analysis of CMOS gates is very similar to that for the CMOS inverter. If both propagation times (high-low and low-high) are to be similar, the overall conductance factors for NMOS and PMOS portions of the gate must be similar. As noted in Chap. 2, typically k'_P is about 40% of k'_N in CMOS circuits. As a result, very wide PMOS devices must be used in the CMOS NOR circuit in which PMOS devices are connected in series, as seen in Fig. 3.14a. On the other

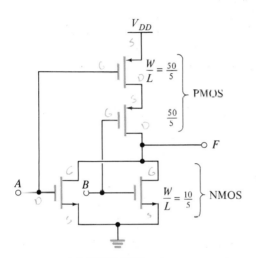

(a) CMOS NOR gate

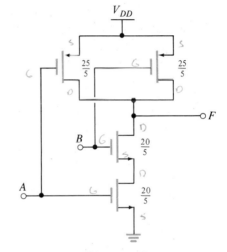

(b) CMOS NAND gate

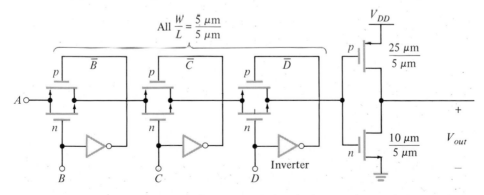

(c) CMOS transmission gates driving a CMOS inverter

FIGURE 3.14
CMOS gates with typical W/L for LSI circuits.

hand, a two- or three-input NAND in CMOS will have approximately equal areas devoted to NMOS and PMOS devices, as seen in the example of Fig. 3.14*b*.

A CMOS transmission gate is shown in Fig. 3.14*c*. Through use of NMOS and PMOS transistors in parallel, driven with complementary signals, the full signal level at *A* is carried through to the inverter input, eliminating the threshold voltage drop of the NMOS transmission gate. However, the timing sensitivity and the parasitic capacitance effects of the NMOS gate remain. Consequently, CMOS transmission gates must be used with care in digital circuit design. They can be very useful in special circuit applications, for instance, to introduce clocking into CMOS static circuits or to selectively close a feedback path. Note that a CMOS transmission gate is a *bilateral* circuit; when on, it conducts equally well in either direction. Sometimes this property is useful.

Conservative results in manual transient analysis are obtained for CMOS if all capacitances are lumped at the output node. Manual analysis and computer simulation of transmission gates is more subject to error than is the case for normal inverters and gates.

3.7.2   4000 Series CMOS Logic Family

This is the oldest standard CMOS logic family. These circuits were introduced in the 1960s, based upon metal-gate CMOS technology. Circuit design for the NAND and NOR gates in the 4000 family follows the schematics shown in Fig. 3.14, with device *W/L* ratios chosen to give approximately equal capability for sourcing and sinking output current. Transmission gates like those shown in Fig. 3.14*c* are available in addition to NAND and NOR gates. Output current is not the same for all circuits in the family, and the current available is not always compatible with requirements of transistor-transistor logic (TTL) circuits. Voltage transfer characteristics of the 4000 Series NAND and NOR gates are similar to those shown in Fig. 3.10.

Propagation delay for these circuits is determined primarily by the off-chip load capacitance and current available from the output node during the high-low and low-high transitions. Measured propagation delay times range from 25 to 100 ns with a 5-V supply and 15 pF of off-chip capacitance.

A useful feature is that these integrated circuits may be operated with supply voltages ranging from 3 to 15 V, while retaining their very low standby power characteristics. (Circuits are often available at lower prices for operation over a narrower range of supply voltage, such as 5 to 10 V.) In addition to basic gates, numerous more complex logic functions such as decoders, latches, registers, adders, etc., are available in this family. Some of these other functions are described in Chap. 8.

3.7.3   4000B and 74C Series

The 4000B and 74C CMOS circuit families are improved versions of the original 4000 Series circuits. The newer circuits are made with silicon-gate CMOS technology, using a minimum feature size of 5 μm and a gate oxide thickness of

1000 Å; this change reduces the chip area and propagation delay of the more complex logic functions. Every output is specified to be capable of driving one TTL series 74LS input (see Sec. 7.3.3) when operating from a 5-V supply. Allowable supply voltage range is from 3 V to as high as 20 V for components from some suppliers.

Every output of every 4000B and 74C CMOS circuit is *double-buffered* in the manner illustrated in Fig. 3.15 for a two-input NAND gate. Devices M_1–M_4 connected to the inputs have relatively small W/L ratios, but with matched capability for sourcing and sinking currents. Instead of taking the gate output directly from the stack of logic devices (as in the original 4000 family), two cascaded inverters are used to isolate the logic circuit from the output node. The final inverter M_7–M_8 is designed with relatively large device W/L ratios in order to drive the off-chip load, possibly including 15 to 50 pF of capacitance and one TTL LS circuit. Intermediate inverter M_5–M_6 has intermediate values of W/L.

Double-buffered CMOS circuits have better voltage transfer characteristics (VTC) and larger noise margins than unbuffered gates. This may be understood by imagining that the two inputs of the NAND gate in Fig. 3.15a are tied together so that the circuit may be represented as a cascade of three inverters, as shown in

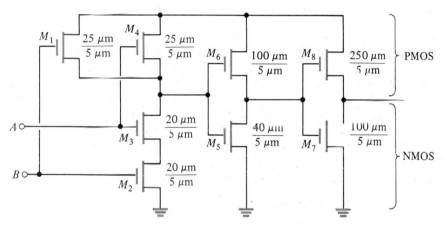

(a) Circuit design with typical device sizes

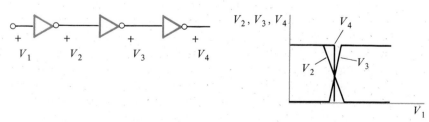

(b) Equivalent circuit for VTC analysis   (c) Equivalent transfer characteristics

FIGURE 3.15
Double-buffered CMOS gates used in small-scaled integrated circuits.

Fig. 3.15b. Suppose that each inverter taken separately has the VTC shown for V_2/V_1 in Fig. 3.15c, with a VTC slope of -2 in the transition region. While in the transition region, this means that for ΔV change in V_1, V_2 changes by $-2\Delta V$, V_3 by $+4\Delta V$, and V_4 by $-8\Delta V$. Clearly V_4 reaches a limit at V_{OH} or V_{OL} for only a small change in V_1.

In fact, the slope in the transition region for a CMOS inverter is much larger than -2, so the overall transfer characteristic of the double-buffered circuits is very steep. It occurs at $V_{in} \approx V_{DD}/2$, regardless of the value of supply voltage V_{DD}.

3.7.4   74HC/HCT Series

A later family of small-scale and medium-scale integrated CMOS logic circuits, the 74HC family, uses more advanced fabrication technology including reduced minimum feature size of 3 μm and a thinner gate oxide of 600 Å to obtain shorter propagation delays and improved load-driving capability compared to older designs. These circuits are also double-buffered. The allowable supply voltage range is from 3 to 6 V.

The principal performance characteristics of the 74HC Series are listed in Table 3.2. Particularly note that with a CMOS load the output levels V_{OH} and V_{OL} are guaranteed to be within 0.1 V of V_{DD} and ground, respectively. This yields a large logic swing of ≥ 4.3 V. The logic swing is the difference between V_{OH} and V_{OL}. With a TTL load the logic swing is reduced a little, since with a TTL load the output current is 4 mA, while for a CMOS load the output current is only 20 μA. The input currents I_{IH} and I_{IL} are simply due to the leakage currents of the input protection diodes (see Fig. 2.10b). The propagation delay time, $t_p = (t_{PHL} + t_{PLH})/2$, is given for a 50-pF capacitor connected to the output node. This is a typical off-chip load capacitance.

A problem with the 74HC Series is that while the pin connections are compatible, a 74HC part cannot be directly substituted for a 74LS Series part. At the time the 74HC Series was introduced there were very many more 74LS parts in use. The 74LS Series of TTL circuits are described in Sec. 7.3.3, but for comparison purposes the characteristics of the 74LS Series are also included in Table 3.2. Notice that while the output voltage levels of the HC part are compatible with the input voltage levels of the LS part, the reverse is not true. The minimum V_{OH} of the LS is less than the minimum V_{IH} of the HC. Hence, with an LS part driving an HC part the output of the HC would be indeterminate. The 74HCT family, a variation on the 74HC, is designed to overcome this problem. The output voltage levels of the HCT are the same as for the HC. But at the input, the minimum V_{IH} has been reduced to 2.0 V. This voltage level shift in the HCT can be achieved by inserting two diodes in series, pointing down, between V_{DD} and the source of the PMOS device at the input inverter of the HCT part (see Fig. 3.10a). With this change the performance characteristics of the 74HCT series are comparable to and compatible with the 74LS series, but with a much lower power consumption.

As noted in Sec. 3.5.3 the power consumption of CMOS circuits varies linearly with operating frequency. The power dissipation per gate of CMOS is usually given for dc or very low frequency of operation. In Fig. 3.16 the power dissipation per gate for a 74HC and a 74LS NAND gate are compared. Notice at low frequency the power dissipation of CMOS is much less than TTL, but there is a crossover at about 5MHz. This figure is typical in comparing CMOS and TTL, or other bipolar, digital circuits.

3.7.5 74AC/ACT Series

The latest development in small-scale and medium-scale integrated CMOS logic circuits, introduced in 1985, is the 74AC and 74ACT Series. These double-buffered circuits use an advanced CMOS technology with minimum feature size of less than 2 μm, a gate oxide thickness of less than 400 Å, and two layers of metal. The performance characteristics of these circuits are listed in Table 3.2. Particularly note that the output transistors of these circuits are capable of sourcing and sinking 24 mA with typical propagation delay times of 5 ns. These characteristics are comparable, and the 74ACT Series is compatible with the 74ALS TTL circuits described in Sec. 7.3.4.

It can be mentioned here that further development of SSI and LSI CMOS circuits with minimum feature sizes of less than 1.5 μm and gate oxides of less than 300 Å has been described. These circuits have typical propagation delay times of 3 ns.

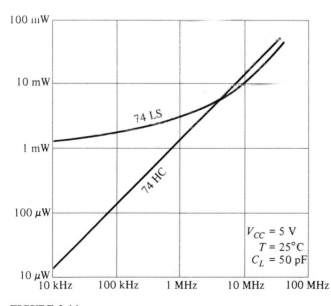

FIGURE 3.16
Power dissipation versus frequency for equivalent NAND gate.

TABLE 3.2
Bipolar and CMOS logic performance characteristics $(T_A = 25°C)$

Parameter	74LS	74HC	74HCT	74AC	74ACT
			Series		
min V_{OH}/max V_{OL}:					
CMOS load	2.7/0.5	4.4/0.1	4.4/0.1	4.4/0.1	4.4/0.1
TTL load	2.7/0.5	4.0/0.3	4.0/0.3	3.9/0.3	3.9/0.3
min V_{IH}/max V_{IL}	2.0/0.8	3.1/0.9	2.0/0.8	3.1/1.3	2.0/0.8
min I_{OH}/min I_{OL}, mA	−0.4/8	±4	±4	±24	±24
max I_{IH}/max I_{IL}, µA	20/−400	±0.1	±0.1	±0.1	±0.1
Typical t_P, ns	10	10	10	5	5
Typical dc P_D/gate	2 mW	2.5 µW	2.5 µW	2.5 µW	2.5 µW

For Series 74LS: $V_{CC} = 5$ V $\quad C_L = 15$ pF
For CMOS Series: $V_{CC} = 4.5$ V $\quad C_L = 50$ pF

3.8 DYNAMIC LOGIC CIRCUITS

All of the circuits described in the previous sections can be used in combinational logic networks without any need of periodic clock signals. Hence they are termed *static* circuits.

All but the very smallest digital systems require sequential as well as combinational logic. The latches and other flip-flop circuits used in sequential logic networks are the subject of Chap. 8. As a practical matter, all systems employing sequential logic require the use of periodic clock signals for correctly synchronized operation. In static circuits, combinational or sequential, clock signals are introduced only at normal gate input nodes, identical to those used for logic inputs. There is no lower limit on clock frequency in static circuits.

Early in the development of MOS logic circuits it was recognized that periodic clock signals could be used advantageously in combinatorial circuits as well as in sequential circuits and memories, particularly for LSI. Clock signals may be introduced at arbitrary circuit nodes. Depending on the application, advantages of *dynamic* logic may include faster operation, reduced power consumption, and greater circuit density. The drawbacks to dynamic operation are:

1. There is a lower limit on clock frequency for proper operation, typically 500 Hz.
2. Design is more difficult, particularly if operation from a single 5-V supply is required.
3. Circuits may be more sensitive to noise and timing errors.

A "family tree" for logic circuits is shown in Fig. 3.17 to clarify the relations among static and dynamic circuits.

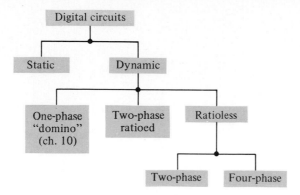

(a) "Family tree" for circuit design

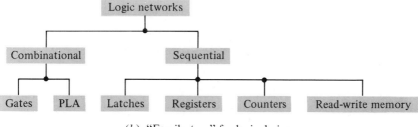

(b) "Family tree" for logic design

FIGURE 3.17
Relations among digital circuits.

Two different dynamic NMOS gate circuits implementing a two-input NOR function are shown in Fig. 3.18, together with the two-phase clock signals needed for operation. In both circuits, certain capacitances as drawn in explicitly on the schematic diagram are essential to operation. These capacitances are not specifically added to the circuit but rather are made up of the inevitable capacitive elements appearing at a MOS circuit node as described in Sec. 3.3. These same capacitances are considered to be undesirable parasitics in static circuits. In dynamic circuits, they are required to store voltages representing binary logic levels during the period of one clock cycle.

The two-phase ratioed NOR gate of Fig. 3.18a derives its name from the fact that a geometry ratio K_R between inverter device and load device, as required for an enhancement load static circuit, must be maintained here also. The ratioless circuit of Fig. 3.18b, on the other hand, can be designed to operate correctly with identical minimum-size transistors in all positions.

Operation of the ratioed logic of Fig. 3.18a is as follows. While clock phase ϕ_1 is high, M_1 and M_2 turn on, transferring the logic input signals to C_1 and C_2 and the gates of M_3 and M_4. When ϕ_1 goes low, voltages representing these logic levels remain stored on capacitances C_1 and C_2.

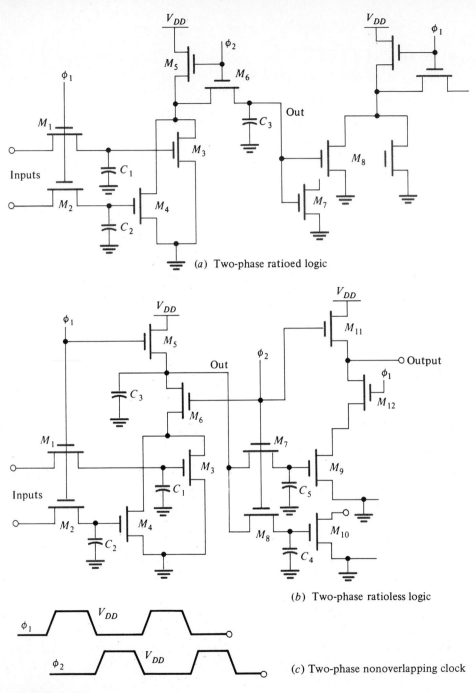

(a) Two-phase ratioed logic

(b) Two-phase ratioless logic

(c) Two-phase nonoverlapping clock

FIGURE 3.18
Dynamic NMOS logic.

A question immediately arises about how long stored high voltages endure before being lost through the junction leakages of M_1 and M_2. The values of C_1 and C_2 are on the order of 0.1 pF, while the junction leakage current for a small transistor is likely to be less than 0.1 pA. A simple calculation using these numbers gives a voltage loss from such a node of 1 V/s. A common, conservative specification is that the clock period should not be longer than 2 ms for dynamic MOS circuits.

To continue the description of this circuit, when ϕ_2 goes high, M_5 and M_6 turn on. If the voltages on the gates of M_3 and M_4 are low, these devices remain nonconducting. Their drains are charged to a high level by M_5, and this high level is transferred to C_3 and the gates of M_7 and M_8 by M_6. Note that the maximum high output level is the threshold drop of M_5 (with body bias!) below the high level of ϕ_2.

If one or both of M_3 and M_4 have high voltages at their gates after ϕ_1, the output will remain low when ϕ_2 goes high. This is because M_3 and M_4 are ratioed to M_5 such that the output voltage under these conditions is only a small fraction—less than V_T—of V_{DD}.

Thus data are taken into the first stage during ϕ_1, and the logical NOR of these inputs is transferred to the next stage during ϕ_2. Strictly speaking, both circuits in Fig. 3.18 should be termed sequential circuits, but because the outputs are a function only of the immediately preceding inputs, they are commonly used and referred to as *combinational* circuits.

Compared to modern NMOS enhancement-depletion static logic circuits, this ratioed two-phase dynamic NOR gate is not particularly advantageous. It still requires area-consuming ratios, it requires more total transistors, and it consumes current from V_{DD} whenever an input is high and M_5 is on. Two-phase ratioed circuits are rarely used for simple gating but do find application in sequential circuits (Sec. 8.6), random-access memories (Sec. 9.4), and shift registers.

The ratioless circuit of Fig. 3.18b can be designed using identical minimum-size transistors throughout, leading to an area saving compared to ratioed circuits. Furthermore, power consumption is low because M_5 and M_6 are never on simultaneously to allow current flow directly from V_{DD} to ground.

Operation of the ratioless circuit proceeds as follows. M_1, M_2, C_1, C_2, and M_3, M_4 function as described above. While ϕ_1 is high, C_3 is charged to a valid high level, one threshold drop below the clock high level. When ϕ_2 goes high, a high level on C_1 or C_2 will turn on M_3 or M_4, discharging C_3 through M_6. This low output level will be transferred through M_7 and M_8 to the next stage.

If both inputs to the first stage are initially low, C_3 is not discharged. When ϕ_2 goes high, the charge on C_3 is shared with C_4 and C_5. The high logic level is thereby further reduced. Care must be taken in design to ensure that it is still adequate in worst-case situations. Obviously it is desirable that C_3 be larger than the sum of C_4 and C_5. Operation from a single 5-V supply is difficult or impossible to achieve. Two-phase ratioless circuits are not widely used today.

By adding two more clock phases, so-called *four-phase ratioless* circuits can be designed which eliminate the drawbacks of the two-phase circuits described above. Unfortunately, four-phase circuits have other constraints. Gates driven with a specific clock phase can fan out only to others driven by other specific clock phases. The resulting design complexities have sharply limited the use of four-phase logic.

The transient performance of dynamic MOS circuits must be analyzed using SPICE or another computer circuit simulator. Special care is needed in modeling device capacitances for circuits such as that of Fig. 3.18b, in which capacitive charge-sharing effects are important. Operating speed of the dynamic circuits is usually determined by the output circuits which must drive signals off the chip and by the quality of clock signals, in terms of rise and fall times and high-level voltage. The RC time constants formed by the channel resistance of one transistor and the capacitance of one internal node rarely are a limiting factor.

3.9   SCALING OF MOS CIRCUITS

Continuing improvements in integrated circuit fabrication technology have made possible steady reductions in the internal dimensions of semiconductor devices. Early integrated circuits (circa 1965) had internal dimensions in the range 10 to 20 μm. Today's most advanced complex integrated circuits employ internal dimensions (*minimum feature size*) no smaller than 2 to 3 μm.

The trend toward progressively smaller internal dimensions is likely to continue for some years until fundamental physical limitations of transistors are approached for internal surface dimensions on the order of 0.25 μm.

In this section we describe the first-order changes in MOS digital circuit characteristics as a function of internal device dimensions. Two cases are considered:

1. All device dimensions (surface and vertical) and all voltages (V_{DD}, V_{TO}) are reduced by the same scaling factor $S(S > 1)$.
2. Only device dimensions, not voltages, are reduced by the factor S. This case is important because we often wish to retain compatibility in supply voltage and logic levels with present 5-V MOS and TTL circuit families.

We will see that several device parameters cannot be scaled as desired and hence that the full performance advantages of scaling cannot be realized in practice.

3.9.1   Full Scaling

Suppose that each surface dimension and vertical dimension, as well as all threshold voltages and supply voltage(s), are reduced by the same factor S. The scaling relations among device and circuit parameters for full scaling are summarized in the left column of Table 3.3. Since t_{ox} is reduced by the factor S, oxide capacitances per unit area are increased by S. However, device area is decreased by $1/S^2$, so that absolute values of oxide capacitance are reduced to $1/S$ of former values.

TABLE 3.3
MOS circuit scaling relationships

Parameter	Full scaling	Constant voltage
W, L, t_{ox}	$1/S$	$1/S$
V_{DD}, V_{T0}	$1/S$	1
Oxide capacitances	$1/S$	$1/S$
k_N, k_P	S	S
I_{DD}	$1/S$	S
DC power consumption	$1/S^2$	S
t_P	$1/S$	$1/S^2$
Power-delay product	$1/S^3$	$1/S$

If we make the reasonable assumption that carrier mobility does not change, the process conduction factor k' is increased by S due to the fact that C_{ox} [per unit area, as defined in Eq. (2.8)] is proportional to S. Maximum and average MOS transistor current is reduced by S because it is a function of $k'V_2 \propto 1/S$; the effect of reducing supply voltage overrides the effect of increasing k'. Power consumption is reduced by $1/S^2 \propto V_{DD}I_{DD}$. Propagation delay is proportional to $(CV/I) \propto 1/S$. Finally, power-delay product based upon consideration of only the factors enumerated above is reduced to $1/S^3$ of its original value. Device W/L ratios may be modified to achieve a different balance of power and delay; to first order, the power-delay product is not affected by these changes.

3.9.2 Constant Voltage Scaling

If supply voltage is held constant while surface and vertical dimensions are reduced by S, device and circuit parameters are related as shown in the right column of Table 3.3. Note that drain currents and power consumption now increase by the factor S. Since oxide capacitances decrease by S, propagation delays are reduced to $1/S^2$ and PDP is reduced to $1/S$ relative to their unscaled values. Of course, circuit designs may be modified (by changing device W/L ratios) to achieve a different balance of power and delay.

3.9.3 Scaling of Potentials and Substrate Doping

Bulk potential ϕ_F, work function difference ϕ_{GC}, and thermal voltage kT/q cannot be scaled. Also, there is a mechanism known as *subthreshold conduction* by which MOS transistor drain current does not go to zero at $V_{GS} = V_T$ but instead falls off exponentially. This puts some limits on scaling the device threshold voltage, particularly for dynamic circuits in which current in the off state must be below a picoampere.

Substrate doping N_A or N_D can be scaled in different ways. If the doping is held constant, depletion-layer capacitances will remain constant per unit area and thus decrease in absolute terms as device areas are reduced. If scaling is done so that maximum depletion-layer width is reduced along with other dimensions, then substrate doping must be increased in proportion to S for full scaling and in proportion to S^2 for constant voltage scaling. Bulk potential ϕ_F is thus increased, although only slowly (logarithmically) with increasing N_A or N_D. If doping is scaled in this way, zero-bias depletion-layer capacitance C_{j0} per unit area will increase approximately as $S^{1/2}$ for full scaling and as S for the constant-voltage case.

3.10   SUMMARY

This chapter provides analysis techniques for the static and dynamic characteristics of NMOS and CMOS inverters and gate circuits.

- Static voltage transfer characteristics of MOS inverter circuits are analyzed.
- Means of calculating the critical voltages V_{OH}, V_{OL}, V_{IH}, and V_{IL} by hand are derived. Results of hand analysis are compared with computer simulation results. Only insignificant differences are found.
- Single-polarity (NMOS) inverters employing four different load devices are compared:

    Resistor load

    Saturated enhancement transistor load

    Linear enhancement transistor load

    Depletion transistor load

The last of these has superior noise margins and requires less chip area than the others.

- The integrated circuit layout of an NMOS inverter is used for calculation of the internal capacitances which limit circuit speed.
- The preparation of SPICE input data for circuit simulation, including both dc parameters and capacitive effects, is described.
- Hand calculation of internal capacitances and of actual propagation delay times are described and illustrated with a detailed example.
- The important concept of *power-delay product* is introduced and explained.
- The dc and transient characteristics of CMOS inverters are described and illustrated with an example.
- The design of NAND and NOR gates based on NMOS and CMOS inverters is explained.
- Standard families of small-scale integrated (SSI) CMOS logic circuits are described, including double-buffered circuits.
- Concepts of dynamic digital circuit design are introduced.
- The trend toward improving the density and performance of all MOS circuits by means of reduced *(scaled)* device dimensions is described.

REFERENCES

1. R. H. Crawford, *MOSFET in Circuit Design*, McGraw-Hill, New York, 1967.
2. W. N. Carr and J. P. Mize, *MOS/LSI Design and Application*, McGraw-Hill, New York, 1972.
3. H. Shichman and D. A. Hodges, "Modeling and Simulation of Insulated-Gate Field-Effect Transistors," *IEEE Journal of Solid-State Circuits*, vol. SC-3, no. 5, September 1968, pp. 285–289.
4. L. W. Nagel, "SPICE2, A Computer Program to Simulate Semiconductor Circuits," *ERL Memorandum ERL-M520*, University of California, Berkeley, May 1975.

DEMONSTRATIONS

D3.1 MOSFET Inverter: Voltage Transfer Characteristic

The objective of this experiment is to examine the voltage transfer characteristic of an NMOS inverter and a CMOS inverter.

1. (*a*) Use the 4007 and connect the NMOS inverter circuit shown in Fig. D3.1*a*. Plot the voltage transfer characteristic (V_{out} versus V_{in}) for V_{in} between 0 and 10 V. Record V_{OH}, V_{OL}, V_{IL}, and V_{IH}.
   (*b*) Repeat part (*a*) but eliminate one of the inverter transistors by connecting its gate to ground.
   (*c*) Connect the CMOS inverter circuit shown in Fig. D3.1*b* and repeat part (*a*) with V_{DD} = 10 and 5 V, limit V_{in} to V_{DD}.
2. (*a*) Use the parameters obtained from Demonstration D2.1 to calculate the VTC for the NMOS inverter, and compare with the observed results in part 1(*a*).
   (*b*) Use the NMOS and PMOS parameters obtained from D2.1 to calculate the VTC for the CMOS inverter with V_{DD} = 10 V, and compare with the observed result in part 1(*c*).

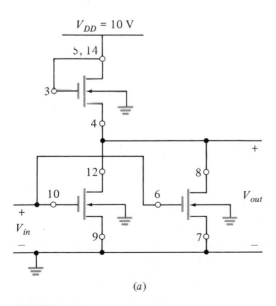
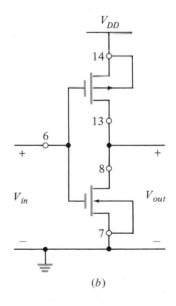

(*a*) (*b*)

FIGURE D3.1

D3.2   CMOS Gates: Propagation Delay Time

In this experiment we will measure the propagation delay time of a CMOS gate, as represented by the 4001 — a two-input NOR gate.

1. (a) The ring oscillator circuit shown in Fig. D3.2 is often used to measure the average propagation delay time of a logic gate. The period of the oscillation frequency is equivalent to the total propagation delay time of the three gates which make up the ring. The average propagation delay time is therefore the period of the oscillation divided by twice the number of gates within the ring. When using high-speed gates it may be necessary to cascade more than three gates to lengthen the period of oscillation and make it easier to measure. An odd number of gates is always required. Keep all connections as short as possible to minimize the effects of stray capacitance. Connect a 1-μF capacitor between V_{DD} and ground to bypass ripples on the V_{DD} line caused by the ring oscillator.

   (b) With the connection as in Fig. D3.2 determine the average propagation delay time with V_{DD} = 10 and 5 V.

   (c) The effective input capacitance of these CMOS gates is about 5 pF. Simulate a fan-out of 10 by adding 50 pF to each output node in Fig. D3.2, and repeat part (b).

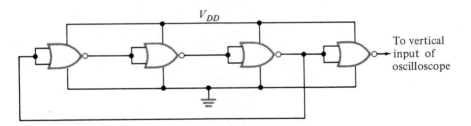

FIGURE D3.2

PROBLEMS

Use the following device parameters unless otherwise specified.

$$k'_{NMOS} = \mu_n C_{ox} = 20 \ \mu\text{A/V}^2 \qquad k'_{PMOS} = \mu_p C_{ox} = 10 \ \mu\text{A/V}^2$$

$$k_N = k'\left(\frac{W}{L}\right)_N \qquad k_P = k'\left(\frac{W}{L}\right)_P$$

$$V_{TN0} = 1 \text{ V} \qquad V_{TP0} = -1 \text{ V} \qquad V_{TD0} = -3 \text{ V}$$

$$\gamma_N = \gamma_P = 0.4 \text{ V}^{1/2} \qquad \lambda = 0$$

$$2|\phi_F| = 0.6 \text{ V}$$

P3.1. For a resistive load inverter with $(W/L)_I$ = 25, R_L = 10 kΩ, and V_{DD} = 5 V:

   (a) Calculate V_{OL}, V_{OH}, NM_L, and NM_H.

   (b) Find the average power dissipated by the inverter and the maximum power dissipated by the MOS transistor if the input varies from V_{OL} to V_{OH} with a 50% duty cycle (i.e., transistor is on 50% of the time).

   (c) Verify the results of parts (a) and (b) using a circuit simulator (e.g., SPICE).

P3.2. For a saturated enhancement load inverter with $(W/L)_I = 5$ and $(W/L)_L = 0.5$ and $V_{DD} = 10$ V:

    (a) Calculate V_{OL}, V_{OH}, NM_L, and NM_H.

    (b) Calculate the average power dissipated by the complete inverter for a 25% duty cycle. (Input is high 25% of the time.)

    (c) Verify the results of (a) using SPICE.

P3.3. For the linear load inverter in Fig. P3.3, the load device has $W/L = 1$.

    (a) Choose the W/L ratio of the inverting transistor for $V_{OL} = 0.3$ V.

    (b) Repeat (a) for $V_{BB} = -5$ V.

    (c) Find V_M (the point at which $V_{in} = V_{out}$) if the W/L of the inverter $= 30$. Neglect body effect.

    (d) Repeat (c) with $\gamma = 0.4$.

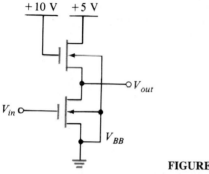

FIGURE P3.3

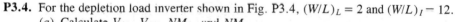

P3.4. For the depletion load inverter shown in Fig. P3.4, $(W/L)_L = 2$ and $(W/L)_I = 12$.

    (a) Calculate V_{OL}, V_{OH}, NM_L, and NM_H.

    (b) Repeat (a) for $V_{DD} = 15$ V.

    (c) Find V_{OL}, V_{OH}, NM_L, and NM_H assuming $\lambda = 0.05$ for $V_{DD} = 15$ V.

    (d) Find V_{OL} and V_{OH} with a 100-kΩ resistor from V_{out} to ground. $V_{DD} = 5$ V.

    (e) Verify your results with SPICE.

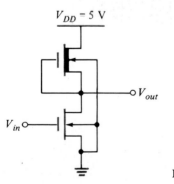

FIGURE P3.4

P3.5. For a depletion load inverter:
   (a) Choose the aspect ratio $K_R = (W/L)_I/(W/L)_L$ for a $V_{OL} = 0.1$ V.
   (b) Repeat (a) to obtain $V_{OL} = 0.01$ V.
P3.6. Calculate the V_{IL} and V_{IH} of the NAND gate shown in Fig. P3.6 for two cases:
   (a) $V_1 = 5$ V and V_2 increases from 0 to 5 V.
   (b) $V_2 = 5$ V and V_1 increases from 0 to 5 V.

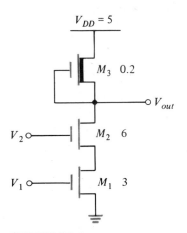

FIGURE P3.6

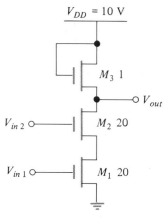

FIGURE P3.7

P3.7. For the NMOS circuit in Fig. P3.7:
   (a) Determine the logic swing $V_{OH} - V_{OL}$ if the two inputs are tied together and driven from another identical circuit.
   (b) If $\gamma = 0.5$, repeat (a).
   (c) If $\gamma = 0$ and $\lambda = 0.025$, repeat (a).
   (d) Verify with SPICE.
P3.8. For the circuit in Fig. P3.8:
   (a) Find the W/L of the load device such that $V_{out} = 0.5$ V for $V_{in} = V_{OH}$.
   (b) With the results of (a), calculate five critical points and sketch the VTC.
   (c) What are t_{PLH}, t_{PHL}, and t_P if the circuit (a) drives an off-chip load of 20 pF?
   (d) Verify with SPICE.

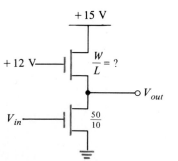

FIGURE P3.8

P3.9. For the PMOS inverter in Fig. P3.9:

(a) Calculate V_{OH}, V_{OL}, NM_H, and NM_L. (V_{OH} is still the most positive output level.) Neglect body effect.

(b) Choose the W/L of the inverting transistor so that $V_{OH} = 9.7$ V.

(c) What are t_{PLH}, t_{PHL}, and t_P if the circuit must drive an internal node capacitance of 0.2 pF?

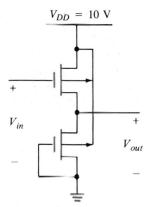

FIGURE P3.9

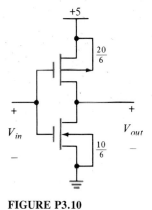

FIGURE P3.10

P3.10. For the CMOS inverter in Fig. P3.10:

(a) Find V_{OH}, V_{OL}, NM_L, and NM_H.

(b) If a 100-kΩ resistor is added from the output node to ground, find V_{OH} and V_{OL}.

(c) Repeat (a) and (b) for $V_{DD} = 15$ V.

(d) What are t_{PLH}, t_{PHL}, and t_P if the circuit drives 0.5 pF?

P3.11. For the CMOS NAND circuit shown in Fig. P3.11, $(W/L)_N = 20$ and $(W/L)_P = 10$.

(a) If $V_A = V_B = 5$ V, sketch the VTC and calculate the five critical points.

(b) Calculate supply current as a function of V_C. ($V_A = V_B = 5$ V.)

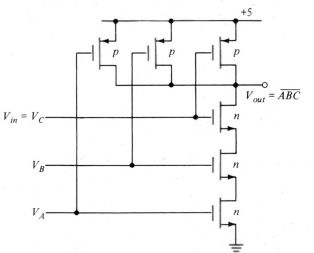

FIGURE P3.11

P3.12. Design a three-input CMOS NOR gate $(F = \overline{A + B + C})$. Specify the W/L of three identical PMOS and three identical NMOS devices such that the maximum output resistance is 500 Ω in either steady state.

P3.13. The circuit in Fig. P3.13 simulates an inverter and load that are far apart on a chip. The inverter device has $W/L = 10$.

(a) Choose a W/L for the load such that $V_{OL} = 0.2$ V.

(b) Calculate five critical points and sketch the VTC as V_{in} varies from 0 to 5 V.

(c) Repeat (b) if the 5-kΩ resistor is in the drain of the load rather than the drain of the inverter.

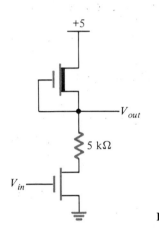

FIGURE P3.13

P3.14. For the CMOS inverter in Fig. P3.14:

(a) Calculate and plot the current that flows through the transistors as a function of $0 \leq V_{in} \leq 5$ V.

(b) Repeat (a) for $V_{DD} = 15$ V.

(c) Verify the results of (a) and (b) using SPICE.

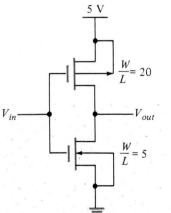

FIGURE P3.14

P3.15. For the NMOS AND gate in Fig. P3.15:

    (*a*) If $V_A = V_B$ and $\gamma = 0$, calculate five critical points of VTC and sketch.

    (*b*) Repeat (*a*) for $\gamma = 0.4$.

    (*c*) If $V_A = V_{DD}$, calculate the VTC with $\gamma = 0$.

    (*d*) Calculate noise margins. What is a suitable definition for noise margin in a case such as this?

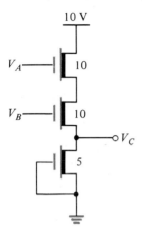

FIGURE P3.15

P3.16. An NMOS process has $V_{T0} = 1.0 \pm 0.5$ V and $t_{ox} = 0.1 \pm 0.02$ μm. (Assume that t_{ox} variations do not cause further variation in V_{T0}.) If both these parameters can vary over the entire range simultaneously, calculate worst-case V_{OL} for the circuit in Fig. P3.16. $(W/L)_L = 1$ and $(W/L)_I = 6$.

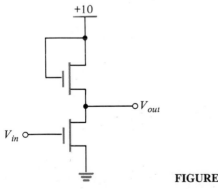

FIGURE P3.16

P3.17. What logical function does the circuit in Fig. P3.17 perform? Design the load transistor so that $V_F = 0.3$ V if the W/L's of other transistors are as shown.

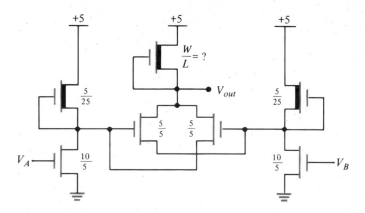

FIGURE P3.17

P3.18. An NMOS inverter with $t_{ox} = 500$ Å has final dimensions $(W/L)_I = 6$ μm/3 μm and $(W/L)_L = 3$ μm/6 μm and operates from $V_{DD} = 5$ V at 27°C. Body doping is 5×10^{14} cm^{-3}. Consider the effects of the following variations: $V_{TE} = 1 \pm 0.3$ V; W, L vary independently ± 0.3 μm.
  (a) Calculate the worst-case range for inverter and load device conductance parameters k_I and k_L.
  (b) Calculate the variation in supply current (output low) for the above process variations and for $\pm$ 10% variation in V_{DD} and 70°C operation. Assume $a = 1.3$ in Eq. (2.15).
  (c) Calculate the worst-case noise margin NM_L by finding the maximum value of V_{OL} and the minimum value of V_{IL} (at 5 V, 27°C).

P3.19. Draw the layout of a two-input NOR gate using the devices specified above. Your layout should be to scale on grid paper, with a level of detail like that of Fig. 3.6. Assume that all dimensions are reduced by a factor of 2 compared to Fig. 3.6.
  (a) Calculate all capacitances and make a hand calculation of t_{PLH} and t_{PHL} assuming nominal dc parameters and a fan-out of 2. Assume input pulses have a rise and fall time of 2 ns.
  (b) Make a SPICE run for the same conditions and compare the results with your hand analysis.

P3.20. Draw the layout of a single-stage CMOS inverter that is able to sink 4 mA with $V_{GSN} = 4$ V and to source 4 mA with $V_{GSP} = -4$ V. In each case $|V_{DS}| = 0.4$ V, $V_{DD} = 5$ V. Use an n well and the simplified design rules described in Sec. 2.8.1. The minimum feature size is 3 μm. Use the SPICE example parameter values given in Table 2.2, but $k'_N = 20$ μA/V^2 and $k'_P = 10$ μA/V^2. Repeat parts (a) and (b) as in P3.19.

CHAPTER
4

SEMICONDUCTOR DIODES

4.0 INTRODUCTION

Semiconductor diodes are fundamental components in digital integrated circuits. Two basic forms of semiconductor diodes are widely used: the *pn junction diode* (or simply *junction diode*) and the *Schottky-barrier diode* (or *Schottky diode*).

The junction diode serves not only as an independent circuit component; bipolar junction transistors require proper interaction of *two pn* junctions in close proximity, and circuit performance is closely related to the properties of these junctions. We have already considered the effects of junction capacitances in MOS transistor circuits. Junctions are also the cause of potentially destructive breakdown and latch-up effects which limit performance and reliability of MOS circuits.

Schottky diodes can be used advantageously in many circuits, but only at the expense of somewhat more complex fabrication processes. Schottky diodes are described in Sec. 4.9.

The intelligent use of a device in circuit design is substantially enhanced by an understanding of the internal physical behavior of the device. Hence we will study briefly potential distributions, carrier distributions, and current transport phenomena in semiconductor diodes. This will lead to a development of the relation between the terminal voltages and currents of the device for both forward and reverse operation. Especially in digital circuits, an important parameter is the *switching speed* of the device, which is the time required to change from a conducting to a nonconducting state. This topic will also be covered in this chapter. In common with most physical phenomena, the semiconductor diode is a temperature-sensitive device. Therefore a short description will be given of the effects of temperature on the circuit properties of the device.

113

Many students studying this text will already have taken a course on the physics of semiconductor devices. For them this chapter will probably simply be a review. However, this chapter is the foundation block for chapters to come, especially those concerning the static and dynamic characteristics of bipolar transistor switching circuits. Hence attention to this chapter is encouraged.

Cross-sectional views of a typical discrete silicon diode and a typical integrated circuit diode are shown in Fig. 4.1 with representative internal dimensions. Both diodes are normally formed by either diffusion or ion implantation processes. (See Chap. 1 for a brief description of fabrication technology.) These techniques inherently create diodes in which one region (the *p*-type region in these examples) is much more heavily doped than the other. The opposite choice of polarities is also common.

The major differences between discrete and integrated circuit (IC) diodes arise from the much smaller dimensions of the latter and the different methods of contacting the lower region (*n*-type in this illustration). Contacting methods

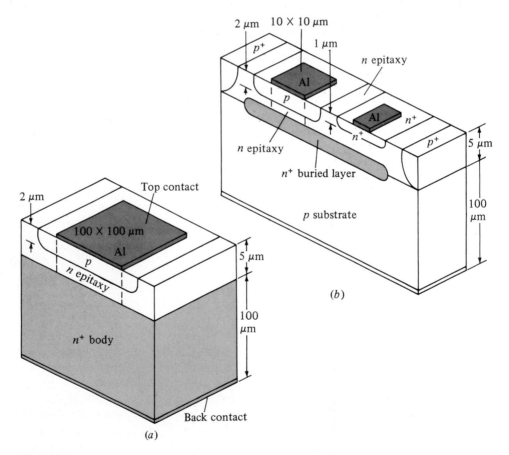

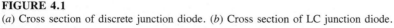

FIGURE 4.1
(*a*) Cross section of discrete junction diode. (*b*) Cross section of LC junction diode.

affect the parasitic resistance in series with the *pn* junction and will be discussed in more detail later. Our initial study will focus on an idealized one-dimensional diode, contacted at opposite ends, as shown in Fig. 4.2*a*. The dashed lines in Fig. 4.1*a* and *b* show where the idealized diode may be considered to exist in the actual diode structures.

4.1   *pn* JUNCTION DIODES

The *pn* junction diode is the simplest of semiconductor devices. In the idealized form sketched in Fig. 4.2*a*, it consists of a semiconductor having a *p*-type region and an *n*-type region separated by a region of transition from one type of doping to the other. For simplicity, we initially consider the *step* (or *abrupt*) junction where this transition region is thin, separating homogeneous regions of *p*- and *n*-type material, as pictured in Fig. 4.2*b*. The *p* region has holes as the dominant, or *majority*, mobile carriers, while in the *n* region electrons are the majority carriers. The standard symbol for the junction diode, shown in Fig. 4.2*a*, shows that the broad end of the arrowhead indicates the *p*-type region.

The *reverse bias* situation has been briefly noted in Secs. 2.5 and 3.3. The applied voltage makes the *p* region negative with respect to the *n* region. The majority carriers are then drawn away from both sides of the junction, and a depletion layer is present. The charge carriers that flow across the junction are the *minority carriers*, electrons from the *p* region and holes from the *n* region. Since in practical diodes the minority carrier concentrations are orders of magnitude lower than the majority carrier concentrations, this reverse "leakage" current flow under reverse bias is orders of magnitude lower than the current that flows under forward bias. Furthermore, the reverse current tends to be constant, almost independent of the applied voltage.

In the *forward bias* situation, the applied voltage makes the *p* region positive with respect to the *n* region. Holes cross the junction from the *p* region to

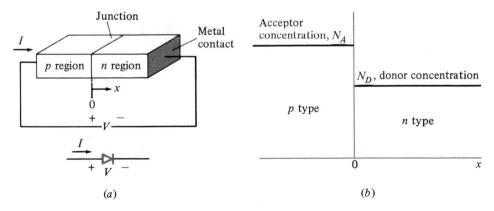

FIGURE 4.2
(*a*) Physical structure of a *pn* diode. (*b*) Idealized impurity distribution in an abrupt *pn* junction diode.

the n region, while electrons cross the junction in the opposite direction. The charge carriers flow from regions of abundant supply, where they are majority carriers, to regions of short supply, where they are minority carriers. Each component of carrier motion contributes to positive current, as indicated by the current arrow in Fig. 4.2a. Even for a small applied voltage this current can be large, and it increases exponentially with the applied voltage.

A typical volt-ampere characteristic of a low-power silicon junction diode is shown in Fig. 4.3a. An expanded scale of the region around the origin is displayed in Fig. 4.3b.

4.2   THE EQUILIBRIUM BARRIER POTENTIAL

We first consider the pn junction with no external voltage bias applied. At a uniform temperature, and with no voltage applied to the terminals, the junction diode is at *equilibrium*. However, we find for this equilibrium condition that there is a potential difference developed across the junction. This is the *equilibrium barrier potential* ϕ_0, also known as the *built-in voltage* V_B. We will first determine the physical basis for this potential and then derive an equation for the potential in terms of the physical properties of the device.

In the p region away from the junction, the hole and electron concentrations are independent of position and are determined solely by the acceptor impurity concentration N_A and the temperature. Normally, with $N_A \gg n_i$, the majority carrier concentration is

$$p_{po} \approx N_A \tag{4.2.1a}$$

and the minority carrier concentration is

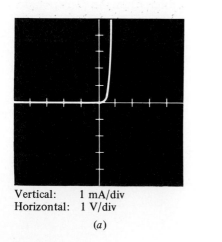

Vertical: 1 mA/div
Horizontal: 1 V/div

(a)

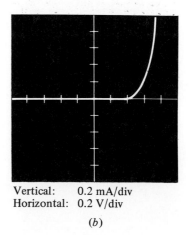

Vertical: 0.2 mA/div
Horizontal: 0.2 V/div

(b)

FIGURE 4.3
Volt-ampere characteristic of a low-power silicon diode.

$$n_{po} \approx \frac{n_i^2}{N_A} \qquad (4.2.1b)$$

The symbols p and n denote the mobile hole and electron concentrations, while the subscript p denotes the p-type region. The second subscript o indicates that the concentration has its thermal equilibrium value. The intrinsic concentration, denoted by n_i, is defined in Sec. 2.3.

Similarly for the n region, the majority carrier concentration is

$$n_{no} \approx N_D \qquad (4.2.2a)$$

and the minority carrier concentration is

$$p_{no} \approx \frac{n_i^2}{N_D} \qquad (4.2.2b)$$

where N_D is the donor impurity concentration.

At equilibrium, the product of hole and electron concentrations is equal to n_i^2 at every point in the semiconductor. From these four equations, note that the hole concentration is much greater in the p region than in the n region, while the electron concentration is much greater in the n region than in the p region.

Although the doping is assumed to change abruptly from p-type to n-type at the junction as seen in Fig. 4.2b, the concentration of mobile carriers does not suddenly change at the junction as might be suggested by Fig. 4.2b. Instead, there is a *depletion region* where, due to the electrostatic forces present, the concentration of mobile carriers is essentially zero, as illustrated in Fig. 4.4a. The depletion region is also known as the *space-charge* region and is typically less than 1 μm wide. Regions where the carrier concentrations are uniform are termed *neutral* regions. Other names for these regions are *ohmic* or *bulk* regions. The boundary of the depletion region is labeled as x_p on the p-region side, and as x_n on the n-region side. The thickness of these boundary regions is much less than the total width of the depletion region, so the boundary regions may be considered abrupt.

The example of Fig. 4.4a is typical of a low power silicon pn junction. In the p region $p_{po} \approx N_A = 10^{18}$ cm^{-3}; therefore $n_{po} = 2 \times 10^2$ cm^{-3}. In the n region, $n_{no} \approx N_D = 10^{16}$ cm^{-3}; therefore $p_{no} \approx 2 \times 10^4$ cm^{-3}. The greater concentration of holes in the p region than in the n region indicates a concentration gradient and hence a *hole diffusion current* from the p region to the n region. Similarly, the greater concentration of electrons in the n region than in the p region leads to an *electron diffusion current* from the n region. The direction of these diffusion currents is indicated in Fig. 4.4b. The diffusion current is always from the area of greater concentration to the area of smaller concentration.

The holes diffusing to the n region leave behind in the p region negatively charged acceptor ions. Likewise, the electrons diffusing to the p region leave behind positively charged donor ions in the n region. Therefore, a potential difference is built up across the depletion region; the n region is positive with respect to the p region. This is the barrier potential ϕ_0 or built-in voltage V_B.

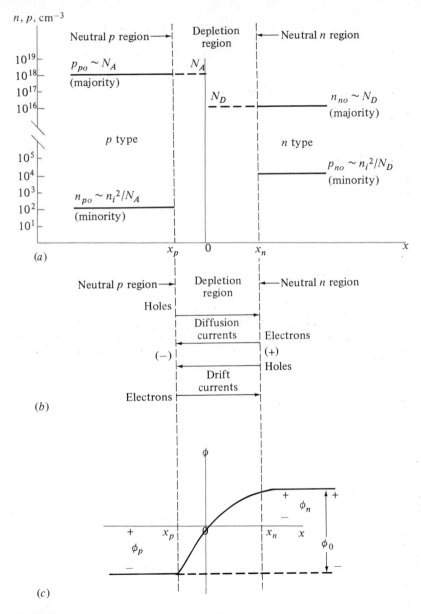

FIGURE 4.4
(a) Mobile carrier concentration at equilibrium. (b) Depletion region current at equilibrium.
(c) Electrostatic potential distribution at equilibrium.

Because of this voltage, an electric field is developed across the depletion region that causes a *hole drift current* from the n region to the p region, and an *electron drift current* from the p region to the n region. The direction of these drift currents is also shown in Fig. 4.4b.

Since at equilibrium the total junction current must be zero, the magnitude of the barrier potential is such that the drift current is just equal to the diffusion current. We will now use some simple ideas to derive a basic equation for this barrier potential. This is done in terms of the hole current, but a similar derivation could be done in terms of the electron current.

The hole diffusion current density is given as

$$J_{p(diff)} = -qD_p \frac{dp(x)}{dx} \qquad (4.2.3)$$

where q = electronic charge, C

D_p = diffusion coefficient for holes, cm^2/s

$dp(x)/dx$ = hole concentration gradient, cm^{-4}

The negative sign in Eq. (4.2.3) is a result of the negative slope of the hole concentration. The hole drift current density is described by

$$J_{p(drift)} = q\mu_p p(x)E(x) \qquad (4.2.4)$$

where μ_p = hole mobility, cm^2/V · s

$p(x)$ = hole concentration as a function of x, carriers/cm^3

$E(x)$ = electric field in the direction of x, V/cm

The total, or net, hole current density is given by

$$J_{p(net)} = J_{p(drift)} + J_{p(diff)} \qquad (4.2.5)$$

$$- q\mu_p p(x)E(x) - qD_p \frac{dp(x)}{dx}$$

But for the equilibrium condition, this current must be zero. Therefore,

$$q\mu_p p(x)E(x) = qD_p \frac{dp(x)}{dx} \qquad (4.2.6)$$

Both and D_p and μ_p are weak functions of the donor concentration N_D, but at any given temperature, the ratio of the two is a constant. This is known as the *Einstein relationship*, namely,

$$\frac{D_p}{\mu_p} = \frac{kT}{q} = V_T \qquad (4.2.7)$$

where k = Boltzmann's constant = 1.38×10^{-23} J/K

T = absolute temperature = 273 + temperature (°C), K

q = electronic charge = 1.60×10^{-19} C

V_T = thermal voltage = 25.9 mV at 27°C

Substituting $V_T{}^*$ in Eq. (4.2.6), we have

$$E(x)\,dx = V_T \frac{dp(x)}{p(x)} \tag{4.2.8}$$

By definition, the electric field in the direction of x is created by the electrostatic potential ϕ:

$$E(x) = -\frac{d\phi(x)}{dx} \tag{4.2.9}$$

Substituting for $E(x)$ in Eq. (4.2.8) we obtain

$$d\phi(x) = -V_T \frac{dp(x)}{p(x)} \tag{4.2.10}$$

We now integrate both sides of this equation from one side (x_p) to the other (x_n), where, as seen in Fig. 4.4c, the corresponding boundary values for the potentials are ϕ_p and ϕ_n, respectively, and p_{po} and p_{no} are the hole concentrations. Then

$$\phi_n - \phi_p = \phi_0 = V_T \ln \frac{p_{po}}{p_{no}} \tag{4.2.11}$$

A further substitution yields the basic equation for the barrier potential, since with $p_{po} = N_A$ and $p_{no} = n_i^2/N_D$:

$$\phi_0 = V_T \ln \frac{N_A N_D}{n_i^2} \tag{4.2.12}$$

from which we obtain the boundary values

$$\phi_n = V_T \ln \frac{N_D}{n_i} \tag{4.2.13a}$$

and

$$\phi_p = -V_T \ln \frac{N_A}{n_i} \tag{4.2.13b}$$

The barrier potential is an electrostatic potential that is developed across the depletion region and, as seen in Eq. (4.2.12), is a function of temperature and the impurity concentrations on either side of the junction. It cannot be measured with any voltmeter because ϕ_0 is exactly canceled out by the difference in contact potentials for probes applied to p-type and n-type semiconductors.

Exercise 4.1. With $N_A = 10^{18}$ cm^{-3} and $N_D = 10^{16}$ cm^{-3}, calculate ϕ_0, ϕ_n, and ϕ_p at 300 K.

---

*The thermal voltage V_T used here should not be confused with the threshold voltage for a MOS transistor, described in Sec. 2.3. Unfortunately, the commonly used symbols for the two are the same.

4.3 DEPLETION REGION CHARGE

For most *pn* junctions the barrier potential ϕ_0 is about 1 V. The width of the depletion region is usually less than 1 μm. Hence the electric field in the depletion region is high (about 10^4 V/cm). Consequently, mobile holes and electrons that enter the depletion region by diffusion from the neutral regions are quickly swept across the depletion region, leaving behind immobile acceptor and donor ions. The distribution of these immobile ions is therefore as sketched in Fig. 4.5a, with the negatively charged acceptor ions on the *p* side of the junction and the positively charged donor ions on the *n* side. Hence we have immobile charged sites in the depletion region.

4.3.1 Charge Density

Here we make use of the *depletion approximation*, which assumes the depletion region is entirely depleted of mobile charge carriers. Furthermore, the immobile

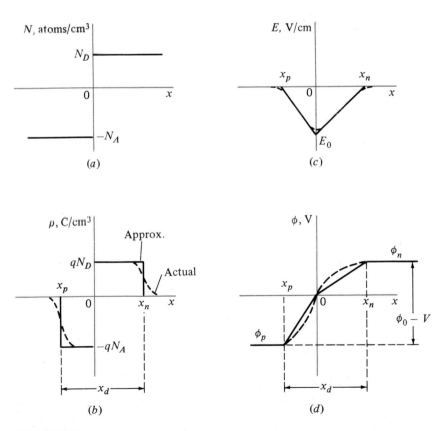

FIGURE 4.5
Approximate distribution of *pn* junction of (*a*) impurity concentration, (*b*) depletion charge density, (*c*) electric field, and (*d*) electrostatic potential.

charged sites in the depletion region form a *dipole layer*; that is, the negative charge on the p side of the junction is exactly equal to the positive charge on the n side. The distribution of the charge on either side of the junction is shown in Fig. 4.5b, where

$$-qN_Ax_p = qN_Dx_n \qquad (4.3.1)$$

4.3.2 Electric Field

With all the voltage being developed across the depletion region, clearly the electric field must be zero at x_n and x_p. Also, with the n side of the junction at a positive potential with respect to the p side and with the direction of x as given, the electric field will be as in Fig. 4.5c, where the maximum value E_o is from Gauss' law,*

$$E_o = \frac{qN_Ax_p}{\epsilon_{si}} = \frac{-qN_Dx_n}{\epsilon_{si}} \qquad (4.3.2)$$

where ϵ_{si} = dielectric constant of silicon = $11.7\epsilon_0$

ϵ_0 = dielectric permittivity of free space = 8.85×10^{-14} F/cm

The electrostatic potential distribution, seen in Fig. 4.5d, is obtained by integrating the electric field seen in Fig. 4.5c. The height of the potential barrier ($\phi_0 - V$) is equal to the area under the curve of Fig. 4.5c. That is,

$$\phi_0 - V = -\frac{1}{2}E_o(x_n - x_p) = -\frac{1}{2}E_oX_d$$

where we now define the total depletion layer width as

$$X_d = x_n - x_p \qquad (4.3.3)$$

4.3.3 Total Charge and Width of Depletion Layer

Solving the Eqs. (4.3.1) to (4.3.3) simultaneously yields:

1. Depletion region charge:

$$Q_j = A\left(2\epsilon_{si}q\frac{N_AN_D}{N_A + N_D}\right)^{1/2}(\phi_0 - V)^{1/2} \qquad C \qquad (4.3.4)$$

2. Maximum electric field:

---

*Gauss' law relates the concentration of the charge in the depletion region to the electric field as $dE/dx = \rho/\epsilon$, where ρ is the charge density and ϵ is the dielectric permittivity of the material.

$$E_o = -\left(\frac{2q}{\epsilon_{si}} \frac{N_A N_D}{N_A + N_D}\right)^{1/2} (\phi_0 - V)^{1/2} \qquad \text{V/cm} \qquad (4.3.5)$$

3. Depletion region width:

$$X_d = \left(\frac{2\epsilon_{si}}{q} \frac{N_A + N_D}{N_A N_D}\right)^{1/2} (\phi_0 - V)^{1/2} \qquad \text{cm} \qquad (4.3.6)$$

In each of these equations, note the sensitivity to the applied voltage. Each of the parameters varies as $(\phi_0 - V)^{1/2}$, increasing with reverse bias ($V < 0$). By convention, the externally applied voltage V is taken to be positive for forward bias.

From Eq. (4.3.6) we can find the boundaries of the depletion region:

$$x_p = X_d \frac{N_D}{N_A + N_D} \qquad (4.3.7a)$$

$$x_n = X_d \frac{N_A}{N_A + N_D} \qquad (4.3.7b)$$

In a practical diode, with the ratio of the impurity concentrations on either side of the junction ≥ 10, the depletion region is almost entirely into the lighter doped side.

4.3.4 Depletion-Layer Capacitance

If the voltage applied to the pn junction diode is changed by a small dV, then the depletion region charge will change by an amount dQ_j. Hence we may define a *depletion-layer (or junction) capacitance* as

$$C_j = \left|\frac{dQ_j}{dV}\right| = A\left(\frac{\epsilon_{si} q}{2} \frac{N_A N_D}{N_A + N_D}\right)^{1/2} \frac{1}{(\phi_0 - V)^{1/2}} \qquad (4.3.8)$$

In this equation the depletion-layer capacitance is defined in terms of the physical parameters of the device. For the circuit designer it is more convenient to express this capacitance in terms of electrical parameters, as follows:

$$C_j = \frac{C_{j0}}{[1 - (V/\phi_0)]^m} \qquad (4.3.9)$$

where C_{j0} is the capacitance at equilibrium, that is, $V = 0$ V, and m is the grading coefficient. For the abrupt junction considered here, $m = \frac{1}{2}$. Shown in Fig. 4.6a is a plot of the junction capacitance as a function of junction voltage for a typical small-area silicon pn junction for which $\phi_0 = 0.94$ V and $C_{j0} = 0.12$ pF. Note the capacitance decreases as the reverse bias increases.

By combining Eqs. (4.3.4) and (4.3.9) we obtain a relationship between $Q_j(V)$ and $C_j(V)$:

$$Q_j(V) = 2C_{j0}(\phi_0)^{1/2}(\phi_0 - V)^{1/2} \qquad (4.3.10)$$

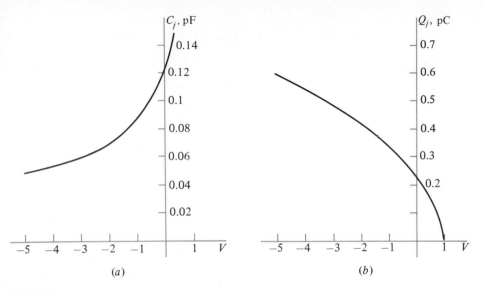

FIGURE 4.6
(*a*) Junction capacitance as a function of junction voltage. (*b*) Depletion region charge as a function of junction voltage.

The plot of $Q_j(V)$ in Fig. 4.6*b* is for the same junction diode as in Fig. 4.6*a*.

It is interesting to note that the junction capacitance may simply be considered as a parallel-plate capacitance, albeit a voltage-sensitive one, having a cross-sectional area ($=A$) and spacing ($X_d = x_n - x_p$). Then,

$$C_j = \frac{\epsilon_{si} A}{X_d} \qquad \text{where } \epsilon_{si} = 1.04 \text{ pF/cm} \qquad (4.3.11)$$

4.3.5 Linear (or Graded) Junction

In the foregoing analysis, the depletion region charge has been derived for a *pn* junction with a step (or abrupt) change of impurity at the junction. Certain older fabrication techniques for transistors did not produce an abrupt junction. Instead the transition from uniform *n*-type to uniform *p*-type doping took place over a finite distance $X_t = 0.5$ to 5 μm. In such cases, a linear (or graded) distribution of the impurities across the junction, as sketched in Fig. 4.7, is a more accurate approximation to the physical situation. For small applied voltages, the depletion layer is confined to the region of varying doping. Its characteristics then become a function of the gradient of the impurity concentration.

An analysis of the graded junction case shows that the junction capacitance is described by Eq. (4.3.9) but with the grading coefficient $m = \frac{1}{3}$. The resulting depletion region charge is related to the capacitance by

$$Q_j(V) = \tfrac{3}{2} C_{j0}(\phi_0)^{1/3}(\phi_0 - V)^{2/3} \qquad (4.3.12)$$

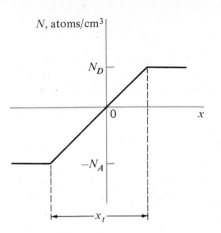

FIGURE 4.7
Idealized distribution of impurity concentration for a graded *pn* junction.

In modern integrated circuit structures, the depletion region extends through the region of graded doping and into the more lightly doped side of the junction with a total voltage across the junction ($\phi_0 - V$) of a few tenths of a volt. For practical purposes, abrupt junction characteristics can be assumed for normal operation of digital circuits. Unless otherwise stated, this is assumed throughout this text.

4.3.6   Large-Signal Equivalent Capacitance

The depletion-layer capacitances defined above are small-signal parameters that vary appreciably over the voltage ranges used in digital circuits. As pointed out in Sec. 3.3.2, it is usually desirable to simplify large-signal circuit analysis by using equivalent voltage-independent capacitances C_{eq} that require the same change in charge as the nonlinear capacitance for a transition between two voltages V_1 and V_2 applied to the junction.

$$C_{eq} = \frac{\Delta Q}{\Delta V} = \frac{Q_j(V_2) - Q_j(V_1)}{V_2 - V_1} \qquad (4.3.13a)$$

where $\Delta Q = \int_{V_1}^{V_2} C_j(V)\, dV = \int_{V_1}^{V_2} C_{j0}\left(1 - \frac{V}{\phi_0}\right)^{-m} dV$

$\Delta V = V_2 - V_1$

Thus

$$C_{eq} = -\frac{C_{j0}\phi_0}{(V_2 - V_1)(1 - m)}\left[\left(1 - \frac{V_2}{\phi_0}\right)^{1-m} - \left(1 - \frac{V_1}{\phi_0}\right)^{1-m}\right] \qquad (4.3.13b)$$

We are now able to define K_{eq}, the dimensionless constant used to relate C_{eq} to C_{j0} for specified values of V_1 and V_2. In the case of an abrupt junction, $m = \frac{1}{2}$, and

$$K_{eq} = \frac{C_{eq}}{C_{j0}} = \frac{-2\phi_0^{1/2}}{V_2 - V_1}\left[(\phi_0 - V_2)^{1/2} - (\phi_0 - V_1)^{1/2}\right] \qquad (4.3.14)$$

Example 4.1

(a) Find ϕ_0 and C_{j0} for an n^+p junction diode with $N_D = 10^{20}$ cm^{-3}, $N_A = 10^{16}$ cm^{-3}, and 20×20 μm$^2 = 400 \times 10^{-8}$ cm^2 in area. From Eq. (4.2.12),

$$\phi_0 = V_T \ln \frac{10^{20} \times 10^{16}}{2 \times 10^{20}} = 0.94 \text{ V}$$

From Eq. (4.3.8),

$$C_{j0} = 4 \times 10^{-6}\left(\frac{1.6 \times 10^{-19} \times 1.04 \times 10^{-12} \times 10^{16}}{2 \times 0.94}\right)^{1/2} = 0.12 \text{ pF}$$

(b) Now find C_j for $V = -5$ V. From Eq. (4.3.9),

$$C_j(-5) = \frac{0.12 \text{ pF}}{(1 + 5/0.94)^{1/2}} = 48 \text{ fF}$$

(c) Find C_{eq} for $V_1 = -5$ V, $V_2 = 0$. From Eq. (4.3.13b),

$$C_{eq} = \frac{(-0.12 \text{ pF})(0.94)}{(0 + 5)(0.5)}\left[\left(1 - \frac{0}{0.94}\right)^{1/2} - \left(1 + \frac{5}{0.94}\right)^{1/2}\right] = 68 \text{ fF}$$

Exercise 4.2. With $N_A = 10^{18}$ cm^{-3}, $N_D = 10^{16}$ cm^{-3}, and $A = 2500$ μm^2, calculate C_j, Q_j, E_o, x_p, and x_n for $V = -5$ and $+0.5$ V. Also find C_{eq} as V changes from -5 to $+0.5$ V.

4.4 pn JUNCTION WITH FORWARD BIAS

No doubt many readers are already aware of the *ideal diode equation* that relates the diode current to the diode voltage, namely,

$$I = I_s(e^{V/V_T} - 1) \qquad (4.4.1)$$

where I_s is a constant, referred to as the *saturation current* of the diode. The thermal voltage V_T is equal to kT/q. In this section we will show the physical basis for the diode equation. Of interest is the concentration of carriers, majority and minority, in the neutral regions adjacent to the depletion region.

By rearrangement of the basic equation for the barrier potential, Eq. (4.2.11), we have

$$p_{no} = p_{po}e^{-\phi_0/V_T} \qquad (4.4.2a)$$

This equation relates the hole concentrations on either side of the junction for the open-circuit equilibrium condition. Similarly, for the electron concentrations,

$$n_{po} = n_{no}e^{-\phi_0/V_T} \qquad (4.4.2b)$$

With forward bias ($V > 0$) applied to the junction diode, the potential barrier is decreased, while with reverse bias ($V < 0$) the barrier is increased. The barrier height then corresponds to the potential $(\phi_0 - V)$. We assume that all the voltage applied to the diode appears across the depletion region; that is, there is no voltage drop across the neutral ohmic regions. Then the two previous equations become

$$p_n(0) = p_p(0)e^{-(\phi_0 - V)/V_T} \tag{4.4.3a}$$

and

$$n_p(0) = n_n(0)e^{-(\phi_0 - V)/V_T} \tag{4.4.3b}$$

The boundary of the depletion region and the neutral regions has been denoted as x_p on the p side and x_n on the n side. Hence, the concentrations at the depletion region boundaries are

$$p_n(x_n) = p_p(x_p)e^{-(\phi_0 - V)/V_T} \tag{4.4.4a}$$

and

$$n_p(x_p) = n_n(x_n)e^{-(\phi_0 - V)/V_T} \tag{4.4.4b}$$

Under the condition of low-level injection (that is, in each of the neutral regions the minority carrier concentration is at all times much less than the majority carrier concentration), we will find that $p_p(x_p) \approx p_{po}$ and $n_n(x_n) \approx n_{no}$. Therefore, Eq. (4.4.4) becomes

$$p_n(x_n) = p_{po}e^{-(\phi_0 - V)/V_T} \tag{4.4.5a}$$

and

$$n_p(x_p) = n_{no}e^{-(\phi_0 - V)/V_T} \tag{4.4.5b}$$

We can eliminate ϕ_0 from these equations if we substitute for $p_{po}e^{-\phi_0/V_I}$ from Eq. (4.4.2a). Then Eq. (4.4.5a) becomes

$$p_n(x_n) = p_{no}e^{V/V_T} \tag{4.4.6a}$$

Similarly, using Eq. (4.4.2b), in Eq. (4.4.5b), we have

$$n_p(x_p) = n_{po}e^{V/V_T} \tag{4.4.6b}$$

These last two equations are commonly called the *law of the junction* in that they relate the minority carrier concentration boundary values to the applied voltage. Especially note the exponential relation. With forward bias the potential barrier is reduced, resulting in a net increase in the flow of mobile carriers across the junction, holes moving to the n region and electrons to the p region, where they become *excess* minority carriers. This is because the concentration is now *greater* than the thermal equilibrium level.

Since powerful electrostatic forces maintain charge neutrality in the p and n regions, the increase in the minority carrier concentrations must be matched by a similar increase in the majority carrier concentrations. This is illustrated in Fig. 4.8, where, for the condition of forward bias, the minority and majority carrier concentrations in the neutral regions are plotted on a linear scale. In the p region, $\Delta p_p = \Delta n_p$, and similarly in the n region, $\Delta p_n = \Delta n_n$. However, note that although the *fractional increase* of the minority carrier concentrations is usually

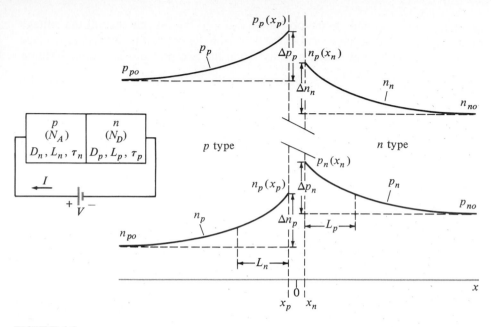

FIGURE 4.8
Carrier concentration in the neutral region near a *pn* junction with forward bias. (Note linear scale.)

large, the *fractional increase* of the majority carrier concentrations is generally infinitesimal. This is because $p_{po} \gg n_{po}$ and $n_{no} \gg p_{no}$. For this reason we were able to replace $p_p(x_p)$ with p_{po} in Eq. (4.4.5). Furthermore, since the fractional increase of the minority carrier concentrations is so large and that of the majority carrier concentrations is so small, we henceforth restrict our interest to just those changes caused by the excess minority carrier concentrations.

As seen in Fig. 4.8, moving away from the depletion region, the concentration of the minority carriers in the neutral regions has an exponential decrease as they recombine with the more copious majority carriers in characteristic distances we call the *diffusion lengths* L_p and L_n. In one diffusion length, the excess concentration is $1/e$ ($= 0.37$) of its original value. Hence, in a few diffusion lengths the concentration reaches its thermal equilibrium level. Associated with the diffusion length is the *mean recombination* time or the *excess minority carrier lifetime*, τ_p and τ_n. The longer the diffusion length the longer will be the carrier lifetime. The two are related by the diffusion coefficient, such that

In the *n* region:
$$D_p = \frac{L_p^2}{\tau_p} \quad \text{cm}^2/\text{s} \tag{4.4.7a}$$

In the *p* region:
$$D_n = \frac{L_n^2}{\tau_n} \quad \text{cm}^2/\text{s} \tag{4.4.7b}$$

Note particularly that these physical parameters are the characteristics of the minority carriers. That is, the diffusion coefficients D_p and D_n are a function of the impurity concentrations N_D and N_A, respectively. In silicon, typical values for L_p and L_n range from 10^{-4} to 10^{-2} cm, and the corresponding values for τ_p and τ_n are in the range 10^{-8} to 10^{-6} s.

With the assumption of no voltage drop in the neutral regions, there can be no drift currents in these regions. However, the concentration gradient for the excess minority carriers, pictured in Fig. 4.8, indicates a hole diffusion current in the neutral n region, and an electron diffusion current in the neutral p region.

Hence in the n region, using the equation for diffusion current density,

$$J_p(x_n) = -qD_p \frac{dp(x)}{dx}\bigg|_{x=x_n} \tag{4.4.8}$$

Our interest is in the excess minority carrier concentration, in particular, for excess holes in the n region, where

$$p_n'(x) = p_n(x) - p_{no} \tag{4.4.9}$$

Due to the exponential decrease of the excess minority carrier concentration,

$$p_n'(x) = p_n'(x_n)e^{-x/L_p} \tag{4.4.10a}$$

$$= [p_n(x_n) - p_{no}]e^{-x/L_p} \tag{4.4.10b}$$

$$= p_{no}(e^{V/V_T} - 1)e^{-x/L_p} \tag{4.4.10c}$$

In the last step we have made use of Eq. (4.4.6a). Now, including this last equation in Eq. (4.4.8),

$$J_p(x_n) = \frac{qD_p p_{no}}{L_p}(e^{V/V_T} - 1) \tag{4.4.11a}$$

Similarly in the p region,

$$J_n(x_p) = \frac{qD_n n_{po}}{L_n}(e^{V/V_T} - 1) \tag{4.4.11b}$$

With the depletion region width $x_n - x_p \ll L_p$ or L_n, we are justified in assuming that there is negligible recombination in this region. Then the total current density is

$$J(0) = J_p(x_n) + J_n(x_p) \tag{4.4.12}$$

$$= q\left(\frac{D_p p_{no}}{L_p} + \frac{D_n n_{po}}{L_n}\right)(e^{V/V_T} - 1)$$

If area is included to obtain the total current instead of the current density,

$$I = qA\left(\frac{D_p p_{no}}{L_p} + \frac{D_n n_{po}}{L_n}\right)(e^{V/V_T} - 1) \tag{4.4.13}$$

which we may write as

$$I = I_s(e^{V/V_T} - 1) \tag{4.4.14}$$

This is the *ideal diode equation*, where the saturation current I_s is, from Eq. (4.4.13),

$$I_s = qA\left(\frac{D_p p_{no}}{L_p} + \frac{D_n n_{po}}{L_n}\right) \tag{4.4.15a}$$

Or alternatively, using Eqs. (4.4.7*a* and *b*),

$$I_s = qA\left(\frac{L_p p_{no}}{\tau_p} + \frac{L_n n_{po}}{\tau_n}\right) \tag{4.4.15b}$$

We note from this equation that the saturation current may be considered as arising from the thermal generation of minority carriers in the neutral regions, within one diffusion length of the depletion region boundary. Also, we note that the saturation current I_s is directly related to the cross-sectional area of the junction. Further, in a practical diode the saturation current is mainly influenced by the physical characteristics of the lighter doped side. That is the side where the minority carrier concentration is the greatest.

With forward bias, the diode current I is a result of the reduction of the potential barrier and the subsequent injection of excess minority carriers into the neutral regions. The exponential relation is a consequence of the law of the junction.

In practice, the diode current I is somewhat less than predicted by Eq. (4.4.14). This is because not all of the applied voltage appears across the junction. There is always some voltage drop across the neutral (ohmic) regions, though the resistance of these regions is usually only a few ohms (1 to 100 Ω). Generally, the departure of the diode equation from the "ideal" is only significant for currents greater than 1 mA.

Finally, in this section the diode equation has been derived for the condition of forward bias, but as we shall see in the next section, the equation is also valid with reverse bias.

Exercise 4.3. Calculate the saturation current I_s for a low-power silicon *pn* junction diode having the following characteristics. Also, solve for the forward voltage V with $I = 1$ mA. Assume $V_T = 26$ mV.

$$N_A = 10^{18} \text{ cm}^{-3} \qquad\qquad N_D = 10^{16} \text{ cm}^{-3}$$

$$D_n = 20 \text{ cm}^2/\text{s} \qquad\qquad D_p = 10 \text{ cm}^2/\text{s}$$

$$L_n = 10 \text{ } \mu\text{m} \qquad\qquad L_p = 5 \text{ } \mu\text{m}$$

$$A = 2500 \text{ } \mu\text{m}^2$$

4.5   *pn* JUNCTION WITH REVERSE BIAS

With reverse bias and $|V| \gg V_T$, we have from the ideal diode equation [Eq. (4.4.14)] that the diode current I approaches a constant, $-I_s$. Let us now see the physical basis for this. With reverse bias the potential barrier is increased, and from the law of the junction [Eq. (4.4.6)], the boundary values for the minority carrier concentration are

$$p_n(x_n) = p_{no}e^{-V/V_T} \qquad (4.5.1a)$$

and

$$n_p(x_p) = n_{po}e^{-V/V_T} \qquad (4.5.1b)$$

Therefore, with $|V| \gg V_T$ the concentration of minority carriers at the depletion region boundaries becomes small ($\to 0$). Some distance away from the depletion region, the concentration is at the thermal equilibrium level. This is sketched in Fig. 4.9.

From the figure, we note a concentration gradient in each of the neutral regions. Consequently there is a diffusion flow of minority carriers toward the junction of holes in the n region and electrons in the p region. Once these mobile carriers approach the depletion region, they are swept across the junction as a drift current, holes to the p region and electrons to the n region. This is because of the electric field in the depletion region caused by the reverse bias voltage, minus on the p side and plus on the n side. However, from Eq. (4.5.1) we note that ideally this drift current is limited, since with $|V| \geq 4V_T$, the concentration gradient does not show much change in slope as the reverse voltage is changed. That is, with reverse bias, each component of the drift current is *saturated* for $|V| \geq 4V_T$.

The two drift currents, one from each side of the junction, constitute the saturation current I_s.

$$I = I_s(e^{V/V_T} - 1) \qquad (4.5.2a)$$

$$= -I_s \qquad \text{with } V \leq -4V_T \qquad (4.5.2b)$$

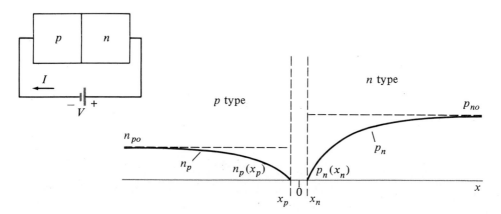

FIGURE 4.9
Minority carrier concentration in the neutral region near a *pn* junction with reverse bias. (Note linear scale.)

In practice, reverse currents much larger than indicated by Eq. (4.5.2b) are generally measured. This is due to the thermal generation of holes and electrons in the depletion region. These mobile carriers are swept out of the region by the reverse voltage as a drift current. For silicon pn junctions, this current from the depletion region is usually much larger than the saturation current. For a low-power silicon pn junction, $I_s \approx 10^{-15}$ A, but the depletion region current is about 10^{-12} A. In addition, unless special precautions are taken, the leakage current across the surface of the device will be in the order of 10^{-9} A. Both the depletion region current and the leakage current increase as the applied voltage is increased.

4.6   DIODE SWITCHING TRANSIENTS

So far, in our study of the electrical and physical roperties of the pn junction diode, we have mostly been concerned with the static, or steady-state, characteristics. As we now turn to consider the switching transients in the pn junction, the action of the diode turning on and turning off, we must be concerned with the dynamic, or time-dependent, characteristics.

The ideal diode equation [Eq. (4.4.14)] is a fundamental equation that relates the diode currents to the diode voltage. The diode current may also be related to the *excess minority carrier charge*. This is another fundamental equation, which introduces the concept of charge control to our study. This concept is a most useful one for analysis of switching transients in both diodes and bipolar junction transistors.

With forward bias applied to the pn junction there is an excess minority carrier concentration in the neutral regions near the depletion region. This is described in Sec. 4.4 and pictured in Fig. 4.8. Each of these excess mobile carriers carries a charge q ($= 1.6 \times 10^{19}$ C). Hence, the area between the curve representing p_n and the level p_{no} in Fig. 4.8 is a measure of the *excess hole charge* Q_p in the neutral n region.* From Fig. 4.8,

$$Q_p = qA \int_0^\infty p_n'(x)dx \qquad (4.6.1)$$

where A = cross-sectional area of the junction

$\quad q$ = electronic charge

From Eq. (4.4.10c) we have

$$p_n'(x) = p_{no}(e^{V/V_T} - 1)e^{-x/L_p}$$

An equation similar to Eq. (4.6.1) can be written for the *excess electron charge* Q_n in the neutral p region. However for a practical diode, one side of the junction

---

*To maintain charge neutrality, the total charge in the neutral region is zero. The *excess* charge refers to the *excess* minority carrier concentration compared to that at equilibrium.

is much more heavily doped than the other. We assume $N_A \gg N_D$, then $p_{no} \gg n_{po}$ and $Q_p \gg Q_n$. Henceforth, therefore, we concentrate our attention on the minority carrier holes in the neutral n region.

In carrying out the integration in Eq. (4.6.1) we assume the depletion region width is much less than the hole diffusion length L_p; then the origin 0 and x_n are essentially synonymous. Including Eq. (4.4.10c), we have from Eq. (4.6.1) that

$$Q_p = qAL_p p_{no}(e^{V/V_T} - 1) \tag{4.6.2}$$

From this equation we see directly that the excess minority carrier charge is related to the diode voltage. Furthermore, this charge is positive for forward bias and negative for reverse bias.

To relate the excess minority carrier charge to the diode current, we first note that the forward current I is the product of the cross-sectional area of the junction A and the current density at the origin $J_p(0)$. Thus from Eq. (4.4.11a),

$$I = AJ_p(0) = \frac{qAD_p p_{no}}{L_p}(e^{V/V_T} - 1) \tag{4.6.3}$$

We may eliminate $p_{no}(e^{V/V_T} - 1)$ in this equation by substitution from Eq. (4.6.2). Then

$$I = Q_p \frac{D_p}{L_p^2} \tag{4.6.4a}$$

But from Eq. (4.4.7a), $D_p/L_p^2 = 1/\tau_p$. Therefore,

$$I = \frac{Q_p}{\tau_p} \tag{4.6.4b}$$

This simple equation relates, on a static basis, the diode current to the excess minority carrier charge and a physical parameter of the device, the excess minority carrier lifetime τ_p. In the steady state, the current I supplies holes to the neutral n region at the same rate as they are being lost by recombination.

To obtain a dynamic, or time-dependent, relation we must include the time rate of change of the excess minority carrier charge. Then we have an equation for the instantaneous, or time-dependent, current as

$$i(t) = \frac{Q_p}{\tau_p} + \frac{dQ_p}{dt} \tag{4.6.5}$$

This is a fundamental charge-control equation for the pn junction diode.* It simply states that current $i(t)$ supplies holes to the neutral n region at the rate at which the stored charge increases plus the rate at which holes are being lost by recombination. We next make use of these charge-control expressions in the calculations of the switching times for a pn junction diode.

---

*For the case where $N_A = N_D$ and $Q_p = Q_n$, the instantaneous current is given as $i = Q_p/\tau_p + dQ_p/dt + dQ_n/\tau_n + dQ_n/dt$. The steady-state current is simply $I = Q_p/\tau_p + Q_n/\tau_n$.

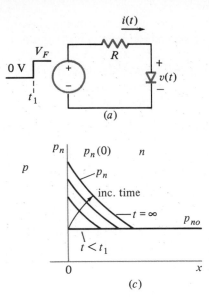

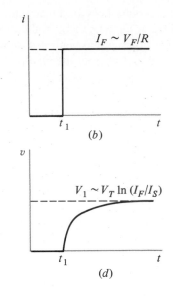

FIGURE 4.10
Turn-on of *pn* junction diode.

4.6.1   Turn-On

Consider the simple circuit of Fig. 4.10a, where at time $t = t_0$, the source voltage is equal to 0 V and the diode is at equilibrium with zero bias. Then at $t = t_1$, the source voltage is suddenly changed to V_F. The diode forward current I_F instantly becomes approximately V_F/R. (This assumes that V_F is much greater than the diode forward voltage V_1.) The current in the circuit as a function of time is seen in Fig. 4.10b. Since the forward current is approximately constant with time, and since $I_F = AJ_p(0)$ and $J_p(0) = -qD_p \, dp/dx|_{x=0}$, the slope of the minority carrier concentration at the depletion region boundary is also approximately constant. The sketch of minority carrier concentration in Fig. 4.10c also represents the increase in the excess minority carrier charge Q_p.

The diode voltage depends upon the absolute value of $p'_n(0)$, since from Eq. (4.4.6a),

$$V_1 = V_T \ln \left[\frac{P'_n(0)}{p_{no}} + 1 \right] \tag{4.6.6}$$

Hence V_1 increases only as $p'_n(0)$ increases. The diode voltage as a function of time is seen in Fig. 4.10d. We see an initial fast rise, followed by a slow settling time to the final steady-state value. This is because initially most of the holes injected into the neutral n region go to increase the charge Q_p. The recombination term, Q_p/τ_p, is small since Q_p is small. As the charge nears its steady-state value, more of the injected holes go to supply recombination and less to increase the charge store. The time taken for V_i to reach its steady-state value is dependent upon the excess minority carrier lifetime τ_p and the forward current I_F. But for the case of low-level injection, this is about 2 times τ_p.

In practice, to ensure a fast turn-on of the diode, the constraint of low-level injection is not observed. Indeed, under high-level injection, the injected carrier concentration may be greater than the equilibrium level of the majority carrier concentration. We then have a phenomenon called *conductivity modulation* taking place. This is because the high level of injected carriers modulates the conductivity of the neutral regions. In a conductivity modulated diode the voltage waveform shows a fast rise, even a peaked response that is followed by a decrease to the steady-state value.

4.6.2 Turn-Off

Now at time $t = t_2$, where $t_2 \gg t_1$, we suddenly change the source voltage from V_F to V_R, where $V_R \ll 0$ V. This is diagrammed in Fig. 4.11a. Hence,

At $t < t_2$: $$i(t) = I_F = \frac{V_F - V_1}{R} \qquad (4.6.7a)$$

At $t = t_2$: $$i(t) = I_R = \frac{V_R - V_1}{R} \qquad (4.6.7b)$$

As seen in Fig. 4.11b, the charge stored in the neutral n region, under the

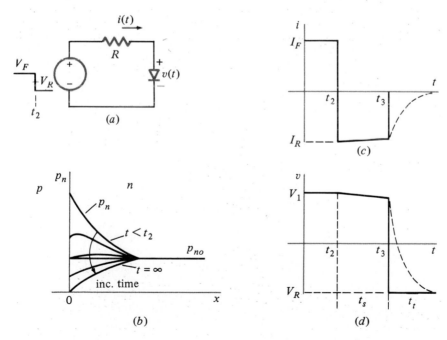

FIGURE 4.11
Turn-off of *pn* junction diode.

forward bias condition ($t < t_2$), must now be removed. This takes time. In fact, the junction cannot be reverse-biased until all this charge is removed. The current I_R is what removes the charge.

Using the charge-control expressions,

At $t < t_2$: $\qquad i_F = \dfrac{Q_p}{\tau_p}$ $\qquad$ the steady state condition $\qquad$ (4.6.8a)

At $t = t_2$: $\qquad i_R = \dfrac{Q_p}{\tau_{p^-}} + \dfrac{dQ_p}{dt}$ $\qquad$ including the transient term $\qquad$ (4.6.8b)

The solution of this simple differential equation is

$$Q_p(t) = i_R \tau_p + C e^{-t/\tau_p} \qquad (4.6.9)$$

To determine the constant C, we know

At $t \le t_2$: $\qquad\qquad Q_p(t) = i_F \tau_p = i_R \tau_p + C \qquad (4.6.10a)$

Therefore,

$$C = \tau_p(i_F - i_R) \qquad (4.6.10b)$$

The complete equation for the excess minority carrier charge is therefore

$$Q_p(t) = \tau_p[i_R + (i_F - i_R)e^{-t/\tau_p}] \qquad (4.6.11)$$

But

At $t = t_3$: $\qquad\qquad\qquad Q_p = 0 \qquad (4.6.12)$

Hence, setting $Q_p(t) = 0$ and solving for $t = t_3$ in Eq. (4.6.11),

$$t_3 = \tau_p \ln \frac{i_F - i_R}{-i_R} \qquad (4.6.13)$$

The time $t_3 - t_2$ we term the *diode storage time* t_s.

At $t = t_3$, $Q_p(t_3) = 0$. Hence, neglecting any depletion-layer capacitance effects, at $t = t_3$ the current $I_R \to 0$ and the voltage across the diode will suddenly switch from V_1 to V_R. This is seen in Fig. 4.11c and d.

Including the effects of the depletion-layer capacitance, we know that the voltage across the diode cannot change instantaneously. To change the charge in the depletion region takes time; this is the *transition time* t_t, indicated by the dashed lines in Fig. 4.11c and d.

The total diode reverse recovery time t_{rr} is the sum of the storage time and the transition time, namely,

$$t_{rr} = t_s + t_t \qquad (4.6.14)$$

Exercise 4.4. In Fig. 4.11a, let $V_F = +10$ V, $V_R = -10$ V, $R = 10$ kΩ, and $\tau_p = 10$ ns. While conducting, the diode voltage is constant at 0.7 V. Solve for the diode storage time t_s.

4.7 DIODE STRUCTURES

We now turn our attention to the electrical characteristics of a particular physical structure of the diode. Consider a *pn* junction, with $N_A = 10^{18}$ cm^{-3} and $N_D = 10^{16}$ cm^{-3}. We have already noted, at the beginning of Sec. 4.6, that with such a ratio for the impurity concentration, the switching time characteristics of the diode will be primarily determined by the minority carriers in the lighter doped side, the *n* region. Hence, as seen in Fig. 4.12*a* we have labeled the length of the *n* region as *W*. The metallurgical junction is at the origin, 0.

We consider the two cases:

1. The *long-base diode*, where $W \gg L_p$
2. The *short-base diode*, where $W \ll L_p$

As described in Sec. 4.4, the distance L_p is the mean diffusion length of the minority carriers (holes) in the *n* region.

4.7.1 Long-Base Diode

Under the condition of forward bias, we picture in Fig. 4.12*b* the concentration of the excess minority carriers in the *n* region as a function of the distance from the junction. Notice for this long-base case that the minority carrier concentration is down to the thermal equilibrium level in a distance less than *W*; that is, $W \gg L_p$. This is the classic case we have been considering so far. For this condition, using the charge-control expressions,

$$i_F = \frac{Q_p}{\tau_p} \tag{4.7.1a}$$

and

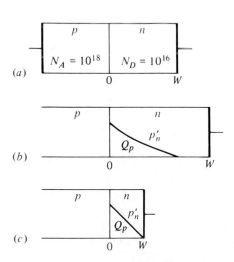

(a)

(b)

(c)

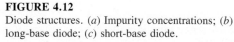

FIGURE 4.12
Diode structures. (*a*) Impurity concentrations; (*b*) long-base diode; (*c*) short-base diode.

$$i_R = \frac{Q_p}{\tau_p} + \frac{dQ_p}{dt} \qquad (4.7.1b)$$

$$t_s = \tau_p \ln \frac{i_F - i_R}{-i_R} \qquad (4.7.1c)$$

For the classic case, the *mean minority carrier lifetime* τ_p is related to the diffusion length L_p and the diffusion constant D_p as

$$\tau_p = \frac{L_p^2}{D_p} \qquad (4.7.2)$$

As an example: with $N_D = 10^{16}$ cm^{-3}, $D_p = 10$ cm^2/s at 300 K. Depending upon the silicon purity and crystal quality, L_p may range from 1 to 100 μm. Taking a typical value of $L_p = 10$ μm, we have

$$\tau_p = \frac{(10 \ \mu m)^2}{10 \ cm^2/s} = 100 \ ns$$

Notice with $W \gg 10$ μm that we make use of the classical equations in the solution of diode reverse recovery time problems.

4.7.2 Short-Base Diode

Now consider the case where $W \ll L_p$. As seen in Fig. 4.12c, recombination is now complete at the ohmic contact that makes connection to the n region. We again make use of the charge-control expressions, but now the minority carrier lifetime τ_p is replaced by the *mean transit time* τ_T. The storage time is given by

$$t_s = \tau_T \ln \frac{i_F - i_R}{-i_R} \qquad (4.7.3)$$

For the short-base case τ_T, the mean transit time of the excess minority carriers through the neutral region, is related to W and D_p by

$$\tau_T = \frac{Q_p}{i_F} = \frac{W^2}{2D_p} \qquad (4.7.4)$$

This is because the excess minority carrier charge Q_p is the product of the electronic charge q, the area of the junction A, and the area under the curve of p_n' in Fig. 4.12c (i.e., a triangle). That is,

$$Q_p = \frac{qAWp_n'(0)}{2} \qquad (4.7.5)$$

But from Eq. (4.4.6a), $p_n'(0) = p_{no}(e^{V/V_T} - 1)$. Also, from Eq. (4.4.13),

$$i_F = \frac{qAD_p p_{no}}{W}(e^{V/V_T} - 1) \qquad (4.7.6)$$

Taking the ratio of Eqs. (4.7.5) and (4.7.6) we have τ_T as in Eq. (4.7.4). The difference between the excess minority carrier transit time, as in Eq. (4.7.4), and the lifetime, as in Eq. (4.7.2), is to be particularly noted.

4.7.3 Discrete *npn* Diode

It is common practice to connect bipolar transistors as junction diodes. Five possible circuit configurations are seen in Fig. 4.13*a*. Each has its own particular advantages and drawbacks.

Sketched in Fig. 4.13*b* is a cross-sectional view of a discrete *npn* transistor. In such a device the collector is generally the back contact and the device thickness is $\approx 100~\mu m$. For this device the base width is $\approx 1~\mu m$.

In the *open-collector diode* of Fig. 4.13*a*, with a base doping of $N_A = 10^{18}$ cm^{-3}, $D_n = 7$ cm^2/s. Assuming a diffusion length $L_n = 10~\mu m$, this connection yields a short-base characteristic, since the base width is only $\approx 1~\mu m$. For this case,

$$\tau_T = \frac{W^2}{2D_n} = \frac{(10^{-4})^2}{2(7)} = 0.7~\text{ns} \tag{4.7.7}$$

The *open-emitter diode* of Fig 4.13*a* is an example of a long-base diode. The width of the lighter doped side of the junction, i.e., the collector region, is $\approx 100~\mu m$. For a collector doping of $N_D = 10^{16}$ cm^{-3}, the diffusion coefficient $D_p = 10$ cm^2/s, and we assume $L_p = 10~\mu m$. For this case,

$$\tau_p = \frac{L_p^2}{D_p} = \frac{(10^{-3})^2}{10} = 100~\text{ns} \tag{4.7.8}$$

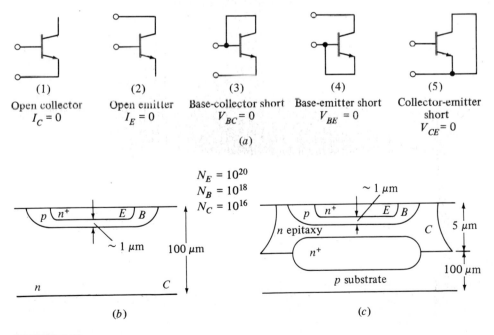

(1)
Open collector
$I_C = 0$

(2)
Open emitter
$I_E = 0$

(3)
Base-collector short
$V_{BC} = 0$

(4)
Base-emitter short
$V_{BE} = 0$

(5)
Collector-emitter short
$V_{CE} = 0$

(*a*)

$N_E = 10^{20}$
$N_B = 10^{18}$
$N_C = 10^{16}$

(*b*)

(*c*)

FIGURE 4.13
(*a*) Diode-connected transistor configuration. (*b*) Cross section of discrete *npn* transistor. (*c*) Cross section of IC *npn* transistor.

Because Eqs. (4.7.1c) and (4.7.3) have identical functional form, one cannot determine from diode storage time measurements whether long-base or short-base behavior dominates. No distinction need be made between these for purposes of circuit design. Measurements on a sample device may be used to determine the value of the time constant for that device. This is the subject of Demonstration D4.3, to be described at the end of this chapter.

4.7.4 Integrated Circuit Diodes

Diodes used in integrated circuits are also generally formed from *npn* transistors in the same manner as described above. The sketch of Fig. 4.13c shows a cross-sectional view of an integrated circuit *npn* transistor. The collector region is formed by an *n*-type epitaxial layer about 5 μm thick on top of the *p*-type supporting substrate. The substrate is about 100 μm thick, and the base is about 1 μm thick.

Exercise 4.5. The IC *npn* transistor of Fig. 4.13c is used in the diode connections of Fig. 4.13a. Determine whether the resulting diode has a long-base or a short-base characteristic. Assume a diffusion length of 10 μm in each of the regions.

4.8 DIODE MODELS FOR CIRCUIT SIMULATION

As noted in Chap. 1 the accurate characterization of an integrated circuit can be done only with a computer and a circuit simulation program. Of the many circuit simulation programs, one of the more successful is SPICE. This requires modeling the circuit in mathematical terms. As well, the data out of the computer are only as good as the data into the computer. This requires developing good techniques to measure the circuit parameters that characterize the circuit model.

4.8.1 SPICE Diode Model

The model used for the *pn* junction diode in SPICE is seen in Fig. 4.14. The dc characteristic of the diode is modeled by the nonlinear current source I_D. From the ideal diode equation [Eq. (4.4.14)],

$$I_D = I_s(e^{V_D/nV_T} - 1) \tag{4.8.1}$$

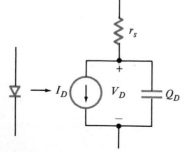

FIGURE 4.14
The SPICE2 diode model.

The parameter n, termed the *emission coefficient*, will be described shortly; it is usually equal to 1, but it can be as high as 2. The resistor r_s models the series resistance contributed by the neutral (ohmic) regions on either side of the junction.

Charge storage in the diode is modeled by the charge storage element Q_D, where

$$Q_D = \tau_T I_s (e^{V_D/nV_T} - 1) + C_{j0} \int_0^{V_D} \left(1 - \frac{V}{\phi_0}\right)^{-m} dV \qquad (4.8.2)$$

Equivalently, this element can be represented by a voltage-dependent capacitor, since

$$C_D = \frac{dQ_D}{dV_D} = \tau_T \frac{I_s}{nV_1} e^{V_D/nV_T} + \frac{C_{j0}}{(1 - V/\phi_0)^m} \qquad (4.8.3)$$

The element Q_D models two totally different charge storage effects in the diode. Charge storage in the depletion region is taken directly from Eq. (4.3.9) and is modeled by the parameters C_{j0}, ϕ_0, and m.

Charge storage due to minority carrier injection across the junction is described by the exponential term in Eq. (4.8.2), which relates the diode current to the minority carrier charge and lifetime as given in Eq. (4.6.4). In SPICE this effect is modeled by the transit time parameter τ_T, which is equal to τ_n, τ_p, or τ_T, depending upon which of these parameters applies.

4.8.2 Measurement of the Diode Parameters

A listing of the parameters required to model the diode is presented in Table 4.1. Included in the table are the default values used in SPICE and an example that might be typical for a low-power switching diode.

I_s, r_s, n. A one-point measurement of the saturation current of the diode can readily be made with the diode forward-biased, since from the ideal diode equation and Eq. (4.8.1),

$$I_s = \frac{I_D}{e^{V_D/nV_T}} \qquad (4.8.4)$$

TABLE 4.1
SPICE diode model parameters

Symbol	Name	Parameter name	Units	Default	Example
I_s	IS	Saturation current	A	1.0E-14	1.0E-14
r_s	RS	Ohmic resistance	Ω	0	10
n	N	Emission coefficient		1	1.0
C_{j0}	CJO	Zero-bias depletion capacitance	F	0	2.0E-12
ϕ_0	VJ	Built-in potential	V	1	0.8
m	M	Grading coefficient		0.5	0.5
τ_T	TT	Transit time	s	0	1.0E-10

Hence measuring V_D and I_D leads to a determination of I_s. This measurement should be made at a low value of I_D so that any voltage drop across r_s is an insignificant part of V_D.

A better method is to plot I_D versus V_D, as seen in Fig. 4.15. Then extrapolating the curve to $V_D = 0$ V, the intercept is at I_s. This is so, since from Eq. (4.8.4),

$$V_D = nV_T \ln \frac{I_D}{I_s} \qquad (4.8.5)$$

where with $I_D = I_s$, $V_D = 0$.

The saturation current of the diode is a function of the junction area, but typically $I_s = 10^{-14}$ A.

The curve in Fig. 4.15 is a straight line over a large range of I_D, from about 10^{-9} to 10^{-3} A. The emission coefficient n is determined from the slope of the diode characteristic. In almost all cases the emission coefficient for diodes made from transistors is unity, leading to $n = 1.0$. For Schottky diodes (next section) and power diodes, n can be somewhat larger than 1.0.

The curvature at high current (>1 mA) is caused by:

1. The bulk ohmic effect of the neutral regions on either side of the junction yielding a larger terminal voltage for V_D at any given value of I_D. Hence an estimate can be made of r_s, since from Fig. 4.15,

$$r_s = \frac{\Delta V_D}{I_D} \qquad (4.8.6)$$

The value of r_s may range from 10 to 100 Ω, but a typical value is 10 Ω.

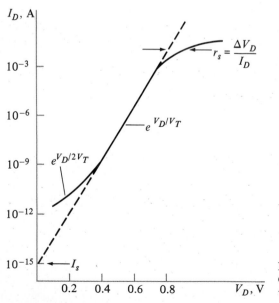

FIGURE 4.15
Graph of diode current versus diode voltage.

2. High-level injection effects. Where the concentration of the minority carriers at the junction boundaries approaches that of the majority carriers, the physical equations derived in Sec. 4.4 are no longer applicable. The diode current no longer follows the simple exponential

$$I_D = I_s e^{V_D/nV_T} \tag{4.8.7a}$$

but approaches

$$I_D = I_s e^{V_D/2nV_T} \tag{4.8.7b}$$

At the low currents (< 1 nA) the change in slope of the diode characteristic is primarily caused by the thermal generation and recombination (trapping) of electrons and holes in the depletion region. This process is active at all current levels, but at low currents the effect is more noticeable since the percentage of carriers lost by trapping is greater.

The measurement of these static parameters (I_s, r_s, and n) is part of Demonstration D4.1 described at the end of this chapter.

C_{j0}, ϕ_0, m.   The measurement of the junction capacitance parameters is readily done with a capacitance bridge and the circuit seen in Fig. 4.16a.

A direct measurement of C_{j0} is difficult because the excitation signal (ac) from the bridge can be as large as 1 V rms. Instead, the junction capacitance is measured as a function of the junction voltage, and plotted as shown in Fig. 4.16b. Then by extrapolation, C_{j0} can be found. Notice, in forward bias the junction capacitance increases rapidly, another reason for using the extrapolation method. By analytic curve fitting the other two parameters may be determined. Some typical values for C_{j0}, ϕ_0, and m are given in Table 4.1. The measurement of these parameters is included in Demonstration D4.2.

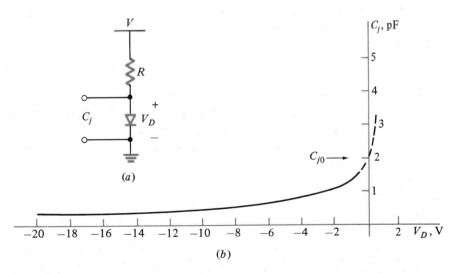

(a)

(b)

FIGURE 4.16
(a) Circuit for measurement of C_j. (b) Graph of junction capacitance versus junction voltage.

τ_T. The transit time τ_T is obtained from a measurement of the diode storage time t_s described in Sec. 4.6.2. From Eq. (4.6.13),

$$\tau_T = \frac{t_s}{\ln[(i_F - i_R)/(-i_R)]} \tag{4.8.8}$$

Further details on the measurement of this parameter are presented in Demonstration D4.3.

4.9 SCHOTTKY-BARRIER DIODE

Very useful diodes can be made by forming a microscopically clean contact between certain metals and a lightly doped ($N_D \leq 10^{16}$ cm^{-3}) n-type semiconductor, as illustrated in Fig. 4.17. Because practical means of forming the necessary clean interface were not developed until about 1970, Schottky-barrier diodes* were not used in early integrated circuits. Included in Fig. 4.17 is the circuit symbol for a Schottky diode.

The principal advantage of Schottky diodes is that all conduction is via electrons flowing in metal or n-type silicon. That is, only majority carriers contribute to current flow. Hence there are no minority carrier storage effects, and their limitations on diode switching speed are completely eliminated.

A detailed understanding of the Schottky diode generally involves energy-band theory, which is beyond the scope of this text. A practical grasp of Schottky diode behavior can be developed through consideration of the electron energy and electrostatic potential distributions in the Schottky diode, as sketched in Fig. 4.18. Figure 4.18b may be compared with the corresponding potential distribution for a pn junction as seen in Fig. 4.4c.

For every metal-semiconductor contact there is a characteristic potential barrier ϕ_B which depends only on the two materials. This is the *Schottky barrier*.

---

*Named for W. Schottky, a pioneer in the study of metal-semiconductor interfaces.

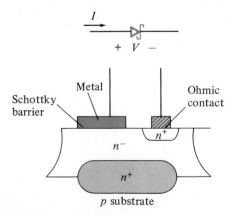

FIGURE 4.17
Cross section of Schottky-barrier diode.

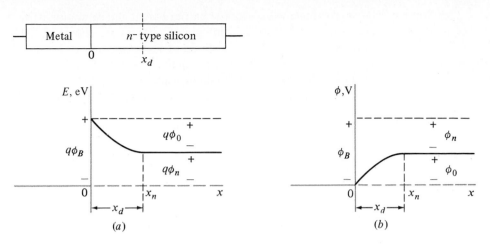

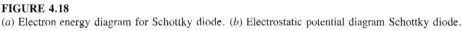

(a)   (b)

FIGURE 4.18
(a) Electron energy diagram for Schottky diode. (b) Electrostatic potential diagram Schottky diode.

It is not a function of semiconductor doping and is only a weak function of temperature. The barrier potential blocks the flow of electrons from the metal toward the semiconductor.

In the neutral semiconductor, far from the barrier, and with no externally applied potentials, the electron potential energy in the semiconductor is, from Eq. (4.2.13a), $\phi_n - V_T \ln (N_D/n_i)$. The potential difference $\phi_B - \phi_n = \phi_0$ is simply the equilibrium difference in energy for electrons in the metal compared to electrons in the semiconductor. It is analogous to ϕ_0 for the *pn* junction diode. Externally applied reverse bias adds to ϕ_0 and increases the width of the depletion region, which extends entirely into the semiconductor. Depletion-layer width, capacitance, etc., are identical to those for an abrupt p^+n junction diode.

Forward bias subtracts from ϕ_0, allowing electrons to flow from the *n*-type semiconductor into the metal. Although holes do not take part in forward conduction, the roles of electrons, ϕ_0, and V are identical to the case of the *pn* junction diode. Hence it is not surprising that detailed analysis, omitted here, yields

$$I = I_0(e^{V/V_T} - 1) \tag{4.9.1}$$

The form of I_0 is somewhat different for the Schottky diode:

$$I_0 = K_{SB} T^2 e^{-\phi_B/V_T} \tag{4.9.2}$$

where $K_{SB} =$ a constant, best evaluated by measurement

$\qquad T =$ absolute temperature, K

$\qquad \phi_B =$ barrier height, V

$\qquad V_T =$ thermal voltage, V

Some typical approximate values for the barrier height and saturation current of metal-n^- silicon barriers often used in integrated circuits are shown in Table 4.2. A typical value of I_s for a pn junction diode of the same area is 2×10^{-16} A. The larger values of I_0 for Schottky diodes result in smaller forward voltage drops at any fixed current density. From the diode equation (4.4.14) we have the useful rule of thumb that forward voltage drop at room temperature decreases by 60 mV for each factor of 10 increase of I_s or I_0.

Not all metal-semiconductor contacts have diode characteristics. When the semiconductor is heavily doped (N_A or $N_D \gg 10^{17}$ cm^{-3}), the depletion layer becomes so narrow that carriers can travel in either direction through the potential barrier by a quantum-mechanical carrier transport mechanism known as *tunneling*. Under these conditions, current flows equally well in either direction, resulting in what is usually described as an *ohmic contact*. In fact, this is the physical explanation for most all ohmic contacts to semiconductor devices.

The diode model used in SPICE, and described in Sec. 4.8.1, applies just as well for the Schottky-barrier diode, but with $I_0 = I_s$, and of course $\tau_T = 0$.

Exercise 4.6. At a forward current I of 2 mA, compare the forward voltage V of a Schottky-barrier and a pn junction diode. Use the values of I_0 and I_s given in this section.

4.10   TEMPERATURE EFFECTS

The ideal diode equation (4.4.14) is affected by temperature in two ways:

1. The saturation current I_s is a function of n_i^2, which increases strongly with temperature.

2. The thermal voltage V_T that appears in the exponent is linearly dependent on temperature.

In the equation for the saturation current [Eq. (4.4.15)], except for $p_{no} = n_i^2/N_D$ and $n_{po} = n_i^2/N_A$, the other terms within the parentheses are not strongly influenced by temperature. The effect of temperature on I_s is therefore mainly the result of the temperature dependence of the square of the intrinsic carrier concentration (n_i^2), and that is

$$n_i^2(T) = KT^3 e^{-V_{go}/V_T} \tag{4.10.1}$$

TABLE 4.2
Schottky-barrier diode characteristic

	Barrier height, V	I_0, 100-μm^2 area, A
Molybdenum	0.58	1×10^{-9}
Titanium silicide	0.59	1×10^{-9}
Aluminum	0.69	2×10^{-11}
Palladium silicide	0.73	2×10^{-12}
Platinum silicide	0.85	2×10^{-14}

where $K = \text{constant}$

$T = \text{absolute temperature (room temperature} \approx 300 \text{ K)}$

$V_{go} = \text{band-gap voltage (for silicon at 300 K, } V_{go} = 1.11 \text{ V)}$

$V_T = \text{thermal voltage (at 300 K, } V_T \approx 26 \text{ mV)}$

The *temperature coefficient* of the saturation current describes the fractional change in I_s per unit change in temperature. This is approximately equal to the fractional change in n_i^2 per unit change in temperature. That is,

$$\frac{1}{I_s}\frac{dI_s}{dT} = \frac{1}{n_i^2}\frac{dn_i^2}{dT} \qquad (4.10.2)$$

$$= \frac{3}{T} + \frac{1}{T}\frac{V_{go}}{V_T}$$

Near room temperature (300 K), the first term is $\approx 1\%/K$, but for silicon the second term is $\approx 14\%/K$. In other words, the saturation current approximately doubles every 5°C, since $(1.15)^5 \approx 2.0$.

Experimentally, in a silicon diode, the reverse current doubles about every 8°C. This is because, for silicon, the reverse current is mainly the result of either the leakage current or the depletion region current, as described in Sec. 4.5. The depletion region current varies directly with the intrinsic carrier concentration n_i instead of n_i^2 for the saturation current.

The temperature coefficient of the *diode current* for a fixed forward bias is given as

$$\left.\frac{1}{I}\frac{dI}{dT}\right|_{dV=0} \approx \frac{1}{I}\frac{dI_s e^{V/V_T}}{dT} \qquad (4.10.3)$$

$$= \left(\frac{1}{I_s}\frac{dI_s}{dT}\right) - \left(\frac{1}{T}\frac{V}{V_T}\right)$$

Obviously, the fractional change in the forward current is less than the fractional change in the saturation current. For example, assume a fixed forward bias of 700 mV at 300 K; then

$$\frac{1}{I}\frac{dI}{dT} = 0.15 - \frac{1}{300}\frac{700}{26} \qquad (4.10.4)$$

$$= 0.15 - 0.09 = 6\%/°C$$

For this example, the forward current doubles about every 12°C.

Finally, to evaluate the temperature dependence of the *diode voltage* at a fixed current,

$$\left.\frac{dV}{dT}\right|_{dI=0} = \frac{d[V_T \ln (I/I_s)]}{dT} \qquad (4.10.5)$$

$$= \frac{V}{T} - V_T\left(\frac{1}{I_s}\frac{dI_s}{dT}\right)$$

Assume now, that at a constant forward current, $V = 700$ mV at 300 K; then

$$\frac{dV}{dT} = \frac{700}{300} - (26 \text{ mV})(0.15) = 2.3 - 3.9 = -1.6 \text{ mV/°C} \qquad (4.10.6)$$

Commonly, at a forward current of about 1 mA, the measured temperature coefficient of a silicon *pn* junction diode is approximately -2 mV/°C. The measurement of the saturation current and diode voltage as a function of temperature is described in Demonstration D4.1.

Derivation of the temperature characteristics of the Schottky diode is similar to the above. The temperature dependence of forward voltage for the Schottky diode is found to be

$$\frac{dV}{dT}\bigg|_{dI=0} = \frac{d[V_T \ln (I/I_s)]}{dT} \qquad (4.10.7)$$

$$= \frac{V}{T} - \frac{V_T}{T}\left(2 + \frac{\phi_B}{V_T}\right)$$

A typical measured value of temperature coefficient of forward voltage for a Schottky diode is -1.2 mV/°C.

Fortunately, these temperature characteristics are generally only of minor importance to the digital IC designer. This is because the signal voltages in most digital circuits are much larger than the variation of the diode voltage over the temperature range of 0 to 70°C, that is, the range of commercial interest. For military usage, where the range of temperature is -55 to 125°C, other effects become more troublesome than the temperature coefficients of either the diode current or diode voltage. However, for the analog IC designer these temperature coefficients are of major importance and become a subtle part of the circuit design.

4.11 BREAKDOWN DIODES

Shown in Fig. 4.19 is the *I-V* characteristic of a silicon diode, and particular attention is drawn to the reverse region. Notice, as expected, that for small reverse voltages the reverse current is essentially zero, but at the breakdown voltage ($BV = -20$ V) the reverse current shows a marked increase. Provided the external circuit limits this current to a safe value, the diode may be operated in this breakdown region, where it acts as a stable voltage reference source.

The breakdown of the reverse characteristic of the diode is the result of one of two distinctively different effects. One of these is called *avalanche breakdown*. Note from Eq. (4.3.5) that with the reverse voltage V_R much greater than the barrier potential ϕ_0, the maximum electric field across the depletion region is given as

$$|E_{max}| = \left(\frac{2qV_R}{\epsilon_{si}} \frac{N_A N_D}{N_A + N_D}\right)^{1/2} \qquad (4.11.1)$$

Hence, with increasing reverse voltage, the magnitude of the electric field

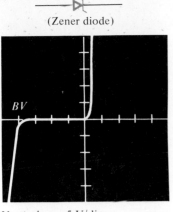

(Zener diode)

BV

Vertical: 5 V/div
Horizontal: 5 mA/div

FIGURE 4.19
I-V characteristic of junction diode, showing reverse voltage
breakdown.

increases. As a consequence, the carriers crossing the depletion region are accelerated to high velocity. At a critical field E_{crit}, the carriers have sufficient energy
that in colliding with other silicon atoms in the depletion region they create new
electron-hole pairs. These secondary carriers can in turn create more carriers
before leaving the depletion region. This is the avalanche breakdown process.
The value of E_{crit} is about 2×10^5 V/cm for impurity concentrations of the order
10^{16} cm^{-3}.

It has been found empirically that if the low-voltage reverse current of the
diode is I_R, in the region of the avalanche breakdown the reverse current becomes

$$I_{RA} = MI_R \qquad (4.11.2a)$$

where M is the multiplication factor, defined by

$$M = \frac{1}{1 - (V_R/BV)^n} \qquad (4.11.2b)$$

where V_R is the applied voltage and n has a value between 3 and 6.

In highly doped diodes, where the impurity concentration is about 10^{18}
cm^{-3}, the value of E_{max} is close to 10^6 V/cm. With this high value of electric
field we have the second mechanism of breakdown in the diode called *Zener
breakdown*. In this process the electric field in the depletion region is so high
that electrons are simply stripped away from the outer orbit of the silicon atoms.
Hence mobile carriers are produced in the depletion region as before, but this
process does not include any multiplication effect.

By close control of the processing, in particular the impurity concentration,
the breakdown voltage of a diode can be controlled to a particular value. When
the breakdown voltage is ≤ 6 V, the process is usually due to the Zener effect,
though commonly all voltage reference diodes are termed Zener diodes even if the
breakdown voltage is > 6 V. The circuit symbol and current-voltage characteristics
for either an avalanche or Zener diode are shown in Fig. 4.19.

4.12 SUMMMARY

- From a simple description of the physical behavior of a *pn* junction diode, we were able to derive equations for some important electrical properties, namely:

    Equilibrium barrier potential ϕ_0, also known as the built-in voltage V_B
    Depletion-layer (or junction) capacitance C_j
    Saturation current I_s

- Further consideration of the physical behavior of the diode with forward and reverse bias led to the fundamental *ideal diode equation*.
- For digital applications, an all-important parameter of the diode is the switching time. Hence, equations were developed for the *turn-on and turn-off* times of the diode.
- Various physical configurations of IC diodes were illustrated and classified as either *short-base* or *long-base* diodes.
- The parameters required to model the diode when using SPICE were listed; also, experimental methods for measuring these parameters were described.
- A simple description of the Schottky-barrier diode has also been included in this chapter.
- Finally, the effects of temperature on both the junction diode and the Schottky diode were presented, and the major temperature coefficients were determined.

REFERENCES

1. P. E. Gray, D. DeWitt, A. R. Boothroyd, and J. F. Gibbons, *Physical Electronics and Circuit Models of Transistors*, Wiley, New York, 1964.
2. P. E. Gray and C. L. Searle, *Electronic Principals—Physics, Models, and Circuits*, Wiley, New York, 1969.
3. R. S. Muller and T. Kamins, *Device Electronics for Integrated Circuits*, Wiley, New York, 1977.
4. I. Getreu, *Modeling the Bipolar Transistor*, Tektronix, Inc., Beaverton, Oreg., 1976.
5. L. W. Nagel, "SPICE2, A Computer Program to Simulate Semiconductor Circuits," *ERL Memorandum ERL-M520*, University of California, Berkeley, May 1975.

DEMONSTRATIONS

D4.1   Si Diode: *I-V* Characteristics

The objective of this experiment is to determine the saturation current I_s of a diode-connected transistor* and demonstrate the variation of I_s and V_D as a function of

---

*Use a low-power *npn* switching transistor, such as the 2N4275 or 2N2222A.

temperature. The open-collector diode of Fig. 4.13a will be used in each of the experiments.

1. (a) At room temperature measure V_D at $I_D = 0.1$, 1.0, and 10 mA. Use a commercial curve tracer or measure directly with a digital voltmeter.
   (b) Repeat part (a) at a temperature of 0 and 100°C. Control the temperature with an ice bath and boiling water as in Demonstration D2.2.
2. (a) Use the data at $I_D = 1$ mA to determine I_s from Eq. (4.4.14). What is your estimate for the temperature coefficient of I_s? Compare your result with Eq. (4.10.2).
   (b) Compute the temperature coefficient of V_D at $I_D = 1$ mA, and compare with Eq. (4.10.5).

D4.2 Si Diode: Depletion Capacitance

In this experiment we will measure the junction capacitance as a function of the junction voltage and determine C_{j0}, ϕ_0, and m.

1. With a capacitance bridge and the circuit seen in Fig. 4.16a, measure C_j versus V_j and graph as in Fig. 4.16b. To prevent erroneous readings $V_j \leq -1$ V, but to prevent voltage breakdown $V_j \geq -5$ V.
2. (a) By extrapolation of the graph determine C_j (0 V), namely, C_{j0}.
   (b) Initially assume an abrupt junction and use Eq. (4.3.9) to determine the barrier potential, ϕ_0.
   (c) Using the values of $C_j = -1$ and -5 V and the method of trial and error, determine better values of ϕ_0 and the grading coefficient m. Is the diode better described as an abrupt or graded junction?

D4.3 Si Diode: Switching Transients

In this experiment we investigate the turn-on and turn-off behavior of the junction diode.

1. (a) Use the circuit of Fig. D4.1 and sketch the turn-on and turn-off voltage waveforms. Compare with Figs. 4.10d and 4.11d.
   (b) Repeat part (a) for each of the diode connections shown in Fig. 4.13a. Comment

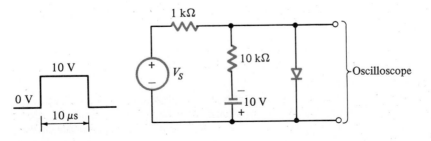

FIGURE D4.1

on how the diode switching transients are affected by each of the diode configurations.

2. (a) Assume the open-collector diode yields a short-base characteristic, and use Eq. (4.7.3) to determine the transit time τ_T of the excess minority carriers in the base region.

   (b) Assume the open-emitter diode yields a long-base characteristic, and use Eq. (4.7.1c) to determine the lifetime τ_p of the excess minority carriers in the collector region.

PROBLEMS

P4.1. For a heavily doped pn junction as would be found at the base-emitter junction of a BJT, $N_A = 2 \times 10^{20}$ cm^{-3} and $N_D = 5 \times 10^{17}$ cm^{-3}. Calculate:
(a) The barrier potential ϕ_0.
(b) The depletion region charge Q_j, depletion region width X_d, and maximum electric field E_0 at $V = -5$ and $+0.5$ V.

P4.2. For a lightly doped pn junction as would be found at the base-collector junction of a BJT, $N_A = 2 \times 10^{15}$ cm^{-3} and $N_D = 5 \times 10^{16}$ cm^{-3}. Calculate:
(a) ϕ_0.
(b) Q_j, X_d, and E_0 for $V = -20$ and $+0.2$ V.

P4.3. For the pn junction of P4.1, in the p region, $D_n = 2$ cm^2/s and $L_n = 10$ μm. In the n region, $D_p = 4$ cm^2/s and $L_p = 5$ μm. The area $= 100$ μm^2 and $m = 0.5$. Find:
(a) The saturation current I_s.
(b) The junction capacitance C_j for $V = -5$ and $+0.5$ V.

P4.4. For the pn junction in P4.2, in the p region, $\mu_n = 1200$ cm^2/V·s and $L_n = 50$ μm. In the n region, $\mu_p = 300$ cm^2/V·s and $L_p = 5$ μm. The area $= 50 \times 50$ μm^2 and $m = 0.5$. Calculate:
(a) I_s.
(b) C_j for $V = -20$ and $+0.2$ V.

P4.5. For the diode described in P4.3, calculate:
(a) The capacitance at equilibrium, C_{j0}.
(b) The equivalent voltage-independent capacitance C_{eq} when V changes from -5 to $+0.5$ V.

P4.6. For the diode described in P4.4, calculate:
(a) C_{j0}.
(b) C_{eq} when V changes from -20 to $+0.2$ V.

P4.7. Two diodes are connected in series, and $I_{s1} = 10^{-16}$ A and $I_{s2} = 10^{-14}$ A. If the applied voltage is 1 V, calculate the diode current I and calculate the voltage across each diode.

P4.8. Two diodes are connected in parallel, and $I_{s1} = 10^{-16}$ A and $I_{s2} = 10^{-14}$ A. Calculate the current in each diode if the total current $I = 1$ mA , and calculate the voltage across each diode.

P4.9. The boundary between low-level and high-level injection may be defined where the excess minority carrier concentration equals the equilibrium majority carrier concentration. Using this criterion find the maximum voltage V and current I for low-level injection in the diodes described in:
(a) P4.3.
(b) P4.4.

P4.10. For the diode logic gate of Fig. P4.10, plot the transfer function with $V_{in} = -5$ to $+5$ V. $V_A = +5$ V. Assume $V_D = 0.6$ V. What problem arises if this gate is used to drive other gates of the same type?

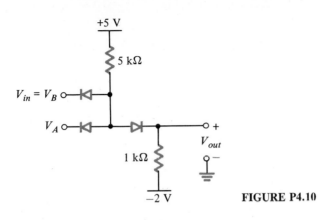

FIGURE P4.10

P4.11. Repeat P4.10 for the circuit in Fig. P4.11. $V_A = 0$ V.

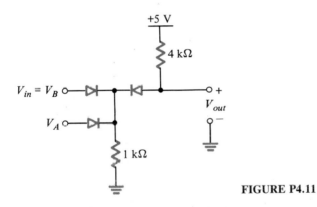

FIGURE P4.11

P4.12. For the circuit in Fig. P4.12, assume a long-base diode and neglect any diode capacitance.
  (a) Sketch $V_D(t)$ and $I_s(t)$ for the input waveform shown. $I_s = 10^{-14}$ A and $\tau_p = 10$ ns.
  (b) Verify the results of (a) with SPICE.

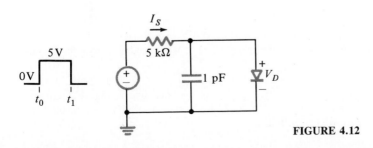

FIGURE 4.12

P4.13. For the circuit in Fig.P4.13 repeat P4.12. Also sketch $V_{out}(t)$.

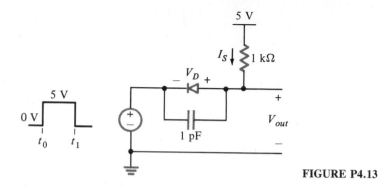

FIGURE P4.13

P4.14. The diode capacitance C_j is measured as a function of the diode voltage V_j. Using the data below extract C_{j0}, ϕ_0, and m_j.

V, V	C_j, pF
0.5	2.18
0.2	1.94
−1	1.19
−5	0.735
−10	0.571

P4.15. What value of I_0 is necessary to create a Schottky diode with $V_D = 0.3$ V at $I_D = 1$ mA?

P4.16. The data below are for a Schottky-barrier diode. Calculate the Schottky-barrier potential of the junction.

Temp., K	V_D, V	I_D, mA
300	0.5	1
400	0.5	4

P4.17. A Schottky-barrier diode and a *pn* junction diode are connected in series and biased at a constant current. ϕ_B is a constant 0.4 V over the temperature range of interest, and K is independent of temperature. For what voltage across the diodes is $\partial V/\partial T = 0$?

P4.18. The voltage applied across a diode is held constant at 0.6 V. The temperature is allowed to vary from −50 to 150°C. Assuming the band-gap voltage V_{go} remains constant over this range, plot $I_D(T)$ if $I_D(27°C) = 1$ mA. Verify your result with the more accurate model in SPICE.

P4.19. Repeat P4.18 if the current is held constant at 1 mA and V_D is allowed to vary. Plot $V_D(T)$.

P4.20. Assuming that voltage breakdown occurs when the maximum electric field in the depletion region reaches 10^6 V/cm, estimate the doping concentration of a symmetrical ($N_A = N_D = N$) silicon diode that exhibits voltage breakdown at -10, -5, and 0 V.

P4.21. Plot the I-V characteristic of an avalanche diode with $I_s = 10^{-16}$ A, $BV = 8$ V, and $n = 4$. Verify with SPICE.

P4.22. Use the diode in P4.21 to find V_{out} in the circuit of Fig. P4.22. How much power is dissipated by the diode?

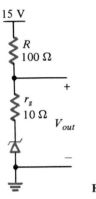

15 V

R
100 Ω

+

r_s
10 Ω

V_{out}

−

FIGURE P4.22

P4.23. Consider a diode like that of Fig. 4.1a, with the vertical dimensions seen in the figure but with an area larger than 100×100 μm. Resistivity is 0.01 Ω ·cm for the n^+ body and for the p-type anode. The n-type epitaxial layer has a resistivity of 0.5 Ω ·cm, corresponding to a donor concentration $N_D = 10^{16}$ cm^{-3}.
(a) Find the diode area needed so that the series resistance of all neutral n and p regions is reduced to 0.1 Ω.
(b) Calculate the total forward voltage drop across this diode at a forward current of 1 A.
(c) Calculate the maximum reverse voltage that can be applied to this diode if the peak electric field in the depletion region is not to exceed 2×10^5 V/cm to prevent avalanche breakdown. Consider the junction to be abrupt.

CHAPTER
5

BIPOLAR JUNCTION TRANSISTOR

5.0 THE BIPOLAR JUNCTION TRANSISTOR

We now progress to discuss the basic device behavior of the bipolar junction transistor (BJT). In this, we will use many of the ideas already developed for the *pn* junction diode in Chap. 4. Again, the aim will be not only to develop the equations but also to describe the physical processes with sufficient detail that the reader is able to relate the equations to the physical process, and vice versa. As well, a knowledge of the physical behavior of these devices will lead to a better understanding of the performance limits of bipolar junction transistor digital circuits. It is also important for an understanding of integrated injection logic (I^2L), latch-up in CMOS circuits, and failure modes in digital integrated circuits.

First, we describe the operation of the BJT in physical terms. Next, we develop the general equations for the terminal currents of the device—the Ebers-Moll equations. Then we find suitable and simplified versions of these equations that are appropriate for a particular region (mode) of operation. This will also include developing simple circuit models for the BJT that will be used later in the analysis and design of bipolar transistor digital circuits. The model for the BJT that is used in the simulation program SPICE will also be described. In addition, procedures for measuring some important electrical parameters of a typical transistor will be given. The chapter concludes with a listing of some simplified design rules appropriate for the layout of a high-yield small-sized BJT in a digital integrated circuit.

5.1 TRANSISTOR OPERATION IN THE FORWARD ACTIVE MODE

By forming two *pn* junctions in a single silicon crystal, such that the *p* region is common and very narrow (≤ 1 μm), an *npn* bipolar junction transistor is made. This is shown in cross section in Fig. 5.1*a*, where the two *n* regions, the emitter *E* and the collector *C*, sandwich the *p* region, the base *B*. The analysis that follows is idealized in that a one-dimensional structure is assumed. The dashed lines in Fig. 5.1*a* show where such a region may be assumed to exist.

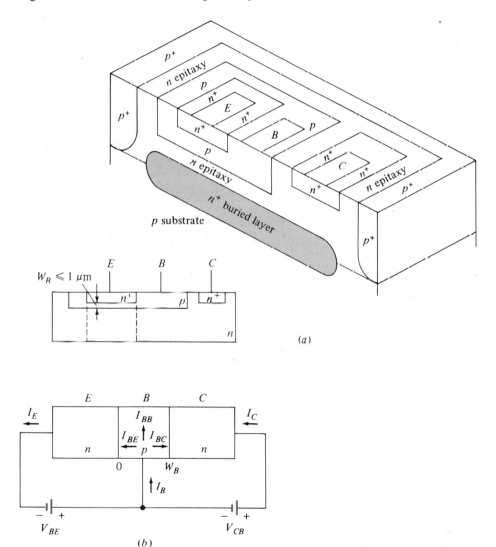

(a)

(b)

FIGURE 5.1
(*a*) Cross section of *npn* bipolar junction transistor. (*b*) *npn* transistor structure biased in the forward active mode.

The biasing batteries as shown in Fig. 5.1*b* cause the transistor to operate in the *forward active mode*. That is, the base-emitter junction is forward-biased by V_{BE} being positive at the base, and the base-collector junction is reverse-biased by V_{CB} being positive at the collector. For the physical consequences of this operating mode we refer to Fig. 5.2.

Figure 5.2 shows a plot on linear coordinates of the concentration of minority carriers as a function of the distance away from the emitter and collector junctions. For a typical transistor at thermal equilibrium, the majority carrier concentrations in the emitter, base, and collector regions are, respectively, 10^{20}, 10^{18}, and 10^{16} atoms/cm^3. Hence the minority carrier concentrations at equilibrium and 300 K are given from Eqs. (4.2.1*b*) and (4.2.2*b*):

In the emitter region:
$$p_{eo} = \frac{n_i^2}{N_D} \approx 2 \text{ cm}^{-3}$$

In the base region:
$$n_{bo} = \frac{n_i^2}{N_A} \approx 2(10^2) \text{ cm}^{-3}$$

In the collector region:
$$p_{co} = \frac{n_i^2}{N_D} \approx 2(10^4) \text{ cm}^{-3}$$

Also shown in Fig. 5.2 are the two depletion regions, one at the emitter junction and the other at the collector junction. The regions outside the depletion regions are termed the *neutral* regions. For simplicity, the dc analyses in this chapter

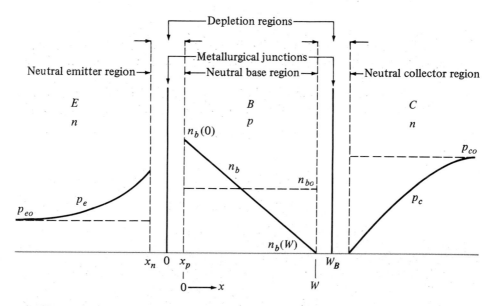

FIGURE 5.2
Minority carrier concentration for *npn* transistor biased in the forward active mode. (Note linear scale.)

assume uniform doping concentrations (uniform equilibrium carrier concentrations) in each of the three regions. In fact, in real transistors the doping concentrations vary within each region. However, analysis based on uniform doping leads to results that are in agreement with measurements made on real devices except for adjustments of some constant terms. This is true because most dc device parameters are determined by the integrals of concentrations within the neutral regions.

The width of the base W_B is the distance between the two metallurgical junctions, but the width of the neutral base region is W, with the origin as shown in the figure. Clearly, if the depletion regions are much narrower than the neutral base region, $W \approx W_B$.

Two basic assumptions are made in the analysis that follows.

1. At all times we assume that in the neutral base region the concentration of minority carriers is much less than the concentration of majority carriers. That is, even with forward bias, for an *npn* transistor $n_b(0) \lll p_{bo}$. This condition is termed *low-level injection* in the base region. Modern digital circuits often violate this condition. The effects of *high-level injection* were described in Sec. 4.8.2.

2. The second assumption is that all terminal voltages appear across the junction depletion regions. That is, we assume negligible voltage drops across the ohmic neutral regions. The effects of such voltage drops were also considered in Sec. 4.8.2.

With forward bias applied to the emitter junction the concentration of minority carriers at the depletion region interface is increased. In particular, in the neutral base region from Eq. (4.4.6b),

$$n_b(0) = n_{bo}e^{V_{BE}/V_T} \qquad (5.1.1)$$

The reverse bias at the collector junction causes the concentration of minority carriers at this interface to be decreased:

$$n_b(W) = n_{bo}e^{V_{BC}/V_T} = n_{bo}e^{-V_{CB}/V_T} \qquad (5.1.2)$$

But with $V_{CB} \ggg V_T, n_b(W) \to 0$.

Typically, the distance $W \lll L_b$, the diffusion length of the minority carriers in the base of the transistor. That is, there is negligible recombination in the neutral base region. The minority carrier distribution across the base then approximates the triangular form as shown in Fig. 5.2. The width of the neutral emitter and collector regions is much greater than the diffusion length for the minority carriers; the minority carrier concentrations therefore show an exponential slope away from the depletion region to the thermal equilibrium level.

5.1.1 The Physical Picture

Physically, what is going on? As we forward-bias the emitter junction the potential barrier at this junction is reduced; electrons (the majority carriers in the emitter of this *npn* example) are thus injected (emitted) into the base, where they become minority carriers since the base is a *p* region. This current, the emitter current I_E, across the emitter-base depletion region is primarily a diffusion current because of the steep concentration gradient due to the electron concentration being majority carriers in the emitter but minority carriers in the base. Across the base there is another concentration gradient from $n_b(0)$ to $n_b(W)$. Hence there is an electron diffusion current across the base region to the collector junction. The collector junction is reverse-biased, so there is an electric field across the collector-base depletion region due to the voltage V_{CB}. Hence the current across this junction, the collector current I_C, is a drift current, because any electrons at W are accelerated across the depletion region (i.e., collected at the collector) owing to the positive voltage at the collector. With negligible recombination in the base region, $I_C \approx I_E$.

We now develop some important device equations for the bipolar junction transistor. But first we rewrite Eqs. (5.1.1) and (5.1.2) in terms of the excess minority carrier concentration, where

$$n_b'(x) = n_b(x) - n_{bo} \tag{5.1.3}$$

Then from Eq. (5.1.1),

$$n_b'(0) = n_{bo}(e^{V_{BE}/V_T} - 1) \tag{5.1.4}$$

and from Eq. (5.1.2),

$$n_b'(W) = n_{bo}(e^{V_{BC}/V_T} - 1) \tag{5.1.5}$$

The distribution of this concentration across the neutral base region is diagrammed in Fig. 5.3. This figure is similar to Fig. 5.2, but now we are plotting the excess minority carrier concentration, denoted as n_b'. Notice that the excess minority carrier concentration at the collector junction $[n'_b(W)]$ is less than 0, but the analysis is simplified by assuming $n'_b(W) = 0$, as shown by the dashed line in the figure.

Now the initial assumption is that indeed $I_C = I_E$, and we note again that across the emitter junction this current is a diffusion current, across the neutral base region the current continues as a diffusion current, but across the collector-base junction the current is a drift current.

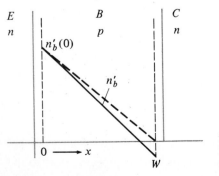

FIGURE 5.3
Excess minority carrier concentration in neutral base region; *npn* transistor biased in forward active mode.

In the neutral base region, therefore, similar to Eq. (4.4.8),

$$I_C = qAD_b \frac{dn_b'}{dx}\bigg|_{x=0} \tag{5.1.6}$$

where D_b represents the diffusion coefficient for the minority carriers in the neutral base region.

From Fig. 5.3, the gradient $dn_b'/dx = [n_b'(0) - n_b'(W)]/W$. Under normal forward bias conditions $n_b'(0) \gg n_b'(W)$; therefore,

$$I_C = \frac{qAD_b n_b'(0)}{W} \tag{5.1.7}$$

Now replacing $n_b'(0)$ from Eq. (5.1.4) we can relate the collector current I_C to the base-emitter voltage V_{BE} using

$$I_C = \frac{qAD_b n_{bo}}{W}(e^{V_{BE}/V_T} - 1) \tag{5.1.8}$$

5.1.2 Excess Minority Carrier Base Charge

Later, when we discuss the dynamic properties of the junction transistor in Chap. 6, we will make use of the charge-control expressions. This will require a knowledge of the *excess minority carrier charge* Q_F stored in the neutral base region.

Now we recognize that each of the electrons injected into the base from the emitter carries a charge q (where $q = 1.6 \times 10^{-19}$ C). Hence the total excess minority carrier base charge, denoted in the forward active mode as Q_F, is represented in Fig. 5.3 by the product of the electronic charge q and the volume contained by the excess minority carrier concentration $[n_b'(x)]$. This is the area of the triangle, with 0 to W as the base and $n_b'(0)$ as the amplitude, multiplied by the cross-sectional area A of the junction at the emitter, namely,

$$Q_F = \frac{qAWn_b'(0)}{2} \tag{5.1.9a}$$

We can combine Eq. (5.1.9a) with Eq. (5.1.7) and describe Q_F in terms of the collector current I_C. That is,

$$Q_F = \frac{W^2}{2D_b}I_C \tag{5.1.9b}$$

Now we introduce a basic BJT parameter in

$$Q_F = \tau_F I_C \tag{5.1.10}$$

The *mean forward transit time* τ_F of the minority carriers in the base is the mean time for the minority carriers to cross (diffuse) the neutral base region from the emitter to the collector.

$$\tau_F = \frac{W^2}{2D_b} \tag{5.1.11}$$

where W is the effective width of the neutral base region and D_b is the diffusion coefficient for the minority carriers in the neutral base region. Notice that Q_F and τ_F for the junction transistor are analogous to Q_p (or Q_n) and τ_T for the short-base junction diode that was described in Sec. 4.7.2.

5.1.3 Forward Transit Time

In Eq. (5.1.11) τ_F is described in what are essentially physical properties of the device. Notice that τ_F is a function of the square of the width W_B of the neutral base region. Now for a high-speed digital circuit we require τ_F to be as short as possible, so an effective way to decrease τ_F is to make W_B smaller. However, there are physical limitations to exactly how small W_B can be.

Some typical values for a low-power transistor might be for the base width, $W = 0.5 \ \mu m$, and for the diffusion coefficient for electrons in the base region, $D_b = 7 \ cm^2/s$. Then,

$$\tau_F = \frac{(0.5 \ \mu m)^2}{2(7 \ cm^2/s)} = 0.18 \ ns$$

Now we note from Eq. (5.1.8) that a change in V_{BE} will lead to a change in I_C, and therefore from Eq. (5.1.9b) a change in Q_F. That is,

$$\Delta V_{BE} \rightarrow \Delta I_C \rightarrow \Delta Q_F$$

5.1.4 The Base Current

The premise so far has been that $I_E = I_C$; that is, in Fig. 5.1 we have assumed that the base current I_B is zero. Now the current gain β_F of the transistor is defined by the ratio of the collector current I_C and the base current I_B, that is,

$$\beta_F = \frac{I_C}{I_B} \tag{5.1.12}$$

Hence, we recognize that I_B is a defect parameter and, for the general case, the smaller it is the better.

For the *npn* transistor of Fig. 5.1, I_B is a majority carrier (hole) current, which is made up of three parts:

$$I_B = I_{BB} + I_{BE} + I_{BC} \tag{5.1.13}$$

1. As the minority carriers (electrons) diffuse across the base from the emitter to the collector, some do not reach the collector, but on the way they recombine with the majority carriers (holes). Now for charge neutrality in the neutral base region, the missing holes must be replaced from the external circuit; this is the part of I_B designated I_{BB}.

$$I_{BB} = qA \int_0^W \frac{n_b(x)}{\tau_b} \, dx \tag{5.1.14}$$

To minimize I_{BB} the base width W is made small; from Eq. (5.1.11) this will reduce τ_F so that the minority carriers spend less time in the base region, therefore reducing the chances of recombination. Also, the mean lifetime τ_b of the minority carriers in the base region should be maximized, again reducing the chances of recombination.

2. From Fig. 5.2 notice that there is a diffusion gradient for holes across the emitter depletion region. Hence there is a hole diffusion current from the base to the emitter, which is designated as I_{BE}. From *pn* junction theory and Eq. (4.4.13), in the emitter region,

$$I_{BE} = \frac{qAD_e p_{eo}}{L_e}(e^{V_{BE}/V_T} - 1) \tag{5.1.15}$$

where D_e and L_e represent the diffusion coefficient and diffusion length of the minority carriers across the neutral emitter region.

To minimize I_{BE}, p_{eo} is made small. Since $p_{eo} = n_i^2/N_D$, N_D in the emitter region is made large. Actually we require that the majority carrier concentration in the emitter be much greater than that in the base. Then we will have a good emitter efficiency, which for the *npn* transistor we define as the ratio of the electron current to hole current across the emitter junction.

3. The final component of I_B is I_{BC}. This is the hole current from the base to the collector. Again, similar to Eq. (4.4.13), but this time for the collector region,

$$I_{BC} = \frac{qAD_c p_{co}}{L_c}(e^{V_{BC}/V_T} - 1) \tag{5.1.16}$$

where the subscript c denotes the neutral collector region.

However, the collector junction is reverse-biased, and the hole current is actually from the collector to the base. To minimize this current, p_{co} is made small, which is done by making N_D in the collector large. For the diffused process by which these devices are made it is not possible to have the impurity concentration in the collector be greater than that in the base. Moreover this is not necessary, since fortunately, with reverse bias ($V_{BC} \leq -0.1$ V), the exponential term in Eq. (5.1.16) is essentially zero, and only the -1 term in parentheses remains. Generally, in a well-designed transistor, $I_{BC} \ll I_{BB} + I_{BE}$.

In Sec. 6.2.1, we show that the current gain β_F can also be equated to the ratio of two time constants, namely,

$$\beta_F = \frac{\tau_{BF}}{\tau_F} \tag{5.1.17}$$

where τ_F describes the *effective minority carrier lifetime* in the neutral base region in the forward mode. Essentially, this is accountable to the combined effects of I_{BB} and I_{BE}. The mean transit time τ_F has been described in Sec. 5.1.2.

5.2   TERMINAL CURRENTS

A simple realization of the *npn* transistor as two *pn* junction diodes connected back to back with a common *p* region making the base region is shown in Fig. 5.4*b*. The *pn* junction diodes formed at the emitter and the collector junctions are designated as the emitter D_E and collector D_C diodes, respectively. The standard circuit symbol for the *npn* transistor is shown in Fig. 5.4*c*, where also are indicated the terminal currents I_E, I_B, and I_C, and the junction voltages V_{BE} and V_{BC}. In Fig. 5.4*d*, the ideas of the back-to-back diodes are combined with the knowledge that a reverse-biased junction exhibits a high impedance. In Fig. 5.4*d*, the current that crosses the emitter junction is shown as I_{DE}. This current is a function of the base-emitter voltage V_{BE}. In the forward active mode we noted that not all the minority carriers injected into the base region from the emitter reach the collector. The large fraction that do are represented in Fig. 5.4*d* by the $\alpha_F I_{DE}$ current generator, where α_F is indicative of the defect factor.

Another mode of operation for the BJT is to forward-bias the collector junction and reverse-bias the emitter junction. For obvious reasons this is known as the *reverse active mode* of operation. This mode is represented in Fig. 5.4*d* by the collector diode current I_{DC}, which is a function of the base-collector voltage V_{BC}, and the current generator, now across the emitter junction, $\alpha_R I_{DC}$. The fraction α_R is due to the ratio of the minority carriers collected at the emitter to the number injected into the base at the forward-biased base-collector junction.

To develop the general circuit equations, we first write the equation for the diode currents I_{DE} and I_{DC}. From Eq. (4.4.14),

$$I_{DE} = I_{ES}(e^{V_{BE}/V_T} - 1) \tag{5.2.1a}$$

$$I_{DC} = I_{CS}(e^{V_{BC}/V_T} - 1) \tag{5.2.1b}$$

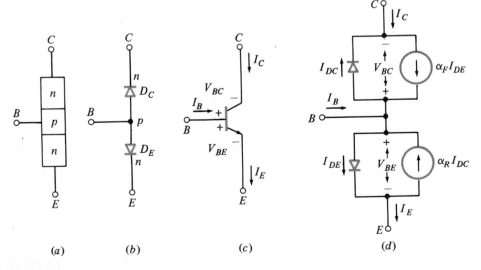

| (a) | (b) | (c) | (d) |

FIGURE 5.4
Development of the Ebers-Moll model for the *npn* transistor.

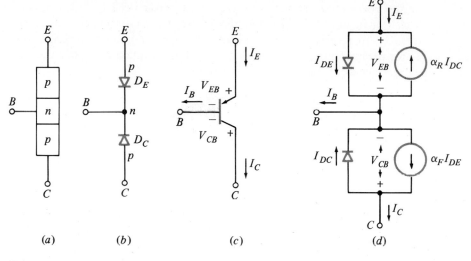

FIGURE 5.5
Development of the Ebers-Moll model for the *pnp* transistor.

where I_{ES} and I_{CS} represent the saturation currents for the emitter and collector junctions, respectively.

Next we express the terminal currents I_E and I_C in terms of the diode currents. From Fig. 5.4d,

$$I_E = I_{DE} - \alpha_R I_{DC} \qquad (5.2.2a)$$

$$I_C = \alpha_F I_{DE} - I_{DC} \qquad (5.2.2b)$$

Now combine Eqs. (5.2.1) and (5.2.2), and we have the general equations for the *npn* transistor due to Ebers and Moll:[*]

$$I_E - I_{ES}(e^{V_{BE}/V_T} - 1) - \alpha_R I_{CS}(e^{V_{BC}/V_T} - 1) \qquad (5.2.3a)$$

$$I_C = \alpha_F I_{ES}(e^{V_{BE}/V_T} - 1) - I_{CS}(e^{V_{BC}/V_T} - 1) \qquad (5.2.3b)$$

These equations are the starting point for all analysis and design of electronic circuits using the bipolar junction transistor in that they are applicable to all modes of operation of the transistor. The terminal current I_B, can simply be obtained from Eq. (5.2.3), since by Kirchhoff's current law in Fig. 5.3d, $I_B = I_E - I_C$.

Exercise 5.1. Use Fig. 5.5 to prove the Ebers-Moll equations for the *pnp* transistor, namely,

$$I_E = I_{ES}(e^{V_{EB}/V_T} - 1) - \alpha_R I_{CS}(e^{V_{CB}/V_T} - 1) \qquad (5.2.4a)$$

$$I_C = \alpha_F I_{ES}(e^{V_{EB}/V_T} - 1) - I_{CS}(e^{V_{CB}/V_T} - 1) \qquad (5.2.4b)$$

---

[*]J. J. Ebers and J. L. Moll, "Large-Signal Behavior of Junction Transistors," *Proceedings of the IRE*, vol. 42, December 1954.

From the Ebers-Moll equations for the *npn* transistor in Eq. (5.2.3) and for the *pnp* transistor in Eq. (5.2.4), notice that the terminal currents for the device are described by two variables: the emitter and collector junction voltages and four transistor parameters I_{ES}, I_{CS}, α_F, and α_R. For the ideal transistor the four parameters are related by the *reciprocity theorem* as

$$\alpha_F I_{ES} = \alpha_R I_{CS} \tag{5.2.5}$$

Typical small-device parameters are

$$\alpha_F \approx 0.99 \qquad I_{ES} \approx 10^{-15}\text{A}$$
$$\alpha_R \approx 0.66 \qquad I_{CS} \approx 10^{-15}\text{A}$$

5.3 MODES OF OPERATION

Mention has already been made of two of the modes of operation, but for completeness, listed below are all the possible modes of operation for the BJT as a function of the bias that is applied to the two junctions.

Emitter junction	Collector junction	Mode of operation
Forward	Reverse	Forward active
Reverse	Forward	Reverse active
Reverse	Reverse	Cutoff
Forward	Forward	Saturation

Generally, in analog or linear circuits the transistors are operated in the forward active mode only. However, in digital circuits all four modes of operation may be involved. This being the case, it is therefore useful to develop from the basic Ebers-Moll equations some simplified equations that are useful for a particular mode of operation. Since digital ICs primarily make use of *npn* transistors only, the starting equations will be from Eq. (5.2.3). A similar development could be made for *pnp* transistors, starting from Eq. (5.2.4).

5.3.1 Forward Active Mode

In the forward active mode the base-emitter junction is forward-biased and the base-collector junction is reverse-biased. Furthermore, assuming operation at 300 K and $V_{BE} \geq 4V_T$ and $V_{BC} \leq -4V_T$, that is, with $|V_{BE}| = |V_{BC}| \geq 100$ mV, then from Eq. (5.2.3) we have the very good approximation

$$I_E = I_{ES}e^{V_{BE}/V_T} + \alpha_R I_{CS} \tag{5.3.1a}$$

$$I_C = \alpha_F I_{ES}e^{V_{BE}/V_T} + I_{CS} \tag{5.3.1b}$$

Notice that with $V_{BE} < 100$ mV the currents are too small to be useful.

A similar simplification can be made of Fig. 5.4d. This is done in Fig. 5.6a, which diagrammatically represents Eq. (5.3.1). In practice, I_{CS} can generally be neglected when compared to I_C. Then the collector current is described in terms of the *diode current* I_{DE}.

An alternative description of I_C can be had by substitution for $I_{ES}e^{V_{BE}/V_T}$ in Eq. (5.3.1b) from Eq. (5.3.1a). Then,

$$I_C = \alpha_F(I_E - \alpha_R I_{CS}) + I_{CS} \qquad (5.3.2a)$$

$$= \alpha_F I_E + I_{CS}(1 - \alpha_F \alpha_R) \qquad (5.3.2b)$$

$$= \alpha_F I_E + I_{CO} \qquad (5.3.2c)$$

where

$$I_{CO} = I_{CS}(1 - \alpha_F \alpha_R) \qquad (5.3.3)$$

This alternative form for I_C is shown diagrammatically in Fig. 5.6b, where with I_{CO} very small, the collector current is described in terms of the *emitter current* I_E.

A final form for the collector current is shown in Fig. 5.6c, where the $(\beta_F + 1)I_{CO}$ term can generally be neglected; then the collector current is directly

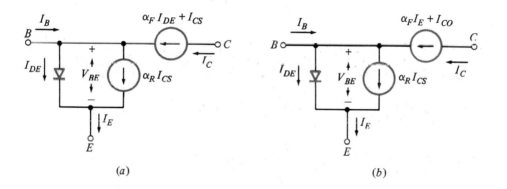

(a) (b)

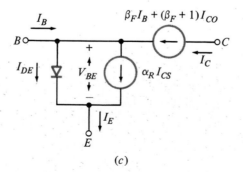

(c)

FIGURE 5.6
Simplified models for the forward active mode.

related to the *base current* I_B. To derive the necessary equations we first note in Fig. 5.4d, that by Kirchhoff's current law (KCL),

$$I_E = I_B + I_C \tag{5.3.4}$$

substituting this for I_E in Eq. (5.3.2c),

$$I_C = \alpha_F I_B + \alpha_F I_C + I_{CO} \tag{5.3.5a}$$

Hence

$$I_C = \frac{\alpha_F}{1 - \alpha_F} I_B + \frac{1}{1 - \alpha_F} I_{CO} \tag{5.3.5b}$$

$$= \beta_F I_B + (\beta_F + 1)I_{CO} \tag{5.3.5c}$$

where

$$\beta_F = \frac{\alpha_F}{1 - \alpha_F} \tag{5.3.6}$$

5.3.2   Reverse Active Mode

In the reverse active mode the base-collector junction is forward-biased and the base-emitter junction is reverse-biased. With the assumptions that $V_{BC} \geq V_T$ and $V_{BE} \leq -4V_T$, it is a simple matter to derive equations for the collector current, now the emitter current I_E, in similar fashion to what was done in the previous section for the collector current I_C.

5.3.3   Cutoff Mode

In the cutoff mode, both junctions are reverse-biased. With the further assumption that V_{BE} and V_{BC} are both $\leq -4V_T$, we may write from Eq. (5.2.3),

$$I_E = -I_{ES} + \alpha_R I_{CS} \tag{5.3.7a}$$

$$I_C = -\alpha_F I_{ES} + I_{CS} \tag{5.3.7b}$$

Then making use of the reciprocity theorem, Eq. (5.2.5), a further simplification results:

$$I_E = -I_{ES}(1 - \alpha_F) \tag{5.3.8a}$$

$$I_C = I_{CS}(1 - \alpha_R) \tag{5.3.8b}$$

These equations, modeled in Fig. 5.7a, show that in the cutoff mode the terminal currents are even less than the saturation currents. However we should point out that these currents are derived from the ideal diode equation, Eq. (4.4.14). That is, the effect of leakage currents and the thermal generation of mobile carriers in the depletion region at the two junctions have not been included. As described in Sec. 4.5, in silicon devices these currents are substantially greater than the ideal saturation currents, though they are still very much smaller (that

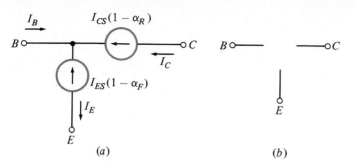

FIGURE 5.7
Simplified models for the cutoff mode.

is, 10^{-9} A) than the usual collector currents of about 10^{-3} A. Hence, effectively there is an open circuit at each of the transistor terminals, as pictured in Fig. 5.7*b*.

5.3.4 Saturation Mode

In the saturation mode both the emitter and the collector junctions are forward-biased. With the assumption that both V_{BE} and V_{BC} are equal to or greater than $4V_T$, from Eq. (5.2.3),

$$I_E = I_{ES}e^{V_{BE}/V_T} - \alpha_R I_{CS}e^{V_{BC}/V_T} \tag{5.3.9a}$$

$$I_C = \alpha_F I_{ES}e^{V_{BE}/V_T} - I_{CS}e^{V_{BC}/V_T} \tag{5.3.9b}$$

This is about as far as the equations for the terminal currents can be simplified. However, when we later discuss the operation of the transistor as a switch in Chap. 6, we will be interested in the voltage that appears across the switch when it is on, that is, the voltage between the collector and emitter when the transistor is saturated. From Fig. 5.8*a* notice by Kirchhoff's voltage

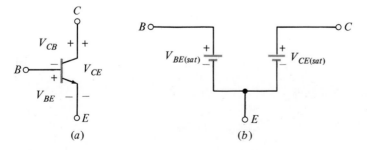

FIGURE 5.8
Simplified model for the saturation mode.

law (KVL) that

$$VCE = V_{CB} + V_{BE} \qquad (5.3.10a)$$

$$= -V_{BC} + V_{BE} \qquad (5.3.10b)$$

That is,
$$V_{CE(sat)} = V_{BE(sat)} - V_{BC(sat)} \qquad (5.3.10c)$$

An equation for $V_{CE(sat)}$ can be derived from the basic saturation equations of Eq. (5.3.9). First, multiply Eq. (5.3.9b) by α_R and subtract from Eq. (5.3.9a) to eliminate the $\alpha_R I_{CSE} e^{V_{BC}/V_T}$ term. The result is

$$I_E - \alpha_R I_C = I_{ES}(1 - \alpha_F \alpha_R)e^{V_{BE}/V_T} \qquad (5.3.11)$$

Now by Kirchhoff's current law, $I_E = I_B + I_C$. Substitute this for I_E in Eq. (5.3.11) to obtain

$$I_B + I_C(1 - \alpha_R) = I_{ES}(1 - \alpha_F \alpha_R)e^{V_{BE}/V_T} \qquad (5.3.12)$$

From this equation we can solve for V_{BE} as

$$V_{BE(sat)} = V_T \ln \frac{I_B + I_C(1 - \alpha_R)}{I_{EO}} \qquad (5.3.13)$$

where

$$I_{EO} = I_{ES}(1 - \alpha_F \alpha_R) \qquad (5.3.14)$$

By a similar method an equation for $V_{BC(sat)}$ can be derived, but this time multiply Eq. (5.3.9a) by α_F, and subtract Eq. (5.3.9b) from the product to eliminate $\alpha_F I_{ES} e^{V_{BE}/V_T}$. The result is

$$\alpha_F I_E - I_C = I_{CS}(1 - \alpha_F \alpha_R)e^{V_{BC}/V_T} \qquad (5.3.15)$$

For $\alpha_F I_E$ substitute $\alpha_F I_B + \alpha_F I_C$, then

$$\alpha_F I_B - I_C(1 - \alpha_F) = I_{CS}(1 - \alpha_F \alpha_R)e^{V_{BC}/V_T} \qquad (5.3.16)$$

Solving for V_{BC} we have

$$V_{BC(sat)} = V_T \ln \frac{\alpha_F I_B - I_C(1 - \alpha_F)}{I_{CO}} \qquad (5.3.17)$$

Here

$$I_{CO} = I_{CS}(1 - \alpha_F \alpha_R) \qquad (5.3.18)$$

Now use Eqs. (5.3.13) and (5.3.17) to solve for $V_{CE(sat)}$ in Eq. (5.3.10c), namely,

$$V_{CE(sat)} = V_T \ln \left[\frac{I_B + I_C(1 - \alpha_R)}{\alpha_F I_B - I_C(1 - \alpha_F)} \frac{I_{CO}}{I_{EO}} \right] \qquad (5.3.19)$$

But from Eqs. (5.3.14) and (5.3.18) and the reciprocity theorem of Eq. (5.2.5),

$$\frac{I_{CO}}{I_{EO}} = \frac{I_{CS}}{I_{ES}} = \frac{\alpha_F}{\alpha_R} \tag{5.3.20}$$

Now in Eq. (5.3.19) divide the numerator by α_R and the denominator by α_F. Also divide both numerator and denominator by I_B. Then

$$V_{CE(sat)} = V_T \ln \frac{\dfrac{1}{\alpha_R} + \dfrac{I_C}{I_B}\left(\dfrac{1 - \alpha_R}{\alpha_R}\right)}{1 - \dfrac{I_C}{I_B}\left(\dfrac{1 - \alpha_F}{\alpha_F}\right)} \tag{5.3.21}$$

Finally substitute β_R for $\alpha_R/(1 - \alpha_R)$ and β_F for $\alpha_F/(1 - \alpha_F)$ to yield

$$V_{CE(sat)} = V_T \ln \frac{\dfrac{1}{\alpha_R} + \dfrac{I_C}{I_B}\left(\dfrac{1}{\beta_R}\right)}{1 - \dfrac{I_C}{I_B}\left(\dfrac{1}{\beta_F}\right)} \tag{5.3.22}$$

Using typical low-power transistor parameters of $\beta_F = 100$, $\beta_R = 2$, and $I_C/I_B = 10$,

$$V_{CE(sat)} = 50 \text{ mV}$$

However this value for $V_{CE(sat)}$ is the *intrinsic saturation voltage* for the transistor. In series with the emitter and the collector junctions there are bulk resistances due to the neutral emitter and collector regions, respectively. Because of the impurity concentrations, the bulk emitter resistance r_e is generally much less than the bulk collector resistance r_c. Representative values might be 1 and 50 Ω, respectively. The terminal currents I_E and I_C develop a voltage across these bulk resistances so that using the above parameters with $I_C = 1$ mA, the *terminal saturation voltage* becomes approximately 100 mV.

A useful model for the transistor in the saturation mode is shown in Fig. 5.8b. Generally the voltages $V_{BE(sat)}$ and $V_{CE(sat)}$ include the bulk resistance effects. As such, typical values are

$$V_{BE(sat)} = 0.8 \text{ V} \qquad \text{and} \qquad V_{CE(sat)} = 0.2 \text{ V}$$

In the saturation mode, the ratio of I_C and I_B [as in Eq. (5.3.22)] is termed the *forced* β, or β_{sat}. Note that for any given value of I_C the value of I_B is greater when the transistor is operating in the saturation mode than when it is in the active mode. Hence,

$$\beta_{sat} < \beta_F$$

For measuring the saturation voltage of a transistor, a typical value of $\beta_{(sat)}$ is 10.

Exercise 5.2. Sometimes, $V_{CE(sat)}$ is simply given as

$$V_{CE(sat)} = V_T \ln \left(\frac{1}{\alpha_R}\right) + I_{EOS} r_c$$

What are the limitations on this approximation?

5.4 BJT MODEL FOR CIRCUIT SIMULATION PROGRAMS

As with the MOS transistor and the *pn* junction diode, accurate characterization of the bipolar transistor requires that the device be modeled in mathematical terms. In this section, the SPICE model for the BJT is described, and techniques for measuring the device parameters are presented.

5.4.1 SPICE BJT Model

The relationship between the Ebers-Moll model and the transistor model used in SPICE is shown in Figs. 5.9 and 5.10.*

The dc characteristics of the intrinsic BJT are determined by the nonlinear current sources I_C and I_E. The equations used in SPICE may be simplified to yield the familiar Ebers-Moll equations presented in Eqs. (5.2.3*a* and *b*), namely,

$$I_E = \frac{I_s}{\alpha_F}(e^{V_{BE}/V_T} - 1) - I_s(e^{V_{BC}/V_T} - 1) \tag{5.4.1a}$$

$$I_C = I_s(e^{V_{BE}/V_T} - 1) - \frac{I_s}{\alpha_r}(e^{V_{BC}/V_T} - 1) \tag{5.4.1b}$$

where, from Eq. (5.2.5),

$$I_s = \alpha_F I_{ES} = \alpha_R I_{CS} \tag{5.4.2}$$

The ohmic resistances of the neutral base, collector, and emitter regions of the BJT are modeled by the linear resistors r_b, r_c, and r_e.

Charge storage in the BJT is modeled by the two nonlinear charge storage elements Q_{BE} and Q_{BC}, which are determined by the equations

$$Q_{BE} = \tau_F I_s(e^{V_{BE}/V_T} - 1) + C_{je0}\int_0^{V_{BE}} \left(1 - \frac{V}{\phi_e}\right)^{-m_e} dV \tag{5.4.3a}$$

$$Q_{BC} = \tau_R I_s(e^{V_{BC}/V_T} - 1) + C_{jc0}\int_0^{V_{BC}} \left(1 - \frac{V}{\phi_c}\right)^{-m_c} dV \tag{5.4.3b}$$

Equivalently, these elements can be represented by voltage-dependent capacitors, since

$$C_{BE} = \frac{dQ_{BE}}{dV_{BE}} = \frac{\tau_F I_s}{V_T} e^{V_{BE}/V_T} + \frac{C_{je0}}{[1 - (V_{BE}/\phi_e)]^{m_e}} \tag{5.4.4a}$$

---

*The models of Figs. 5.9 and 5.10 and all the equations in this section relate to an *npn* transistor. The models and equations for a *pnp* device may simply be obtained by analogy of the Ebers-Moll model for a *pnp* transistor, shown in Fig. 5.5, to the *npn* model of Fig. 5.4.

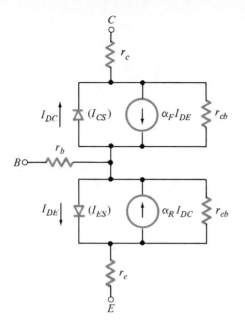

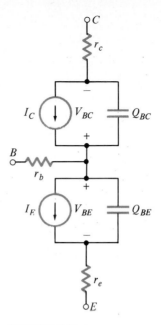

FIGURE 5.9
The static Ebers-Moll model, including second-order effects.

FIGURE 5.10
The SPICE BJT model.

$$C_{BC} = \frac{dQ_{BC}}{dV_{BC}} = \frac{\tau_R I_s}{V_T} e^{V_{BC}/V_T} + \frac{C_{jc0}}{\left[1 - (V_{BC}/\phi_c)\right]^{m_c}} \qquad (5.4.4b)$$

The charge storage elements Q_{BE} and Q_{BC} model the charge stored in the neutral base region Q_F and Q_R as well as the charge stored in the depletion regions Q_{je} and Q_{jc}.

Charge storage in the depletion regions is from Eq. (4.3.9) and is represented by V_{BE} and the model parameters C_{je0}, ϕ_e, and m_e for the emitter junction and V_{BC}, C_{jc0}, ϕ_c, and m_c for the collector junction.

Charge storage due to minority carrier injection across the junctions is described by the exponential terms in Eqs. (5.4.3a and b), which are related to Eqs. (5.1.8) and (5.1.10). The effect is modeled by the transit time parameters τ_F for the emitter junction and τ_R for the collector junction.

The model shown in Fig. 5.10 is complete for a discrete BJT. For an integrated circuit transistor we must add one more component. As illustrated in Fig. 5.1 at the beginning of the chapter, isolation of transistor collectors from each other and from the substrate is achieved using a reverse-biased diode. Assuming it remains reverse-biased, we find that this diode may be modeled by a current source (representing the diode leakage current) in parallel with a voltage-dependent depletion-layer capacitance.

For an IC *npn* transistor, this isolation diode is effectively connected between the collector and substrate, as shown in Fig. 5.11a. The substrate is common to all the components in the integrated circuit. In Fig. 5.11b, this diode is modeled as a capacitance between the collector and substrate (C_{CS}) and is

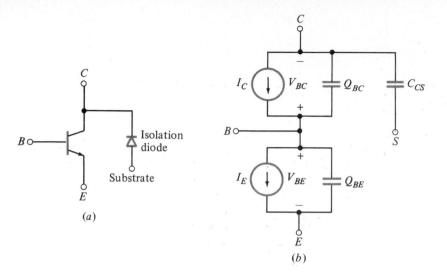

FIGURE 5.11
The integrated circuit *npn* transistor.

added to the basic model for the transistor in the forward active mode. From Eq. (4.3.9),

$$C_{SC} = \frac{C_{js0}}{\left[1 - (V_{SC}/\phi_s)\right]^{m_s}} \tag{5.4.5}$$

Exercise 5.3. Use the transistor data given as an example in Table 5.1 to solve for I_E, I_C, C_{BE}, and C_{BC}. Let $V_{BE} = 0.7$ V, $V_{CE} = 5.0$ V, and $V_T = 26$ mV.

5.4.2 Measurement of the BJT Parameters

A listing of the parameters required to model the BJT is shown in Table 5.1. This is a partial listing, but it is adequate to introduce the reader to computer-aided analysis and design of digital ICs. Included in the table are the default values used in SPICE and an example that might be typical for a low-power switching transistor.

A most useful instrument to measure the BJT parameters is the transistor curve tracer, which plots the collector current I_C as a function of the collector-emitter voltage V_{CE} with either the base-emitter voltage V_{BE} or the base current I_B as an independent variable.

I_s. Saturation current I_s is the extrapolated intercept of the graph of the natural logarithm (ln) of I_C as a function of V_{BE} with $V_{CE} = V_{BE}$, so that $V_{BC} = 0$. With $V_{BC} = 0$, the second term in Eq. (5.4.1b) is eliminated. V_{BC} is set to zero as illustrated in the curve-tracer picture of Fig. 5.12.

β_F, β_R. The forward current gain β_F of the transistor may also be determined from the curve tracer. From Eq. (5.3.5c),

TABLE 5.1
SPICE BJT model parameters

Symbol	Name	Parameter	Units	Default	Example
I_S	IS	Saturation current	A	1.0E-16	1.0E-15
β_F	BF	Forward current gain		100	100
β_R	BR	Reverse current gain		1	1
r_b	RB	Base resistance	Ω	0	100
r_c	RC	Collector resistance	Ω	0	50
r_e	RE	Emitter resistance	Ω	0	1
C_{je0}	CJE	B-E zero-bias depletion capacitance	F	0	1.0E-12
ϕ_e	VJE	B-E built-in potential	V	0.75	0.8
m_e	MJE	B-E junction grading factor		0.33	0.5
C_{jc0}	CJC	B-C zero-bias depletion capacitance	F	0	0.5E-12
ϕ_c	VJC	B-C built-in potential	V	0.75	0.7
m_c	MJC	B-C junction grading factor		0.33	0.5
C_{js0}	CJS	Zero-bias collector-substrate capacitance	F	0	3.0E-21
ϕ_s	VJS	Substrate junction built-in potential	V	0.75	0.6
m_s	MJS	Substrate junction grading factor		0	0.5
τ_F	TF	Forward transit time	s	0	1.0E-10
τ_R	TR	Reverse transit time	s	0	1.0E-8

$$\beta_F = \frac{I_C}{I_B} \tag{5.4.6}$$

At any given operating point (I_C, V_{CE}), β_F can be calculated. Care should be taken in this measurement since β_F does vary with I_C, as shown in Fig. 5.13. The current gain falls at high currents because of high-level injection effects and at low currents because of trapping effects. These two effects have been described in Sec. 4.8.2. β_F should be measured in the region of I_C, where the current gain is approximately constant.

A measurement of β_R, the reverse current gain, can similarly be made with the collector and emitter leads interchanged. In digital ICs β_F can have values from 20 to 200, but a typical value is 100. β_R varies from 0.1 to 10, with a typical value of 1.

r_b, r_c, r_e. The resistors r_b, r_c, and r_e model the resistance effects due to the bulk ohmic regions of the transistor.

As described in Sec. 5.3.4, for a transistor in saturation with a large collector current, most of $V_{CE(sat)}$ is due to the voltage drop across the collector ohmic region. Actually, the voltage drop across the emitter ohmic region is also a contributor. But since the emitter is the most heavily doped region of the transistor, we have from the I_C-V_{CE} characteristic of Fig. 5.14, with heavy saturation,

$$r_c = \frac{\Delta V_{CE}}{\Delta I_C} = \frac{V_{CE(sat)}}{I_{C(sat)}} \tag{5.4.7}$$

Typical values for r_c are 10 to 100 Ω, and for r_e, 0.5 to 5 Ω.

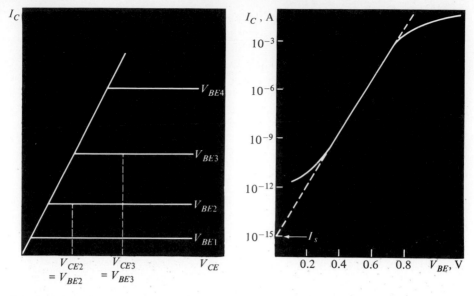

FIGURE 5.12

I_C-V_{CE} characteristic for constant V_{BE}.

The base resistance r_b is something of a problem to measure accurately, and several methods are used. For digital circuits the exact value of r_b is not critical. Hence, an estimate of r_b can be made with the simple circuit of Fig. 5.15a. Then, with the transistor operating in the forward active mode, the value of r_b is obtained from the graph as illustrated in Fig. 5.15b,

$$r_b = \frac{\Delta V_{BE}}{\Delta I_B} \tag{5.4.8}$$

Typically, r_b is in the range of 20 to 200 Ω.

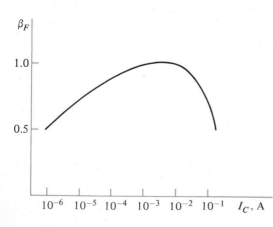

FIGURE 5.13

Normalized current gain β_F as a function of I_C.

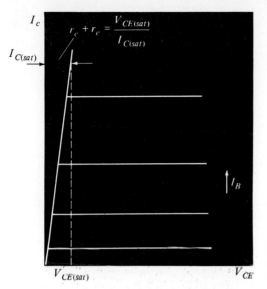

FIGURE 5.14
I_C-V_{CE} characteristic for a low-power *npn* transistor.

The BJT parameters that have so far been described are all *static parameters*. That is, they all relate to the steady-state, or dc, characteristics of the transistor. The measurement of these parameters is the objective of Demonstration D5.2, included at the end of this chapter.

The objective of Demonstration D5.3 is to measure the *dynamic parameters* of the BJT. These are the time-dependent, or ac, characteristics, and their measurement will now be described.

C_{je0}, ϕ_e, m_e. The measurement of the emitter junction capacitor parameters is done with a capacitance bridge and the circuit shown in Fig. 5.16a. The procedure is similar to that for the diode junction capacitance described in Sec. 4.8.2. It should be noted here that because of the heavy doping in the emitter region,

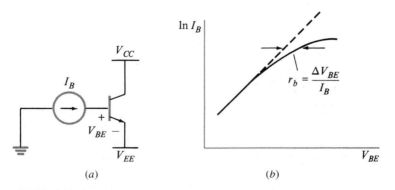

FIGURE 5.15
(a) Circuit for measurement of r_b. (b) Graph of ln I_B versus V_{BE}.

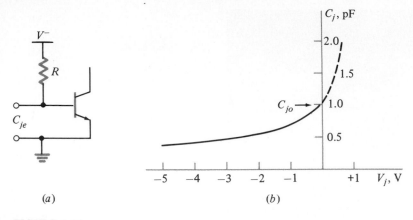

(a)   (b)

FIGURE 5.16
(a) Circuit for measurement of C_{je}. (b) Graph of junction capacitance versus junction voltage.

and consequently the low breakdown voltage BV_{EBO}, the reverse junction voltage in an *npn* transistor is generally limited to approximately 5 V.

Also notice from Eq. (4.3.9), assuming an abrupt junction, $m_e = 0.5$, and $V_{BE} = -\phi_e$,

$$C_{je} = \frac{C_{je0}}{\sqrt{2}} = 0.7C_{je0} \tag{5.4.9}$$

That is, at $C_{je} = 0.7C_{je0}$, $V_{BE} = -\phi_e$. Similarly, for a graded junction where $m_e = 0.3$, at $C_{je} = 0.8C_{je0}$, $V_{BE} = -\phi_e$. These ideas are helpful in the curve-fitting procedure. Some typical values for an IC *npn* transistor are $C_{je0} = 1.0$ pF, $\phi_e = 0.8$ V, $m_e = 0.5$.

By interchanging the collector and emitter leads in Fig 5.16a, the parameters associated with the collector junction capacitance may similarly be determined. Typically, $C_{jc0} = 0.5$ pF, $\phi_c = 0.7$ V, and $m_c = 0.5$.

Finally, a similar method may be used to measure the substrate junction capacitance, and some typical values are $C_{js0} = 3.0$ pF, $\phi_s = 0.6$ V, and $m_s = 0.5$.

τ_F, τ_R. With the transistor operating in the forward active mode, τ_F is the mean transit time of the minority carriers across the neutral base region from the emitter to the collector. Typically, it is not normally measured directly but is obtained from a measurement of the mean recombination time τ_{BF}, since from Eq. (5.1.17), $\tau_{BF} = \beta_F \tau_F$. A simple experiment for the measurement of τ_{BF} is described in Demonstration D5.3.

With the transistor operating in the reverse active mode, τ_R is the mean transit time of the minority carriers from the collector to the emitter. A similar measurement for τ_{BR} as for τ_{BF} can be made but with the transistor biased in the reverse active mode; then $\tau_{BR} = \beta_R \tau_R$. The reverse transit time τ_R, is normally much longer than the forward transit time τ_F because of the different junction areas for the emitter and the collector and also because of the doping gradient in the base region and other effects in the structure.

For a low-power switching transistor, typically, $\tau_F = 0.2$ ns and $\tau_R = 20$ ns.

5.5 LAYOUT DESIGN OF BJT

The cross section and plan view of an integrated circuit low-power *npn* switching transistor is presented in Fig. 5.17. This transistor has one base stripe and one collector stripe, and junction isolation is used. A minimum of six masks are required to fabricate this device; they are:

1. n^+ buried-layer diffusion mask
2. p^+ isolation diffusion mask
3. p base region diffusion mask
4. n^+ emitter and n^+ collector contact region diffusion mask
5. Contact windows mask
6. Metallization mask

There are many rules for the layout of each of the above masks. Below we give a few basic rules for the design of a minimum-sized high-yield *npn* transistor.

5.5.1 Simplified Design Rules

The minimum feature size for the layout of this device is given as M units.

A. Isolation mask
    1.   Width of isolation wall   M
    2.   Spacing from wall edge to buried layer   2.5M
B. Base mask
    1.   Spacing between base and isolation region   2.5M
C. Emitter mask
    1.   Area   2M $\times$ 2M
    2.   Spacing between emitter and base diffusion   M
    3.   Spacing between multiple emitter diffusions   1.5M
D. Collector contact mask
    1.   Area   M $\times$ 5M
    2.   Spacing between n^+ and base diffusion   M
E. Contact windows mask
    1.   Emitter contact area   M $\times$ M
    2.   Base contact area   M $\times$ 2M
    3.   Collector contact area   M $\times$ 2M
    4.   Spacing between base contact and emitter diffusion   M
F. Metallization mask
    1.   Width   1.5M
    2.   Metal to metal spacing   M

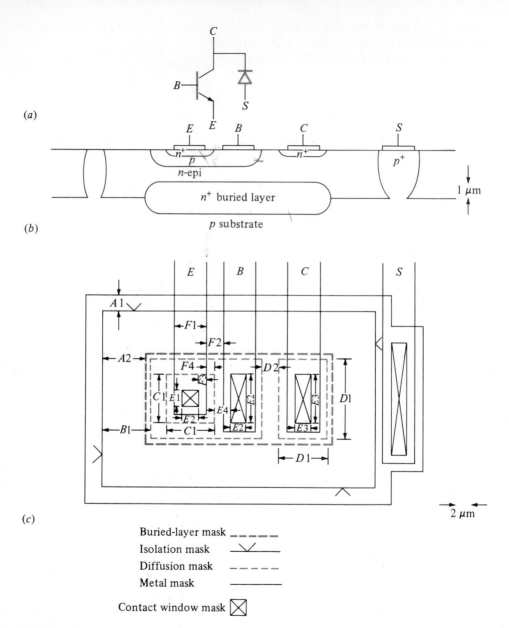

FIGURE 5.17
Integrated circuit *npn* transistor structure.

3. Contact window overlap   0.5M
4. Metal to diffusion edge spacing   0.5M

The example shown in Fig. 5.17c has a 3-μm epitaxial region and a minimum feature size of 2 μm.

5.6 SUMMARY

- In this chapter the operation of the bipolar junction transistor was first described in physical terms.
- This led to the development of equations for the terminal currents (I_B, I_C, I_E) of the device, the fundamental Ebers-Moll equations.
- From the basic Ebers-Moll equations, simpler equations were derived to describe the terminal currents in each of the modes of operation:

  Cutoff
  Forward active
  Reverse active
  Saturation

- The parameters used to model the BJT when using SPICE were listed, and experimental methods for measuring these parameters were given.
- The basic design rules for the layout of an IC *npn* transistor were presented.

REFERENCES

1. P. E. Gray, D. DeWitt, A. R. Boothroyd, and J. F. Gibbons, *Physical Electronics and Circuit Models of Transistors*, Wiley, New York, 1964.
2. P. E. Gray and C. L. Searle, *Electronic Principals—Physics, Models, and Circuits*, Wiley, New York, 1969.
3. D. J. Hamilton and W. G. Howard, *Basic Integrated Circuit Engineering*, McGraw-Hill, New York, 1975.
4. I. Getreu, *Modeling the Bipolar Transistor*, Tektronix, Inc., Beaverton, Oreg., 1976.
5. R. S. Muller and T. Kamins, *Device Electronics for Integrated Electronics*, Wiley, New York, 1977.
6. L. W. Nagel, "SPICE2, A Computer Program to Simulate Semiconductor Circuits," *ERL Memorandum No. ERL-M520*, University of California, Berkeley, May 1975.

DEMONSTRATIONS

D5.1 BJT: Electrical Parameters I

The objective of this experiment is to demonstrate the relationship of β_F, $V_{BE(on)}$, $V_{BE(sat)}$, and $V_{CE(sat)}$ to the collector current I_C in a bipolar junction transistor.

1. (a) With the use of a commercial curve tracer, or the simple circuit of Fig. D5.1, determine the I_C-V_{CE} characteristics of a low-power *npn* switching transistor.*
   (b) Determine and then plot the following parameters as a function of the collector current. Use I_C values of 0.1, 0.2, 0.5, 1, 2, 5, 10, 20, and 50 mA.
   (1) $V_{BE(on)}$ at $V_{CE} = 1$ V.
   (2) β_F at $V_{CE} = 1$ V.
   (3) $V_{BE(sat)}$ at $I_B = 0.1I_C$.
   (4) $V_{CE(sat)}$ at $I_B = 0.1I_C$.
   (c) Interchange the emitter and collector leads of the transistor to determine and plot for $I_E = 0.1$, 1.0, and 10 mA and β_R at $V_{EC} = 1$ V.

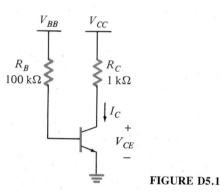

FIGURE D5.1

2. (a) Use the measured values of β_F and β_R to compute $V_{CE(sat)}$ from Eq. (5.3.22). Compare with the measured value of $V_{CE(sat)}$. From the difference in the measured and calculated values determine the ohmic collector resistance r_c.
   (b) Use the measured values of $V_{BE(on)}$ at $I_C = 1$ mA to calculate I_{ES} from Eq. (5.3.1). Now use this value of I_{ES} to calculate $V_{BE(on)}$ at $I_C = 10$ mA. Compare with the measured value of $V_{BE(on)}$. From the difference of the measured and calculated values determine the ohmic base resistance r_b.

D5.2   BJT: Electrical Parameters II

In this experiment we will measure the following *static* parameters of the BJT: I_{ES}, β_F, r_b, I_{CS}, and β_R.

(a) Use the curve tracer, or the circuit of Fig. 5.12, and Eq. (5.4.1b) to determine I_{ES}. Let $I_C = 1$ mA and $V_{CE} = 1$ V.
(b) Use the curve tracer, or the circuit of Fig. D5.1, and Eq. (5.4.6) to determine β_F. Let $I_C = 1$ mA and $V_{CE} = 1$ V.
(c) Reverse the emitter and collector leads of the transistor and repeat the operations of parts (a) and (b) to determine I_{CS} and β_R.

---

*Such as the 2N4275 or 2N2222A.

(d) With the transistor operating in the saturated region ($I_C = 10I_B = 10$ mA), measure $V_{CE(sat)}$ and from Eq. (5.4.7) calculate the ohmic collector resistance r_c.

(e) With the transistor operating in the active region ($I_C = 10$ mA, $V_{CE} = 1$ V), measure V_{BE} (call it $V_{BE(M)}$). Use the value of I_{ES} from part (a) in Eq. (5.3.1b) to calculate the value of V_{BE} at $I_C = 10$ mA (call it $V_{BE(C)}$). Determine a value for the ohmic base resistance r_b from Eq. (5.4.8).

D5.3   BJT: Electrical Parameters III

In this experiment we will measure the following *dynamic parameters* of the BJT:

$$\tau_F \qquad C_{je0} \qquad \phi_e \qquad m_e$$

$$\tau_R \qquad C_{jc0} \qquad \phi_c \qquad m_c$$

(a) Connect the circuit shown in Fig. D5.2, and operate the transistor in the forward active mode with $I_C = 10$ mA, $V_{CE} = 4$ V. With the output pulse ≤ 0.5 V, measure the transition time (the time between the 90 and 10% points on the output voltage waveform). From Eq. (6.2.25b), $t_r = 2.2\tau_{BF}$, and from Eq. (5.1.17), $\tau_F = \tau_{BF}/\beta_F$.

(b) Repeat part (a) with the emitter and collector leads interchanged so that the transistor operates in the reverse active mode with $I_E = 10$ mA and $V_{EC} = 4$ V, and determine τ_{BR} and hence τ_R.

(c) With a capacitance bridge and the circuit shown in Fig. 5.16a, measure C_{je} versus V_{BE} and graph as in Fig. 5.16b. To prevent erroneous readings $V_{BE} \leq -1$ V, but to prevent voltage breakdown $V_{BE} \geq -5$ V. (This assumes $BV_{EBO} > 5$ V.) By extrapolation of the graph determine C_{je0}. Initially assume an abrupt junction and use Eq. (4.3.9) to determine the emitter barrier potential ϕ_e. Using values of $C_{je} = -1$ and -5 V and the method of trial and error, determine better values for ϕ_e and the emitter grading coefficient m_e.

(d) Repeat part (c) with the emitter and collector leads of the transistor reversed to determine C_{jc0}, ϕ_c, and m_c. Since the maximum voltage V_{CBO} that can be applied to

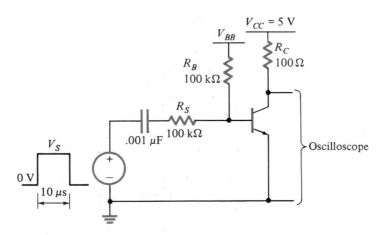

FIGURE D5.2

the collector junction is generally much greater than V_{BEO}, the bias voltage V_{BC} can likewise be increased.

PROBLEMS

For all problems use the following data unless otherwise specified.

$$I_{ES} = 10^{-16} \text{ A} \qquad I_{CS} = 2 \times 10^{-16} \text{ A}$$
$$\alpha_F = 0.98 \qquad \alpha_R = 0.49$$
$$C_{je0} = 1 \text{ pF} \qquad C_{jc0} = 0.5 \text{ pF}$$
$$\phi_e = 0.9 \text{ V} \qquad \phi_c = 0.8 \text{ V}$$
$$m_e = 0.5 \qquad m_c = 0.33$$
$$\tau_F = 0.2 \text{ ns} \qquad \tau_R = 10 \text{ ns}$$
$$C_{js0} = 3.0 \text{ pF} \qquad r_b = 50 \ \Omega$$
$$\phi_s = 0.7 \text{ V} \qquad r_c = 20 \ \Omega$$
$$m_s = 0.33 \qquad r_e = 1 \ \Omega$$

P5.1. Use the Ebers-Moll equations to show that:

(a) $V_{BE} = V_T \ln \left(1 + \dfrac{I_E - \alpha_R I_C}{I_{EO}} \right)$

(b) $V_{BC} = V_T \ln \left(1 - \dfrac{I_C - \alpha_F I_E}{I_{CO}} \right)$

P5.2. For the open-collector diode-connected transistor shown in Fig. P5.2, use the Ebers-Moll equations to find:
(a) I_E as a function of V_{BE}.
(b) V_{BE} if $I_E = 1$ mA.

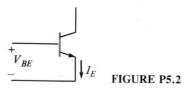

FIGURE P5.2

P5.3. For the shorted base-collector diode-connected transistor shown in Fig. P5.3, use the Ebers-Moll equations to find:
(a) I_E as a function of V_{BE}.
(b) V_{BE} if $I_E = 1$ mA.

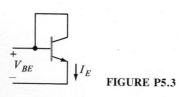

FIGURE P5.3

P5.4. For the circuit in Fig. P5.4 find V_{BE}, V_{BC}, and I_C if:
  (a) $V_{CE} = +10$ V.
  (b) $V_{CE} = -10$ V.

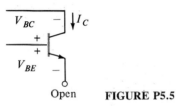

FIGURE P5.4

P5.5. For the circuit in Fig. P5.5 find V_{BE} and I_C for:
  (a) $V_{BC} = 0.7$ V.
  (b) $V_{BC} = -5$ V.

FIGURE P5.5

P5.6. For a silicon transistor, if the collector current I_C is held constant while the temperature is increased, the base current I_B decreases in magnitude, passes through zero, and then becomes negative. What physical effects account for this behavior?

P5.7. For a silicon transistor biased in the forward active mode with a constant V_{CE}, discuss how the collector current varies with an increase in temperature:
  (a) With a constant V_{BE}.
  (b) With a constant I_E.

P5.8. For the circuit in Fig. P5.8:
  (a) Calculate I_C, V_{CE} if $\alpha_F = 0.98$, $\alpha_R = 0.49$, and $I_{ES} = 10^{-15}$ A.
  (b) Repeat (a) if $r_B = 10$ kΩ.

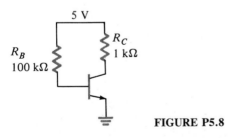

FIGURE P5.8

P5.9. Invert the transistor in Fig. P5.8:
  (a) Calculate I_E, V_{EC} if $\alpha_F = 0.98$, $\alpha_R = 0.49$, and $I_{ES} = 10^{-15}$ A with $r_B = 10$ kΩ and $r_C = 1$ kΩ.
  (b) Repeat (a) if $R_B = 1$ kΩ.

P5.10. For the inverter in Fig. P5.10 find V_{BE}, V_{CE}, I_B, I_C, Q_{je}, Q_{jc}, and Q_F if $V_{in} = -5$ V.

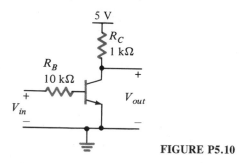

FIGURE P5.10

P5.11. Repeat P5.10 if $V_{in} = +5$ V.

P5.12. In the inverter circuit of Fig. P5.10 interchange the emitter and collector, i.e., reverse the transistor and solve for V_{BC}, V_{EC}, I_B, I_E, Q_{jc}, Q_{je}, and Q_R:
(a) For $V_{in} = -5$ V.
(b) For $V_{in} = +5$ V.

P5.13. Compute values for the elements in Fig. 5.11b if $I_C = 0.5$ mA and $V_{CE} = V_{CS} = 5$ V. Use the standard data.

P5.14. Repeat P5.13 for $I_C = 5$ mA and $V_{CE} = V_{CS} = 0.5$ V.

P5.15. Compare the standard transistor data with a similar device in which all the linear dimensions are doubled (i.e., the area is quadrupled).

P5.16. Repeat P5.13 using the 2X transistor of P5.15.

P5.17. For the layout of Fig. 5.17c, assume the following doping concentrations; for the emitter region, $N_E = 10^{20}$ cm^{-3}; for the base region, $N_B = 10^{18}$ cm^{-3}; for the epitaxial layer, $N_{EPI} = 10^{16}$ cm^{-3}; for the buried layer, $N_{BL} = 10^{19}$ cm^{-3}; and for the substrate, $N_{SUB} = 10^{17}$ cm^{-3}. Calculate the approximate values for the zero-bias capacitances C_{je0}, C_{jc0}, and C_{js0}. Assume the junction area is the same as that of the mask area; i.e., neglect sidewall capacitance.

P5.18. Sketch a minimum-sized layout of an *npn* transistor that is characterized by the standard data, given that $N_E = 10^{18}$ cm^{-3}. Use the simplified design rules described in Sec. 5.5.1. and a minimum feature size of 2 μm. Your layout should be to scale on grid paper with a level of detail like that of Fig. 5.17.

CHAPTER 6

BIPOLAR
TRANSISTOR
INVERTER

6.0 INTRODUCTION

Well prepared by the previous chapters on the characteristics of *pn* junctions and the bipolar junction transistor in an integrated circuit, we now come to make use of these devices in digital circuits. We start with a basic element of most digital systems—the logic inverter.

The logic inverter requires that with the input in one logic state, the output is in the opposite state. That is, with a logic 0 at the input, the output is a logic 1, or vice versa. Using voltage levels and positive logic, this means that with a low voltage at the input, the output is at a high voltage level, or vice versa. The symbol for a logic inverter is shown in Fig. 6.1*a*.

In this chapter we implement this logic inverter with a bipolar junction transistor (BJT). We first review the static (or steady-state) characteristics of a saturating inverter circuit. Then we introduce a new model for the BJT—the charge-control model—and use this model to determine the dynamic (or transient) characteristics of the inverter circuit. The transient characteristics of the inverter can be improved by including a Schottky-barrier diode in the circuit to prevent the inverter from saturating, so we will study the effect of this *clamp diode* on the static and dynamic characteristics of the inverter circuit. Finally, we make use of the computer simulation program SPICE and compare the steady-state and transient characteristics obtained by hand analysis with the computer-aided results for both the saturating and the Schottky-clamped inverter circuit.

6.1   STATIC CHARACTERISTICS

A simple but practical configuration of a bipolar transistor logic inverter is shown in Fig. 6.1*b*, and a perspective view of a typical IC realization is shown in Fig. 6.1*c*. Listed on Fig. 6.1*b* are some typical data for the transistor in this circuit. We assume that these data are independent of collector current, but with $I_C = 0.01I_{C(EOS)}$, $V_{in} = V_{BE(on)}$, where (*EOS*) stands for "edge of saturation." It can be readily seen from the circuit that with the input voltage V_{in} less than the turn-on voltage $V_{BE(on)}$ for the transistor, the collector current will essentially be zero and the output voltage V_{out} will be equal to V_{CC}; that is, the transistor will be cut off. When the input voltage is increased above $V_{BE(on)}$, the transistor turns on and enters the forward active region, where the collector current is related to the base current as $I_C = \beta_F I_B$. Now $V_{out} = V_{CC} - I_C R_C$. Therefore as the

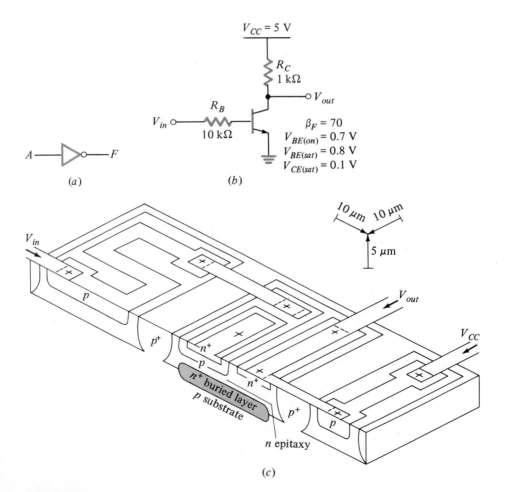

FIGURE 6.1
Bipolar transistor inverter.

input voltage increases, the output voltage decreases; the direction of the voltage change is inverted. With sufficient input voltage, when the output voltage (which is in fact V_{CE} of the transistor) has fallen sufficiently, the transistor enters the saturation region. In the saturation region the output voltage shows little, if any, change as the input voltage is further increased.

6.1.1 Voltage Transfer Characteristics

One of the principal properties of interest for any digital circuit is the *voltage transfer characteristic*, which relates the output voltage to the input voltage under static (or steady-state or low-frequency) conditions. Such a characteristic curve for the transistor inverter is shown in Fig. 6.2, where straight-line asymptotes are used to join the two main breakpoints of the characteristic. At BP 1, the input voltage is just at the point of turning the transistor on, but the output voltage is still very close to the cutoff value; i.e., the collector current is very small. At BP 2, the input voltage is now sufficient enough so that the transistor is at the edge of the saturation region. The collector current is now at, or very nearly at, its maximum value, since any further increase in the input voltage results in hardly any change in the output voltage. Observe that the two breakpoints separate the following three regions of operation for the transistor:

1. Cutoff
2. Active
3. Saturation

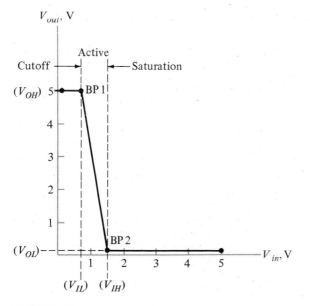

FIGURE 6.2
Voltage transfer characteristic.

The coordinates of BP 1 are V_{IL} and V_{OH}, and for BP 2 they are V_{IH} and V_{OL}, where:

V_{IL} represents the input low voltage, which is the maximum value of V_{in} to guarantee that $V_{out} = V_{OH}$.

V_{IH} represents the input high voltage, which is the minimum value of V_{in} to guarantee that $V_{out} = V_{OL}$.

V_{OL} represents the guaranteed output low voltage. With positive logic this is the 0 state.

V_{OH} represents the guaranteed output high voltage. With positive logic this is the 1 state.

Numeric values for these quantities are quickly obtained, as follows. From the circuit of Fig. 6.1b:

1. V_{OH}, which is equivalent to V_{CE} with the transistor at the edge of the cutoff region, that is, V_{CC}. For the given example it is 5 V. Hence,

$$V_{OH} = V_{CC}$$

But note that for an inverter circuit V_{OH} is usually measured with $V_{in} = V_{OL}$.

2. V_{OL}, which is equivalent to V_{CE} with the transistor at the edge of the saturation region, that is, $V_{CE(sat)}$. In the example it is 0.1 V. Thus

$$V_{OL} = V_{CE(sat)}$$

Again note that for an inverter circuit V_{OL} is commonly measured with $V_{in} = V_{OH}$.

3. V_{IL}, which is the input voltage just sufficient to turn the transistor on. In this example it is $V_{BE(on)}$, which is given as 0.7 V. Therefore,

$$V_{IL} = V_{BE(on)}$$

4. V_{IH}, which is the input voltage just sufficient to saturate the transistor. Now with the transistor just at the edge of saturation, $I_C = I_{C(EOS)}$, where

$$I_{C(EOS)} = \frac{V_{CC} - V_{CE(sat)}}{R_C} \tag{6.1.1}$$

But since the transistor is also at the edge of the forward active region,

$$I_{C(EOS)} = \beta_F I_{B(EOS)} \tag{6.1.2}$$

where, with the input at V_{IH},

$$I_{B(EOS)} = \frac{V_{IH} - V_{BE(sat)}}{R_B} \tag{6.1.3}$$

Therefore, solving for V_{IH} using Eqs. (6.1.2) and (6.1.1),

$$V_{IH} = V_{BE(sat)} + \frac{R_B}{R_C} \frac{V_{CC} - V_{CE(sat)}}{\beta_F} \tag{6.1.4}$$

Using the numeric values,

$$V_{IH} = 0.8 + \frac{10 \text{ k}\Omega}{1 \text{ k}\Omega} \frac{5 - 0.1}{70} = 1.5 \text{ V}$$

Hence the coordinates of BP 1 are $V_{in} = 0.7$ V and $V_{out} = 5.0$ V; for BP 2 they are $V_{in} = 1.5$ V and $V_{out} = 0.1$ V.

Note that the slope of the characteristic in Fig. 6.2 relates to the voltage gain of the transistor inverter, since the voltage gain is given as

$$a_v = \frac{\Delta V_{out}}{\Delta V_{in}} \tag{6.1.5}$$

In the cutoff and saturation regions the slope is zero as is, of course, the voltage gain. But in the active region the slope directly indicates the voltage gain. Now straight-line asymptotes have been used in the voltage transfer characteristic to uniquely identify the breakpoints. More accurately, these breakpoints define the points on the characteristic curve where the small-signal voltage gain is 1.0; they are then identified as the *unity-gain points*. In particular, we note from Fig. 6.2 that to the left of BP 1 and to the right of BP 2 the voltage gain is <1.0, but to the right of BP 1 and to the left of BP 2, that is, in the active region, the voltage gain is >1.0.

6.1.2   Logic-Level Diagram

The concept of the voltage levels in a digital circuit can more readily be understood with the use of the *logic-level diagram* shown in Fig. 6.3.

On the input side, $V_{IL} = 0.7$ V and $V_{IH} = 1.5$ V. From the diagram V_{IL} is the maximum allowed voltage at the input for a logic low level, and V_{IH} is the minimum input voltage for a logic high level. Between the two levels the transistor is in the active region, where because of the loose control on the transistor parameters, in particular β_F, the output level is not uniquely determined. Hence this is a forbidden region, and the difference between V_{IH} and V_{IL} is known as the *transition width*, i.e.,

$$TW = V_{IH} - V_{IL} \tag{6.1.6}$$

$$= 1.5 - 0.7 = 0.8 \text{ V}$$

On the output side of Fig. 6.3, $V_{OH} = 5$ V and $V_{OL} = 0.1$ V. Notice that V_{OH} is sufficiently greater than V_{IH} and V_{OL} is sufficiently smaller than V_{IL} so that the input of a similar inverter may safely be connected to the output of this inverter. Indeed the difference between V_{OH} and V_{IH} designates the *high noise margin* and that between V_{IL} and V_{OL} the *low noise margin* for a digital circuit. Thus

$$NM_H = V_{OH} - V_H \tag{6.1.7}$$

$$= 5.0 - 1.5 = 3.5 \text{ V}$$

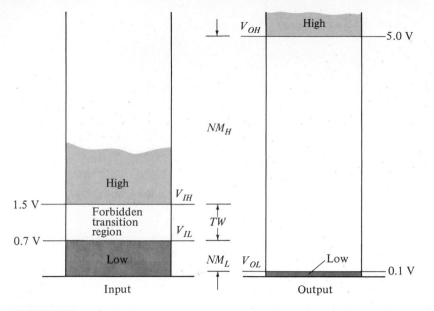

FIGURE 6.3
Logic-level diagram.

$$NM_L = V_{IL} - V_{OL} \tag{6.1.8}$$
$$= 0.7 - 0.1 = 0.6 \text{ V}$$

The difference between the two output voltage levels designates the *logic swing*:

$$LS = V_{OH} - V_{OL} \tag{6.1.9}$$
$$= 5.0 - 0.1 = 4.9 \text{ V}$$

6.1.3   Fan-Out

The term *fan-out* as used with digital circuits describes the maximum number of load gates (circuits), of similar design as the driver gate, that can be connected to the output of a logic circuit, i.e., the driver gate. More generally it simply describes the number of load gates, of similar design, connected to the output of a driver gate, but not necessarily the maximum number. Figure 6.4 shows only one load gate (Q_1); hence the fan-out from the driver (Q_0) is 1.

We have already noted that V_{OH} for the individual inverter of Fig. 6.1 is $V_{CC} = 5$ V and the high noise margin is $NM_H = 3.5$ V. In that example the load is zero (LOAD = 0).

Now with LOAD = 1, as in Fig. 6.4, note that V_{OH} at F is due to the voltage divider action of R_C and R_B. In particular,

$$V_{OH} = V_{BE(sat)} + \frac{R_B}{R_C + R_B}(V_{CC} - V_{BE(sat)}) \tag{6.1.10}$$

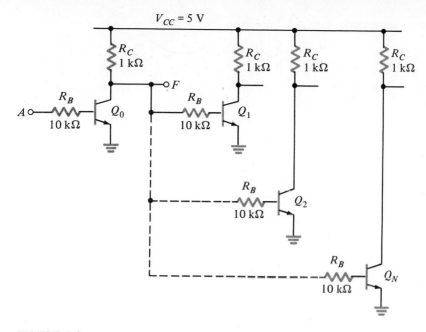

FIGURE 6.4
Fan-out.

$$= 0.8 + \frac{10}{1 + 10}(5.0 - 0.8) = 4.6 \text{ V}$$

That is, due to the application of one load gate, V_{OH} has been reduced from 5.0 to 4.6 V. As a consequence, NM_H has also been reduced. Hence, from Eq. (6.1.7),

$$NM_H = V_{OH} - V_{IH} = 4.6 - 1.5 = 3.1 \text{ V}$$

The question that now arises is what is the maximum number N of load gates that can be connected to the output F in Fig. 6.4? That is, what is the (maximum) fan-out? The answer to the question is that the absolute limit is when $NM_{H} = 0$. Then, from Eq. (6.1.7),

$$V_{OH} = V_{IH} \tag{6.1.11}$$

Now we note from Fig. 6.4 that with inverter Q_0 off and with N number of load gates, there are N base resistors R_B all effectively connected in parallel to $V_{BE(sat)}$, represented in the example by a 0.8-V battery. Hence modifying Eq. (6.1.10) to account for N load gates and then combining it with Eq. (6.1.11), we have

$$V_{BE(sat)} + \frac{R_B/N}{R_C + R_B/N}(V_{CC} - V_{BE(sat)}) = \frac{R_B}{R_C}\frac{V_{CC} - V_{CE(sat)}}{\beta_F} \tag{6.1.12}$$

Solving for N in this equality, we have the maximum number of load gates

$$N \le \beta_F \frac{V_{CC} - V_{BE(sat)}}{V_{CC} - V_{CE(sat)}} - \frac{R_B}{R_C}$$

Substituting the numeric values from Fig. 6.1,

$$N \leq 70 \frac{5 - 0.8}{5 - 0.1} - \frac{10}{1} = 50$$

Since a fractional load gate is an impossibility in practice, in all fan-out problems we conservatively round down, that is, $53.8 \to 53$.

Since $NM_H = 0$, this calculation yields the absolute maximum number of load gates for the fan-out. A more practical solution might be to set $NM_H = NM_L$ and then use the modified form of Eq. (6.1.10) to solve for N. This is left as an exercise for the reader.

Further, it should be noted that even with the qualification of $NM_H = NM_L$, no allowance has been made for variations in the parameters of the passive and active devices with processing, temperature, and time. This further complication is in the realm of statistical and worst-case design, which is beyond the scope of the coverage here.

Exercise 6.1. For the logic inverter circuit of Fig. 6.1, calculate the fan-out (N_{max}) with $NM_H = NM_L$.

6.2   CHARGE-CONTROL ANALYSIS

In the description of the static characteristics of the transistor inverter in Sec. 6.1, we noted that as the transistor changes state from off to on, it starts in the cutoff region, passes through the forward active region, and finishes up in the saturation region. That is, with each change of logic state the transistor will operate for some time in each of the three regions of operation. In Sec. 5.3 simplified models were derived for the BJT, but each of these models is applicable only in one particular operating region of the transistor. Moreover, the model in the forward active mode is applicable only at one particular operating point—a particular collector current I_C and collector-emitter voltage V_{CE}. As we now attempt to characterize the dynamic or transient properties of the transistor inverter, this approach presents some difficulties.

An analysis of the bipolar transistor inverter is possible using the static models in the so-called *piecewise-linear approximation* method. Then, the operation of the inverter is described at a particular operating point with a particular model; the model is then changed to accommodate a subsequent operating point. The method continues until a complete change of logic state has been effected. The pieces are then put together, and a piecewise-linear approximation of the characteristic is described.

Another, and better way, is to make use of the charge-control model described by Beaufoy and Sparkes.* In our description of this model we will

---

*R. Beaufoy and J. J. Sparkes, "The Junction Transistor as a Charge Controlled Device," *ATE Journal*, vol. 13, October 1957, pp. 310–327. J. J. Sparkes, "A Study of Charge Controlled Parameters," *Proceedings of the IRE*, vol. 48, October 1960, pp. 1696–1705.

first restrict our attention to the neutral base region of the transistor. Following that, an extension of the model will be made to include the effects of the depletion region at the emitter and collector junctions of the transistor.

6.2.1 Neutral Base Region

In this section we will initially determine a suitable charge-control model to describe the operation of the bipolar transistor in the forward active mode. We then extend the model to include operation in the reverse active mode.

FORWARD ACTIVE MODE. As we consider only forward injection of carriers from the emitter into the neutral base region, we are concerned with the excess minority carrier base charge Q_F, first described in Sec. 5.1. Figure 5.3 from that section, which diagrams the excess minority carrier concentration across the neutral base region, is shown again here as Fig. 6.5.

The excess minority carrier concentration is given, similar to Eq. (5.1.3), as

$$n'_b(x) = n_b(x) - n_{bo} \tag{6.2.1}$$

With forward injection only, we have, similar to Eq. (5.1.4),

$$n'_b(0) = n_{bo}(e^{V_{BE}/V_T} - 1) \tag{6.2.2}$$

Also, in the forward active mode, the collector current is related to the minority carrier concentration by the diffusion equation. Similar to Eq. (5.1.6),

$$I_C = qAD_b \frac{dn'_b(x)}{dx} \tag{6.2.3a}$$

$$\approx qAD_b \frac{n'_b(0)}{W} \tag{6.2.3b}$$

The excess minority carrier base charge, from Eq. (5.1.9a), is

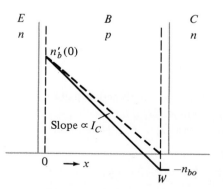

E
n

B
p

C
n

$n'_b(0)$

Slope $\propto I_C$

$0 \quad \longrightarrow x$

W

$-n_{bo}$

FIGURE 6.5
Excess minority carrier concentration in the neutral base region for the forward active mode.

$$Q_F = \frac{qAWn_b'(0)}{2} \tag{6.2.4}$$

Substituting for $n'_b(0)$ from Eq. (6.2.2) we have

$$Q_F = \frac{qAWn_{bo}}{2}(e^{V_{BE}/V_T} - 1) \tag{6.2.5a}$$

$$= Q_{F0}(e^{V_{BE}/V_T} - 1) \tag{6.2.5b}$$

where Q_{F0} is a physical parameter that relates the base charge to the base-emitter voltage in the forward active mode. From Eq. (6.2.5b),

$$Q_{F0} = \frac{qAWn_{bo}}{2} \tag{6.2.6}$$

Combining Eq. (6.2.5a) with Eq. (6.2.3b) we describe Q_F in terms of I_C, as in Eq. (5.1.9b),

$$Q_F = \frac{W^2}{2D_b}I_C \tag{6.2.7a}$$

$$= \tau_F I_C \tag{6.2.7b}$$

where the mean forward transit time of the minority carriers in the neutral base region is

$$\tau_F = \frac{W^2}{2D_b} \tag{6.2.8}$$

From Eq. (6.2.7b),

$$I_C = \frac{Q_F}{\tau_F} \tag{6.2.9}$$

Here the collector current is described by the excess minority carrier base charge and the mean forward transit time of the minority carriers in the neutral base region.

Transistor base current I_B is due to recombination in the neutral base region and hole injection from base to emitter. These two components of the base current are linearly proportional to Q_F. Therefore we may write the base current as

$$I_B = \frac{Q_F}{\tau_{BF}} \tag{6.2.10}$$

where τ_{BF} is a time constant which represents the combined effects of base recombination current I_{BB} and the current I_{BE} injected by the base into the emitter (see Sec. 5.1.4). Since $\beta_F = I_C/I_B$, we have from Eqs. (6.2.9) and (6.2.10)

$$\beta_F = \frac{\tau_{BF}}{\tau_F} \tag{6.2.11}$$

Note from Eq. (6.2.11) that to obtain a high value of β_F we must design a transistor for a large ratio of τ_{BF} and τ_F. That is, we desire a large ratio of minority carrier lifetime to base transit time, and a high emitter injection efficiency.

In Eq. (6.2.9) the static collector current I_C is described by the excess minority carrier base charge Q_F and a physical property of the device (τ_F). Hence, to determine the collector current, the *base charge* may be viewed as the independent variable in place of the *base current*.

Previously the bipolar junction transistor has been described as a *current-controlled device*, that is, $I_C = \beta_F I_B$, where β_F is a constant and I_B is the variable. Now the bipolar transistor is described as a *charge-controlled device*, with $I_C = Q_F/\tau_F$, where τ_F is a constant and Q_F is the variable.

So far we have been describing the static, or steady-state, currents, the collector current I_C and the base current I_B. However, in an analysis of the dynamic or transient properties of the bipolar transistor inverter, we will be interested in the instantaneous values of these currents as a function of time. Hence, for the collector current, similar to Eq. (6.2.9),

$$i_C(t) = \frac{Q_F(t)}{\tau_F} \qquad (6.2.12)$$

But for the base current,

$$i_B(t) = \frac{Q_F(t)}{\tau_{BF}} + \frac{dQ_F}{dt} \qquad (6.2.13)$$

The first term in Eq. (6.2.13) includes the loss of minority carriers in the base region because of recombination with the majority carriers and the majority carrier current injected by the base into the emitter. The second term is introduced to account for the time rate of change of Q_F in the neutral base region of the transistor. The first part of the equation is the steady-state term; the second part is the transient term. Thus in the steady state, $dQ_F/dt = 0$.

By combining Eqs. (6.2.12) and (6.2.13) we have

$$i_E(t) = i_C(t) + i_B(t) \qquad (6.2.14)$$

$$= \frac{Q_F(t)}{\tau_F} + \frac{Q_F(t)}{\tau_{BF}} + \frac{dQ_F}{dt}$$

Following the practice of Chap. 5 we can represent these simple charge-control equations with a model. This is done in Fig. 6.6a; note especially that since Q_F is a nonlinear function of V_{BE} [see Eq. (6.2.5b)] we represent the excess minority carrier base charge as stored on a nonlinear capacitor. For the nonlinear capacitor we introduce a new symbol, shown in Fig. 6.6b, which is used simply to represent that the charge stored is a nonlinear function of the voltage across the capacitor. In addition, note that the charge stored is positive when the voltage across the nonlinear capacitor is also positive. Likewise, the charge stored is negative when the voltage is negative.

Example 6.1. As an example of the use of this elementary charge-control model, consider the simple circuit of Fig. 6.7a, where we desire to know the collector

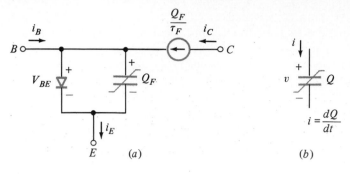

FIGURE 6.6
(a) Simple charge-control model for BJT, due to forward injection in the neutral base region only.
(b) Symbol for nonlinear capacitance, where $i = dQ/dt$.

current as a function of time. The input is a step function at $t = t_0$ that changes instantaneously from 0 V to V, where $V \gg V_{BE(on)}$.

Initially we neglect the capacitor C and note that for $t > 0$ the base current of the transistor is simply given as:
From the circuit:

$$i_B(t) = I_B = \frac{V - V_{BE(on)}}{R} \tag{6.2.15}$$

For the transistor:

$$i_B(t) = I_B = \frac{Q_F(t)}{\tau_{BF}} + \frac{dQ_F}{dt} \tag{6.2.16}$$

Solving this simple differential equation, we obtain

$$Q_F(t) = I_B \tau_{BF} + C e^{-t/\tau_{BF}} \tag{6.2.17}$$

To solve for the constant C, we make use of the initial conditions that at $t = 0$ the collector current is zero and so also is the excess minority carrier base charge. Therefore,

$$Q_F(0) = 0 = I_B \tau_{BF} + C \tag{6.2.18}$$

Hence,
$$C = -I_B \tau_{BF} \tag{6.2.19}$$

Substituting Eq. (6.2.19) in Eq. (6.2.17),

$$Q_F(t) = I_B \tau_{BF}(1 - e^{-t/\tau_{BF}}) \tag{6.2.20}$$

Now from Eq. (6.2.12), $i_C(t) = Q_F(t)/\tau_F$; therefore,

$$i_C(t) = I_B \frac{\tau_{BF}}{\tau_F}(1 - e^{-t/\tau_{BF}}) \tag{6.2.21}$$

From Eq. (6.2.11), $\beta_F = \tau_{BF}/\tau_F$; hence,

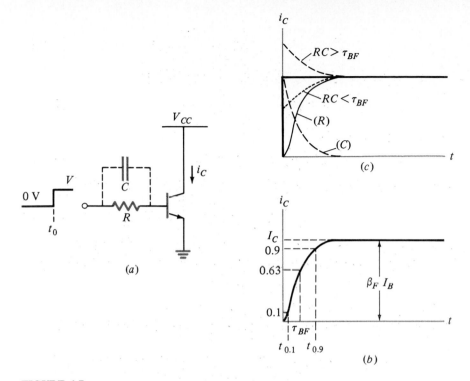

FIGURE 6.7
(a) Simple common-emitter amplifier with speedup capacitor. (b) Collector current waveform for circuit without speedup capacitor. (c) Collector current waveform illustrating effect of speedup capacitor.

$$i_C(t) = \beta_F I_B (1 - e^{-t/\tau_{BF}}) \qquad (6.2.22)$$

A plot of $i_C(t)$ using this equation is shown in Fig. 6.7b. In the figure, and from the equation, note that

$$i_C(\infty) = I_C = \beta_F I_B \qquad (6.2.23)$$

Also note that with $t = \tau_{BF}$ the collector current is 63% of its final value. Hence, this is a simple way of measuring the mean lifetime of the minority carriers in the neutral base region of the transistor.

Because of the exponential form of the collector current, it is difficult to determine just when the collector current has reached its final value. Hence it is customary to describe for a transistor operating in the forward active region the time taken for the collector current to change by the rise time t_r. This is the time taken between the 10 and 90% points of the collector current waveform (collector current versus time).

Transposing Eq. (6.2.22) in terms of time t,

$$t = \tau_{BF} \ln \frac{\beta_F I_B}{\beta_F I_B - i_C(t)} \qquad (6.2.24)$$

Hence,
$$t_{0.9} = \tau_{BF} \ln \frac{\beta_F I_B}{\beta_F I_B - 0.9 I_C}$$

and
$$t_{0.1} = \tau_{BF} \ln \frac{\beta_F I_B}{\beta_F I_B - 0.1 I_C}$$

Therefore
$$t_r = t_{0.9} - t_{0.1}$$

$$= \tau_{BF} \ln \frac{\beta_F I_B - 0.1 I_C}{\beta_F I_B - 0.9 I_C} \qquad (6.2.25a)$$

$$\approx \tau_{BF} \ln 9 = 2.2 \tau_{BF} \qquad (6.2.25b)$$

That is, in the simple circuit of Fig. 6.7a, it takes 2.2 time constants (τ_{BF}) for the collector current to change from 0.1 to 0.9 of its final value. Hence with a typical value of $\tau_{BF} = 20$ ns, $t_r = 44$ ns.

SPEED-UP CAPACITOR. Ideally of course, with a step of voltage at the input, we would like to have a step change of collector current. However, we have just noted that the change in collector current takes time. The performance of the circuit can be improved if the resistor R is bypassed by a capacitor C, as shown dashed in Fig. 6.7a.

Now with the voltage across the capacitor unable to change instantaneously, the step input of voltage V will cause an injection into the base of the transistor of a charge of majority carriers given by the capacitance C multiplied by the voltage across C of $V - V_{BE(on)}$. As a consequence there will appear in the neutral base region a charge of minority carriers:

$$Q_F = C(V - V_{BE(on)}) \qquad (6.2.26)$$

Hence,
$$i_C = \frac{Q_F}{\tau_F} = \frac{C(V - V_{BE(on)})}{\tau_F} \qquad (6.2.27)$$

but
$$i_C = \beta_F I_B = \frac{\tau_{BF}}{\tau_F} \frac{V - V_{BE(on)}}{R} \qquad (6.2.28)$$

Equating these last two equations, we solve for the optimum value of the time constant RC as

$$RC = \tau_{BF} \qquad (6.2.29)$$

hence with $RC = \tau_{BF}$ and a step voltage at the input there is a step change of collector current, as indicated by the solid line in Fig. 6.7c.

Also shown in Fig. 6.7c is $i_C(t)$ due to the resistor R alone—the slow exponential rise of current—and due to the capacitor C alone—the step rise of current followed by the slow exponential fall. Also notice that if $RC \ll \tau_{BF}$, the collector current shows an initial step and then a slow rise to its final value. With $RC \gg \tau_{BF}$, there is an overshoot in the collector current waveform.

The use of a speedup capacitor to improve the transient characteristics is a viable technique with discrete transistors; it was used in early integrated circuits. It is not used in modern ICs since the area taken up by the capacitor compared with the area of the rest of the circuit is simply too great.

Exercise 6.2. In Fig.6.7a, $V = 10$ V, $V_{BE(on)} = 0.7$ V, and $R = 20$ kΩ. The final value of the collector current is 20 mA. When $C = 10$ pF the collector current rises to its final value essentially instantaneously. Determine the transistor parameters τ_F, τ_{BF}, and β_F.

REVERSE ACTIVE MODE. Similar to the forward injection case, we now account for operation in the reverse active mode. That is, the base-collector junction is now forward-biased, and the base-emitter junction is reverse-biased. Then there is injection of carriers from the collector into the base and collection of these carriers at the emitter junction. For this reverse injection case, the distribution of excess minority carriers in the neutral base region is shown in Fig. 6.8. For comparison purposes, also shown in Fig. 6.8 as a dashed line is the distribution of excess minority carriers due to forward injection only. In this one-dimensional figure, the excess minority carrier forward base charge Q_F is represented by the triangle with the dashed gradient from the emitter to the collector; the excess minority carrier reverse base charge Q_R is represented by the triangle with the solid-line gradient from the collector to the emitter. In general, $Q_F \gg Q_R$.

Listed below are the charge-control equations for the collector, base, and emitter currents in the forward active mode from Eqs. (6.2.12), (6.2.13), and (6.2.14), respectively. Following them are listed the comparable equations for these currents in the reverse active mode, where the collector current is now the emitted current and the emitter current is the collected current. The reverse-mode equations can be obtained by an analogy with the forward-mode equations. The subscripts F and R have been added to differentiate between the forward and reverse components of current.

Forward active mode:

$$i_{CF} = \frac{Q_F}{\tau_F} \tag{6.2.30a}$$

$$i_{BF} = \frac{Q_F}{\tau_{BF}} + \frac{dQ_F}{dt} \tag{6.2.30b}$$

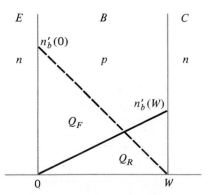

FIGURE 6.8
Distribution of excess minority carriers in the neutral base region for forward (dashed line) and reverse (solid line) modes of operation.

$$i_{EF} = \frac{Q_F}{\tau_F} + \frac{Q_F}{\tau_{BF}} + \frac{dQ_F}{dt} \qquad (6.2.30c)$$

$$Q_F = Q_{FO}(e^{V_{BE}/V_T} - 1) \qquad (6.2.30d)$$

where
$$Q_{FO} = \frac{qA_E W n_{bo}}{2} \qquad (6.2.30e)$$

Reverse active mode:

$$i_{CR} = -\frac{Q_R}{\tau_R} - \frac{Q_R}{\tau_{BR}} - \frac{dQ_R}{dt} \qquad (6.2.31a)$$

$$i_{BR} = \frac{Q_R}{\tau_{BR}} + \frac{dQ_R}{dt} \qquad (6.2.31b)$$

$$i_{ER} = -\frac{Q_R}{\tau_R} \qquad (6.2.31c)$$

$$Q_R = Q_{RO}(e^{V_{BC}/V_T} - 1) \qquad (6.2.31d)$$

where
$$Q_{RO} = \frac{qA_C W n_{bo}}{2} \qquad (6.2.31e)$$

Note that in general $Q_{FO} \neq Q_{RO}$, since the emitter junction area A_E is usually much smaller than the collector junction area A_C, for one reason.

These two sets of equations can be combined to obtain a set of charge-control equations for the neutral base region that are applicable in either the forward or reverse active mode.

$$i_C = i_{CF} + i_{CR} = \frac{Q_F}{\tau_F} - Q_R\left(\frac{1}{\tau_F} + \frac{1}{\tau_{BR}}\right) - \frac{dQ_R}{dt} \qquad (6.2.32a)$$

$$i_B = i_{BF} + i_{BR} = \frac{Q_F}{\tau_{BF}} + \frac{dQ_F}{dt} + \frac{Q_R}{\tau_{BR}} + \frac{dQ_R}{dt} \qquad (6.2.32b)$$

$$i_E = i_{EF} + i_{ER} = -\frac{Q_R}{\tau_R} + Q_F\left(\frac{1}{\tau_F} + \frac{1}{\tau_{BF}}\right) + \frac{dQ_F}{dt} \qquad (6.2.32c)$$

From these equations a charge-control model for the bipolar transistor can be derived, as shown in Fig. 6.9. In comparing the model to the equations, note, for example, in the collector circuit of the model that from the equation for the collector current:

The Q_F term is represented by the current generator.
The Q_R term is represented by the diode and its exponential characteristic.
The dQ_R/dt term is represented by the nonlinear capacitor.

A similar relationship can be made for the emitter circuit in the model and the emitter current in the equations.

The observant reader will already have noticed a similarity between the

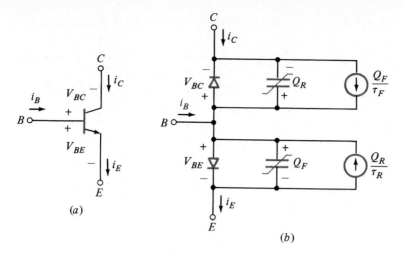

FIGURE 6.9
(a) Symbol for *npn* transistor showing terminal currents and voltages. (b) Charge-control model for charge stores in the neutral base region, including effects of forward and reverse operation.

charge-control model for the *npn* transistor in Fig. 6.9 and the Ebers-Moll model for a similar transistor in Fig. 5.4. This similarity is the subject of the next section.

COMPARISON WITH EBERS-MOLL EQUATIONS. It is interesting and informative to make a comparison of the Ebers-Moll equations derived in Sec. 5.2 with the charge-control equations just obtained. Since the basic Ebers-Moll equations are for the static terminal currents I_C, I_B, and I_E, then the comparison with the charge-control equations must be on the same basis.

Thus in the charge-control equation for the collector current, Eq. (6.2.32a), for the static, or steady-state, condition the term $dQ_R/dt \rightarrow 0$, then

$$I_C = \frac{Q_F}{\tau_F} - Q_R\left(\frac{1}{\tau_R} + \frac{1}{\tau_{BR}}\right) \tag{6.2.33a}$$

$$= \frac{Q_{FO}}{\tau_F}(e^{V_{BE}/V_T} - 1) - Q_{RO}\left(\frac{1}{\tau_R} + \frac{1}{\tau_{BR}}\right)(e^{V_{BC}/V_T} - 1) \tag{6.2.33b}$$

From the Ebers-Moll equation, namely, Eq. (5.2.3b),

$$I_C = \alpha_F I_{ES}(e^{V_{BE}/V_T} - 1) - I_{CS}(e^{V_{BC}/V_T} - 1) \tag{6.2.34}$$

In comparing these two equations, particularly note:

1. $\alpha_F I_{ES} = \dfrac{Q_{FO}}{\tau_F}$ \hfill (6.2.35)

2. $I_{CS} = Q_{RO}\left(\dfrac{1}{\tau_R} + \dfrac{1}{\tau_{BR}}\right)$ \hfill (6.2.36)

Also, by charge control, the static emitter current is, from Eq. (6.2.32c),

$$I_E = Q_F\left(\frac{1}{\tau_R} + \frac{1}{\tau_{BF}}\right) - \frac{Q_R}{\tau_R} \tag{6.2.37a}$$

$$= Q_{FO}\left(\frac{1}{\tau_F} + \frac{1}{\tau_{BF}}\right)(e^{V_{BE}/V_T} - 1) - \frac{Q_{RO}}{\tau_R}(e^{V_{BC}/V_T} - 1) \tag{6.2.37b}$$

And from the Ebers-Moll equations, in Eq. (5.2.3a),

$$I_E = I_{ES}(e^{V_{BE}/V_T} - 1) - \alpha_R I_{CS}(e^{V_{BC}/V_T} - 1) \tag{6.2.38}$$

Comparing the two equations for the emitter current, note that

1. $I_{ES} = Q_{FO}\left(\dfrac{1}{\tau_F} + \dfrac{1}{\tau_{BF}}\right)$ \hfill (6.2.39)

2. $\alpha_R I_{CS} = \dfrac{Q_{RO}}{\tau_R}$ \hfill (6.2.40)

Now from Eq. (6.2.35),

$$\frac{I_{ES}}{Q_{FO}} = \frac{1}{\alpha_F \tau_F} \tag{6.2.41}$$

But in Eq. (6.2.39),

$$\frac{I_{ES}}{Q_{FO}} = \frac{1}{\tau_F} + \frac{1}{\tau_{BF}} = \frac{\tau_F + \tau_{BF}}{\tau_F \tau_{BF}} \tag{6.2.42}$$

Combining Eqs. (6.2.41) and (6.2.42),

$$\alpha_F = \frac{\tau_{BF}}{\tau_F + \tau_{BF}} \tag{6.2.43}$$

Now, by definition,

$$\beta_F = \frac{\alpha_F}{1 - \alpha_F}$$

Hence, from Eq. (6.2.43),

$$\beta_F = \frac{\tau_{BF}}{\tau_F} \tag{6.2.44}$$

This is in agreement with Eq. (6.2.11) presented earlier.
Similarly, from Eqs. (6.2.36) and (6.2.40),

$$\alpha_R = \frac{\tau_{BR}}{\tau_R + \tau_{BR}} \tag{6.2.45}$$

But \hfill $\beta_R = \dfrac{\alpha_R}{1 - \alpha_R}$

therefore,
$$\beta_R = \frac{\tau_{BF}}{\tau_R} \qquad (6.2.46)$$

Further, from the reciprocity theorem of Eq. (5.2.5),

$$\alpha_F I_{ES} = \alpha_R I_{CS}$$

Hence, from Eqs. (6.2.35) and (6.2.40),

$$\frac{Q_{FO}}{\tau_F} = \frac{Q_{RO}}{\tau_R} \qquad (6.2.47a)$$

or
$$\frac{Q_{FO}}{Q_{RO}} = \frac{\tau_F}{\tau_R} \qquad (6.2.47b)$$

Another interesting relation is obtained from Eq. (6.2.35),

$$Q_{FO} = \alpha_F I_{ES} \tau_F \qquad (6.2.48)$$

And from Eq. (6.2.40),

$$Q_{RO} = \alpha_R I_{CS} \tau_R \qquad (6.2.49)$$

In Eq. (6.2.48), Q_{FO} is described in terms of the electrical parameters of the transistor, whereas in Eq. (6.2.30e), Q_{FO} is described in terms of the physical properties of the device. A similar relationship holds for Q_{RO}.

6.2.2 Depletion Region

So far, attention has been focused on the consequences of the excess minority carrier charge storage in the neutral base region of the transistor. We have developed charge-control equations and models that cover both forward and reverse modes of operation. But there is also charge stored in the depletion regions of the transistor at the emitter and collector junctions. Consequently, whenever the voltage across either of these junctions changes, there are charging currents to these depletion regions which must be accounted for. These regions of charge storage are indicated in Fig. 6.10. The depletion region charge, for an abrupt junction is, from Eq. (4.3.10),

$$Q_j = 2C_{jo}\phi_0^{1/2}(\phi_0 - V)^{1/2} \qquad (6.2.50)$$

Note in this equation that Q_j is zero when $V = \phi_0$. Also the charge is always positive, increasing as the applied voltage is made more negative. Now from the circuit designer's point of view, it is simpler to consider the depletion region charge to be zero when the applied voltage is zero and for the charge to be positive for positive bias and negative for negative bias. Hence we introduce a new variable for the depletion region charge, namely,

$$Q_V(V) = Q_j(0) - Q_j(V) \qquad (6.2.51)$$

Then, from Eq. (6.2.50),

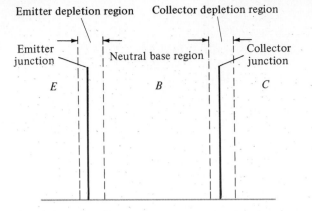

FIGURE 6.10
Region of charge storage in a BJT.

$$Q_V(V) = K_a[\phi_0^{1/2} - (\phi_0 - V)^{1/2}] \qquad (6.2.52)$$

where

$$K_a = 2C_{j0}\phi_0^{1/2}$$

is a constant for any particular abrupt junction.

Similarly, for a linear graded junction, from Eqs. (4.3.12) and (6.2.51),

$$Q_V(V) = K_g[\phi_0^{2/3} - (\phi_0 - V)^{2/3}] \qquad (6.2.53)$$

where

$$K_g = \frac{3}{2}C_{j0}\phi_0^{1/3}$$

We now extend the charge-control equations and model to include the charge stored in the depletion regions.

From Eq. (6.2.52), assuming an abrupt junction, the charge Q_V is zero when the applied voltage V is zero. Hence for the emitter junction,

$$Q_{VE}(0) = 0 \qquad (6.2.54a)$$

and the charging current

$$i_E = \frac{dQ_{VE}}{dt} \qquad (6.2.54b)$$

Similarly, for the collector junction,

$$Q_{VC}(0) = 0 \qquad (6.2.55a)$$

but, from the circuit,

$$i_C = \frac{-dQ_{VC}}{dt} \qquad (6.2.55b)$$

The negative sign is owing to the direction of i_C being to oppose forward current in the collector diode of the transistor.

Listed below are the complete charge-control equations for an *npn* transistor, where we have incorporated the above charging currents into the basic equations of Eq. (6.2.32). A model that pictures these equations is shown in Fig. 6.11. For an *npn* transistor:

$$i_C = \frac{Q_F}{\tau_F} - Q_R\left(\frac{1}{\tau_F} + \frac{1}{\tau_{BR}}\right) - \frac{dQ_R}{dt} - \frac{dQ_{VC}}{dt} \tag{6.2.56a}$$

$$i_B = \frac{Q_F}{\tau_{BF}} + \frac{dQ_F}{dt} + \frac{Q_R}{\tau_{BR}} + \frac{dQ_R}{dt} + \frac{dQ_{VC}}{dt} + \frac{dQ_{VE}}{dt} \tag{6.2.56b}$$

$$i_E = -\frac{Q_R}{\tau_R} + Q_F\left(\frac{1}{\tau_F} + \frac{1}{\tau_{BF}}\right) + \frac{dQ_F}{dt} + \frac{dQ_{VE}}{dt} \tag{6.2.56c}$$

$$Q_F = Q_{FO}(e^{V_{BE}/V_T} - 1) \tag{6.2.56d}$$

$$Q_R = Q_{RO}(e^{V_{BC}/V_T} - 1) \tag{6.2.56e}$$

Finally, to complete the picture, we list the complete charge-control equations for a *pnp* transistor, with the corresponding model shown in Fig. 6.12*b*. For reference purposes, Fig. 6.12*a* shows the schematic of a *pnp* transistor labeled with the terminal voltages and currents.

For a *pnp* transistor:

$$i_C = \frac{Q_F}{\tau_F} - Q_R\left(\frac{1}{\tau_R} + \frac{1}{\tau_{BR}}\right) - \frac{dQ_R}{dt} - \frac{dQ_{VC}}{dt} \tag{6.2.57a}$$

$$i_B = \frac{Q_F}{\tau_{BF}} + \frac{dQ_F}{dt} + \frac{Q_R}{\tau_{BR}} + \frac{dQ_R}{dt} + \frac{dQ_{VC}}{dt} + \frac{dQ_{VE}}{dt} \tag{6.2.57b}$$

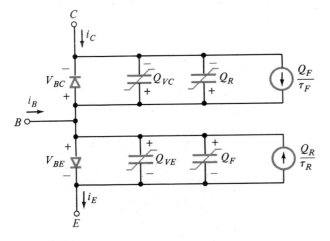

FIGURE 6.11
Complete charge-control model for *npn* transistor.

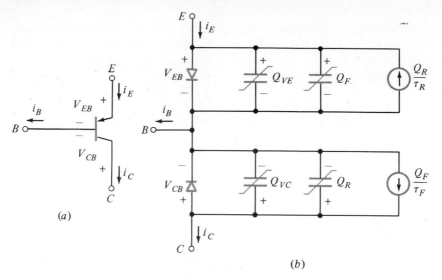

(a)

(b)

FIGURE 6.12
(a) *pnp* transistor. (b) Charge-control model.

$$i_E = -\frac{Q_R}{\tau_R} + Q_F\left(\frac{1}{\tau_F} + \frac{1}{\tau_{BF}}\right) + \frac{dQ_F}{dt} + \frac{dQ_{VE}}{dt} \qquad (6.2.57c)$$

$$Q_F = Q_{FO}(e^{V_{EB}/V_T} - 1) \qquad (6.2.57d)$$

$$Q_R = Q_{RO}(e^{V_{CB}/V_T} - 1) \qquad (6.2.57e)$$

Exercise 6.3. Given $I_S = 10^{-15}$ A, $\tau_R = 0.1$ ns, $V_{BE} = 0.7$ V, $V_{CE} = 5.0$ V, and $V_T = 26$ mV. Solve for the excess minority carrier base charges Q_F and Q_R.

6.2.3 Modes of Operation

Following the initial development of the Ebers-Moll equations, in Sec. 5.3 we derived simpler equations for each of the operating modes of the transistor. Similarly, we now determine a set of simple charge-control equations that are applicable for a particular mode of operation. The derivation will be from the *npn* equations of Eq. (6.2.56), but a similar development could be made starting with the *pnp* equations of Eq. (6.2.57).

CUTOFF MODE. In the cutoff mode both the emitter and the collector junctions are reverse-biased; i.e., for an *npn* transistor both V_{BE} and V_{BC} are negative. Then from Eqs. (6.2.56d and e), both Q_F and Q_R are negative. Furthermore with both V_{BE} and $V_{BC} \ll -4V_T$, it will generally be found that the magnitude of both Q_F and Q_R is much less than either Q_{VE} or Q_{VC}. That is,

$$|Q_F, Q_R| \ll |Q_{VE}, Q_{VC}|$$

Introducing this simplification into Eq. (6.2.56), we obtain very simple expressions for the collector and base currents in the cutoff mode, namely,

$$i_C = -\frac{dQ_{VC}}{dt} \qquad\qquad (6.2.58a)$$

$$i_B = \frac{d}{dt}(Q_{VE} + Q_{VC}) \qquad\qquad (6.2.58b)$$

These equations for the cutoff mode are modeled in Fig. 6.13a. Henceforth we will omit the equation for i_E since it is simply equal to $i_B + i_C$.

FORWARD ACTIVE MODE. In the forward active mode, the base-emitter junction is forward-biased and the base-collector junction is reverse-biased. Now Q_F is positive and Q_R is negative, but with $V_{BE} \gg 4V_T$ and $V_{BC} \ll -4V_T$,

$$|Q_F| \gg |Q_R|$$

With this simplification, the charge-control equations of Eq. (6.2.56) reduce to

$$i_C = \frac{Q_F}{\tau_F} - \frac{dQ_{VC}}{dt} \qquad\qquad (6.2.59a)$$

$$i_B = \frac{Q_F}{\tau_{BF}} + \frac{d}{dt}(Q_F + Q_{VE} + Q_{VC}) \qquad\qquad (6.2.59b)$$

These simple forms of the charge-control equations in the forward active mode are modeled in Fig. 6.13b.

SATURATION MODE. In the saturation mode, both the emitter and the collector junctions are forward-biased. Moreover, both V_{BE} and V_{BC} are essentially

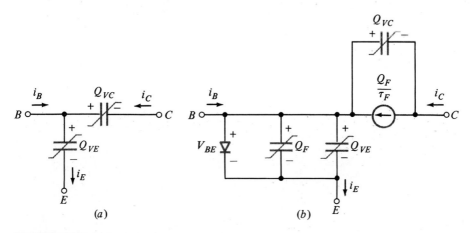

FIGURE 6.13
Simplified charge-control models. (a) Cutoff mode; (b) forward active mode.

constant. Consequently, in saturation there is little change in either Q_{VE} or Q_{VC}; we may therefore neglect these components in Eq. (6.2.56). Hence,

$$i_C = \frac{Q_F}{\tau_F} - Q_R\left(\frac{1}{\tau_F} + \frac{1}{\tau_{BR}}\right) - \frac{dQ_R}{dt} \tag{6.2.60a}$$

$$i_B = \frac{Q_F}{\tau_{BF}} + \frac{Q_R}{\tau_{BR}} + \frac{d}{dt}(Q_F + Q_R) \tag{6.2.60b}$$

Note that each of these equations contains the parameter for the mean lifetime of the minority carriers in the neutral base region in the forward mode (τ_{BF}) and reverse mode (τ_{BR}). The dynamic characteristics of the transistor in saturation are better handled with a new parameter, the *saturation time constant* τ_S. As can be shown, this new parameter is a weighted mean of both τ_{BF} and τ_{BR}.

In general, the total excess minority carrier charge stored in the neutral base region is the superposition of the two charges Q_F and Q_R. That is, $Q_B = Q_F + Q_R$.

Two ways of representing the total base charge Q_B are shown in Fig. 6.14. In Fig. 6.14a, we sum the forward and reverse components—the two triangles— and obtain

$$Q_B = Q_F + Q_R \tag{6.2.61a}$$

Another way is represented in Fig. 6.14b, where we sum a triangle Q_A and a rectangle Q_S. Then

$$Q_B = Q_A + Q_S \tag{6.2.61b}$$

In each case the total area, being representative of Q_B, is the same. In the latter equation:

1. Q_A represents the base charge that brought the transistor to the edge of the saturation region. The collector current at the edge of saturation ($I_{C(EOS)}$) is represented by the slope of the triangle containing Q_A. Hence,

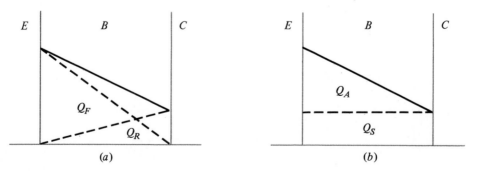

(a)   (b)

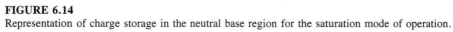

FIGURE 6.14
Representation of charge storage in the neutral base region for the saturation mode of operation.

$$Q_A = \tau_F I_{C(EOS)} = \tau_{BF} I_{B(EOS)} \qquad (6.2.62a)$$

where
$$\frac{I_{C(EOS)}}{I_{B(EOS)}} = \frac{\tau_{BF}}{\tau_F} = \beta_F \qquad (6.2.62b)$$

2. Q_S represents the overdrive base charge.

$$Q_S = \tau_S I_{BS}$$

where I_{BS} is the base current, over and above $I_{B(EOS)}$, that really drives the transistor into saturation and gives rise to Q_S. The saturation time constant τ_S has been defined as the apparent lifetime that characterizes the recombination of the overdrive base charge Q_S.

With the transistor in saturation, the static equation for the base current is

$$I_B = I_{B(EOS)} + I_{BS} \qquad (6.2.63)$$

But the instantaneous value of the base current is given by

$$i_B(t) = \frac{Q_A}{\tau_{BF}} + \frac{Q_S}{\tau_S} + \frac{dQ_S}{dt} \qquad (6.2.64a)$$

Note, using Eq. (6.2.62a), that

$$\frac{Q_A}{\tau_{BF}} = \frac{\tau_F I_{C(EOS)}}{\tau_{BF}} = \frac{I_{C(EOS)}}{\beta_F}$$

then
$$i_B(t) - \frac{I_{C(EOS)}}{\beta_F} = \frac{Q_S}{\tau_S} + \frac{dQ_S}{dt} \qquad (6.2.64b)$$

This last equation equates the instantaneous base current to the dynamics of the overdrive base charge and contains only one transistor time constant (τ_S). As such, we will find this equation simpler to use than that given earlier in Eq. (6.2.60b), when we calculate later the saturation storage time t_s.

*DERIVATION OF SATURATION TIME CONSTANT τ_s.** With the transistor in the saturation mode, the excess minority carrier base charge is, from Fig. 6.14 and Eq. (6.2.61),

$$Q_B = Q_F + Q_R = Q_A + Q_S \qquad (6.2.65)$$

where
$$Q_F = I_{CF} \tau_F$$

with I_{CF} = current that crosses collector junction in forward mode
$\quad\tau_F$ = mean transit time for minority carriers to cross base region from emitter to collector

---

*Optional.

$$Q_R = I_{ER}\tau_R$$

with I_{ER} = current that crosses emitter junction in reverse mode
τ_R = mean transit time for minority carriers to cross base region from collector to emitter

$$Q_A = I_{C(EOS)}\tau_F$$

with $I_{C(EOS)}$ = collector current with transistor just at edge of saturation

$$Q_S = I_{BS}\tau_S$$

with I_{BS} = overdrive base current with transistor in saturation
τ_s = saturation time constant

The general charge-control equations for the transistor in saturation are, from Eq. (6.2.60),

$$i_C = \frac{Q_F}{\tau_F} - Q_R\left(\frac{1}{\tau_R} + \frac{1}{\beta_R\tau_R}\right) - \frac{dQ_R}{dt} \qquad (6.2.66a)$$

$$i_B = \frac{Q_F}{\beta_F\tau_F} + \frac{Q_R}{\beta_R\tau_R} + \frac{d}{dt}(Q_F + Q_R) \qquad (6.2.66b)$$

But in saturation, the static collector current is simply

$$i_C = I_{C(EOS)} = \frac{Q_A}{\tau_F} \qquad (6.2.67)$$

Substituting this for i_C in Eq. (6.2.66a), and also noting from

$$\beta_R = \frac{\alpha_R}{1 - \alpha_R}$$

that

$$\frac{1}{\alpha_R} = 1 + \frac{1}{\beta_R}$$

we obtain

$$\frac{Q_F - Q_A}{\tau_F} = \frac{Q_R}{\alpha_R\tau_R} \qquad (6.2.68)$$

But from Eq. (6.2.65),

$$Q_F - Q_A = Q_S - Q_R \qquad (6.2.69)$$

Using Eq. (6.2.69) in Eq. (6.2.68) and solving for Q_R, we have

$$Q_R = \frac{Q_S}{\tau_F}\frac{1}{1/\alpha_R\tau_R + 1/\tau_F} = \frac{\alpha_R\tau_R}{\tau_F + \alpha_R\tau_R}Q_S \qquad (6.2.70)$$

From Eq. (6.2.65) we also note that

$$Q_F = Q_A + Q_S - Q_R \qquad (6.2.71)$$

Hence substituting this for Q_F in Eq. (6.2.66b),

$$i_B = \frac{Q_A + Q_S - Q_R}{\beta_F \tau_F} + \frac{Q_R}{\beta_R \tau_R} + \frac{d}{dt}(Q_A + Q_S) \qquad (6.2.72)$$

(In saturation, $dQ_A/dt \to 0$.)

Now the instantaneous overdrive base current i_{BS} is related to i_B; from Eq. (6.2.63),

$$i_{BS} = i_B - I_{B(EOS)}$$

where
$$I_{B(EOS)} = \frac{Q_A}{\beta_F \tau_F} \qquad (6.2.73)$$

Therefore, from Eq. (6.2.72),

$$i_{BS} = \frac{Q_S}{\beta_F \tau_F} - \frac{Q_R}{\beta_F \tau_F} + \frac{Q_R}{\beta_R \tau_R} + \frac{dQ_S}{dt} \qquad (6.2.74a)$$

Then substituting for Q_R from Eq. (6.2.70),

$$i_{BS} = \frac{Q_S}{\beta_F \tau_F} - \frac{Q_S}{\beta_F \tau_F} \frac{\alpha_R \tau_R}{\tau_F + \alpha_R \tau_R} + \frac{Q_S}{\beta_R \tau_R} \frac{\alpha_R \tau_R}{\tau_F + \alpha_R \tau_R} + \frac{dQ_S}{dt} \qquad (6.2.74b)$$

$$= Q_S \left[\frac{\alpha_R \tau_R}{\tau_F + \alpha_R \tau_R} \left(\frac{1}{\beta_R \tau_R} - \frac{1}{\beta_F \tau_F} + \frac{\tau_F + \alpha_R \tau_R}{\alpha_R \tau_R \beta_F \tau_F} \right) \right] + \frac{dQ_S}{dt} \qquad (6.2.74c)$$

After further manipulation we obtain

$$i_{BS} = \frac{1 - \alpha_F \alpha_R}{\alpha_F (\tau_F + \alpha_R \tau_R)} Q_S + \frac{dQ_S}{dt} \qquad (6.2.74d)$$

That is,
$$i_{BS} = \frac{Q_S}{\tau_S} + \frac{dQ_S}{dt} \qquad (6.2.75)$$

where
$$\tau_S = \frac{\alpha_F (\tau_F + \alpha_R \tau_R)}{1 - \alpha_F \alpha_R} \qquad (6.2.76)$$

As a typical example, let

$$\alpha_F = 0.98 \qquad \tau_F = 0.2 \text{ ns}$$
$$\alpha_R = 0.5 \qquad \tau_R = 20 \text{ ns}$$

then
$$\tau_S = \frac{0.98(0.2 + 10)}{1 - 0.49} = 19.6 \text{ ns}$$

Before concluding this section, one important point should be noted. The excess minority carrier charge profile shown in Fig. 6.14a, and repeated in Fig. 6.15a, implies that both the emitter and collector regions are more heavily doped than the base region. That is, for an *npn transistor*, the majority carrier concentration is such that

$$n_{ne} \gg p_{pb} \ll n_{nc}$$

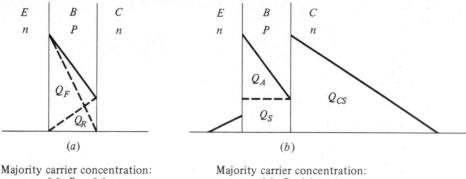

(a)

(b)

Majority carrier concentration:
$$n_{ne} \gg P_{pb} \ll n_{nc}$$

Majority carrier concentration:
$$n_{ne} \gg P_{pb} \gg n_{nc}$$

FIGURE 6.15
Excess minority carrier charge distribution in the neutral base region for the saturation mode of operation.

With the transistor in saturation, both junctions are forward-biased, and these figures show charge stored in the base region only. In particular, we have neglected, or assumed very small, the injection of carriers from the base.

However, the more normal case for a planar-diffused *npn* transistor is that the majority carrier concentration is given as

$$n_{ne} \gg P_{pb} \gg n_{nc}$$

Hence with the collector junction forward-biased there can be appreciable injection of carriers into the collector from the base. A typical charge profile for a saturated-diffused *npn* transistor is shown in Fig. 6.15*b*. Consequently, the charge Q_{CS} stored in the collector region must be added to the overdrive base charge Q_S to obtain the total excess minority carrier charge stored in saturation.

Because of this extra charge storage, calculations using the equation for τ_S from Eq. (6.2.76) will result in an underestimation of the saturation storage time t_S. Fortunately, τ_S can be measured indirectly from observed t_S data. Then the effect of the extra charge Q_{CS} is included in τ_S. That is, $\tau_{S(meas)} > \tau_{S(calc)}$.

Exercise 6.4. An alternative equation for the saturation time constant is

$$\tau_S = \frac{\tau_{BF}(\beta_R + 1) + \tau_{BR}\beta_F}{\beta_F + \beta_R + 1} \tag{6.2.77}$$

Show how this can be derived from Eq. (6.2.76).

6.3   BJT INVERTER SWITCHING TIMES

We will now make use of the charge-control model in the calculation of the switching times for the BJT inverter. The inverter circuit of Fig. 6.1 is repeated here as Fig. 6.16*a*, along with additional transistor data that pertain particularly

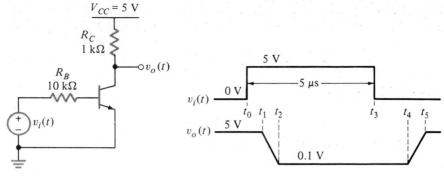

$V_{BE(on)} = 0.7$ V $\quad\tau_F = 0.2$ ns

$V_{BE(sat)} = 0.8$ V $\quad\tau_{BF} = 14$ ns

$V_{CE(sat)} = 0.1$ V $\quad\tau_S = 20$ ns

$C_{jeo} = 0.30$ pF $\quad C_{jco} = 0.15$ pF

$\phi_e = 0.9$ V $\qquad \phi_c = 0.7$ V

$m_e = 0.50$ $\qquad m_c = 0.33$

(a)

Delay time $\quad = t_d = t_1 - t_0$
Fall time $\qquad = t_f = t_2 - t_1$
Saturation time $= t_s = t_4 - t_3$
Rise time $\qquad = t_r = t_5 - t_4$

(b)

FIGURE 6.16
(a) BJT inverter. (b) Input and output voltage waveforms.

to the transient characteristics of the circuit. These parameters are typical of a low-power *npn* switching transistor.

The input voltage is a rectangular pulse that at time t_0 abruptly changes from 0 to 5 V. To describe the turn-on sequence briefly: with the input at 0 V, the transistor is cut off, so the output voltage is equal to V_{CC} (= 5 V). Following the step input at time t_0 there is no change at the output until t_1, the time when the collector current causes a noticeable decrease in the output voltage. This *delay time* $(t_1 - t_0)$ is caused by the voltage across the emitter and collector junctions being unable to change instantaneously due to the junction capacitances at these depletion regions. From t_1 to t_2 there is a *fall time* $(t_2 - t_1)$. This time delay is again due to the junction capacitive effects but also includes the effects of the finite transit time τ_F of the transistor. At time t_2, the transistor is at the edge of saturation, so the output voltage is essentially constant at $V_{CE(sat)}$ (= 0.1 V).

At some time t_3, much later than the turn-on time (in this example at 5 μs), there is another step change in the input voltage back to 0 V. The delay time from t_3 to t_4 is the *saturation time* $(t_4 - t_3)$, when the overdrive charge is being removed from the neutral base region (as in Fig. 6.14b), or the base and collector regions (as in Fig. 6.15b). The *rise time* of the output waveform is designated by $t_5 - t_4$, which is similar to the fall time, except that the transistor is now turning off. At time t_5, the transistor is at the edge of cutoff, so the output voltage is back to V_{CC}. Subsequent to t_5, there is another delay time not apparent at the output waveform, *the final recovery time* $(t_6 - t_5)$, as the base voltage changes from $V_{BE(on)}$ to 0 V, the quiescent input voltage.

CUTOFF REGION ($t_0 \rightarrow t_1$). In the cutoff region we make use of the simplified charge-control equations of Eq. (6.2.58); that is, we neglect both the forward and reverse charges Q_F and Q_R. Hence we are concerned only with the depletion region charges Q_{VE} and Q_{VC}.

During the time t_0 to t_1 the base voltage is changing to the turn-on base-emitter voltage $V_{BE(on)}$. That is,

$$V_{BE}(t_0) = 0 \text{ V}$$

$$V_{BE}(t_1) = 0.7 \text{ V}$$

With the collector voltage practically unchanged at V_{CC}, the voltage across the collector junction is given as

$$V_{BC}(t_0) = 0 - 5 = -5 \text{ V}$$

$$V_{BC}(t_1) = 0.7 - 5.0 = -4.3 \text{ V}$$

This change in junction voltage implies a change in charge at these junctions. The change in charge is caused by the current i_B which flows through the resistor R_B under the influence of the input voltage v_i. Hence, the initial base current at t_0 is

$$i_B(t_0) = \frac{5 - 0}{10 \text{ k}\Omega} = 0.5 \text{ mA}$$

but, at t_1,

$$i_B(t_1) = \frac{5 - 0.7}{10 \text{ k}\Omega} = 0.43 \text{ mA}$$

Therefore the average base current during the time t_0 to t_1 is given simply as

$$i_B = \frac{0.5 + 0.43}{2} = 0.465 \text{ mA}$$

We can equate this average current flowing for a time $(t_1 - t_0)$ to a change in junction charge through the charge-control equations of Eq. (6.2.58b):

$$i_B(t) = \frac{d}{dt}(Q_{VE} + Q_{VC})$$

Integrating this equation from t_0 to t_1, we have an equation for the average base current during this period:

$$\int_{t_0}^{t_1} i_B(t) \, dt = \left[Q_{VE}(t_1) - Q_{VE}(t_0) \right] + \left[Q_{VC}(t_1) - Q_{VC}(t_0) \right] \tag{6.3.1}$$

To determine the depletion region charge for a particular junction voltage, we could make use of Eq. (6.2.52) or (6.2.53) as appropriate. But simpler yet is to use the large-signal equivalent depletion region capacitance described in Sec. 4.3.6 and given in Eq. (4.3.13b).

For the emitter junction (with $m = 0.5$), let*

$$\Delta Q_E = Q_E(t_1) - Q_E(t_0)$$

but

$$\Delta Q_E = C_{eq} \Delta V_{BE}$$

where

$$C_{eq} = - C_{je0} \frac{\phi_e}{(\Delta V_{BE})(1 - m_e)} \left\{ \left[1 - \frac{V_{BE}(t_1)}{\phi_e} \right]^{1-m_e} - \left[1 - \frac{V_{BE}(t_0)}{\phi_e} \right]^{1-m_e} \right\}$$

$$= \frac{-0.3(0.9)}{(0.7 - 0)(\frac{1}{2})} \left[\left(1 - \frac{0.7}{0.9} \right)^{1/2} - \left(1 - \frac{0}{0.9} \right)^{1/2} \right] = 0.408 \text{ pF}$$

Therefore

$$\Delta Q_E = (0.408 \text{ pF})(0.7 \text{ V}) = 0.285 \text{ pC}$$

For the collector junction (where, in this example, $m = 0.33$), let

$$\Delta Q_C = Q_C(t_1) - Q_C(t_0)$$

but

$$\Delta Q_C = C_{eq} \Delta V_{BC}$$

where

$$C_{eq} = - \frac{0.15(0.7)}{(-4.3 + 5.0)(\frac{2}{3})} \left[\left(1 + \frac{4.3}{0.7} \right)^{2/3} - \left(1 + \frac{5.0}{0.7} \right)^{2/3} \right] = 0.076 \text{ pF}$$

Therefore

$$\Delta Q_C = (0.076 \text{ pF})(0.7 \text{ V}) = 0.053 \text{ pC}$$

Hence from Eq. (6.3.1),

$$\int_{t_0}^{t_1} (0.456 \text{ mA}) \, dt = 0.285 + 0.053 = 0.338 \text{ pC}$$

Then, the delay time

$$t_d = t_1 - t_0 = \frac{0.338 \text{ pC}}{0.465 \text{ mA}} = 0.73 \text{ ns}$$

ACTIVE REGION $(t_1 \rightarrow t_2)$. In the active region the base voltage is changing from $V_{BE(on)}$ to $V_{BE(sat)}$ and the collector voltage from V_{CC} to $V_{CE(sat)}$. Therefore,

$$V_{BE}(t_2) = 0.8 \text{ V}$$

and

$$V_{BC}(t_2) = 0.8 - 0.1 = 0.7 \text{ V}$$

---

*The change in the depletion region charge (ΔQ_j) is exactly equal to ΔQ_V. That is, $\Delta Q_E = \Delta Q_{VE}$.

Since $i_B(t_1) = 0.43$ mA, the average base current during the time t_1 to t_2 is 0.425 mA.

From the simplified charge-control equations in the forward active mode, from Eq. (6.2.59b),

$$i_B(t) = \frac{Q_F}{\tau_{BF}} + \frac{d}{dt}(Q_F + Q_{VE} + Q_{VC})$$

Integrating from t_1 to t_2,

$$\int_{t_1}^{t_2} i_B(t)\, dt = \frac{1}{\tau_{BF}} \int_{t_1}^{t_2} Q_F(t)\, dt + \Delta Q_F + \Delta Q_{VE} + \Delta Q_{VC} \qquad (6.3.2)$$

where $$\Delta Q_F = Q_F(t_2) - Q_F(t_1)$$

and $$Q_F(t_2) = I_{C(EOS)}\tau_F = (4.9\ \text{mA})(0.2\ \text{ns}) = 0.98\ \text{pC}$$

$$Q_F(t_1) = 0 \qquad \text{since } I_C(t_1) = 0$$

therefore $$\Delta Q_F = 0.98\ \text{pC}$$

Since $$\Delta Q_E = \Delta Q_{VE} \qquad \text{and} \qquad \Delta Q_E = C_{eq}\,\Delta V_{BE}$$

where

$$C_{eq} = -\frac{0.3(0.9)}{(0.8 - 0.7)(\frac{1}{2})}\left[\left(1 - \frac{0.8}{0.9}\right)^{1/2} - \left(1 - \frac{0.7}{0.9}\right)^{1/2}\right] = 0.745\ \text{pF}$$

therefore $$\Delta Q_E = (0.745\ \text{pF})(0.1\ \text{V}) = 0.075\ \text{pC}$$

Also, since

$$\Delta Q_C = \Delta Q_{VC} \qquad \text{and} \qquad \Delta Q_C = C_{eq}\,\Delta V_{BC}$$

where

$$C_{eq} = -\frac{0.15(0.7)}{(0.7 + 4.3)(\frac{2}{3})}\left[\left(1 - \frac{0.7}{0.7}\right)^{2/3} - \left(1 + \frac{4.3}{0.7}\right)^{2/3}\right] = 0.117\ \text{pF}$$

therefore $$\Delta Q_C = (0.117\ \text{pF})(5.0\ \text{V}) = 0.584\ \text{pC}$$

Neglecting for the moment the recombination term in Eq. (6.3.2),

$$\int_{t_1}^{t_2} (0.425\ \text{mA})\, dt = 0.98 + 0.075 + 0.584 = 1.64\ \text{pC}$$

Then the fall time

$$t_f = t_2 - t_1 = \frac{1.64\ \text{pC}}{0.425\ \text{mA}} = 3.9\ \text{ns}$$

Now to include the recombination term in Eq. (6.3.2), we assume a linear increase in Q_F from t_1 to t_2 as in Fig. 6.17. Then

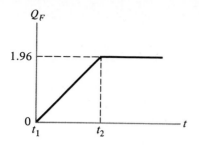

FIGURE 6.17
Forward base charge as a function of time while transistor is in the forward active mode.

$$\int_{t_1}^{t_2} (0.425 \text{ mA}) \, dt = \frac{1}{14 \text{ ns}} \frac{(0.98 \text{ pC})(t_2 - t_1)}{2} + 1.64 \text{ pC}$$

Solving for $t_2 - t_1$,

$$t_f = t_2 - t_1 = \frac{1.64 \text{ pC}}{(0.425 - 0.035) \text{ mA}} = 4.2 \text{ ns}$$

Note that the loss of carriers by recombination results in an increase in the fall time.

SATURATION REGION ($t_3 \rightarrow t_4$). Inasmuch as the input pulse is 5 μs wide and $t_d + t_f = 4.9$ ns, there is ample time for the transistor to reach steady-state saturation. Hence from the simplified charge-control equations for the transistor in the saturation mode of Eq. (6.2.64b),

$$i_{BF} - \frac{I_{C(EOS)}}{\beta_F} = \frac{Q_S}{\tau_S} \tag{6.3.3}$$

where i_{BF} is defined as in Fig. 6.18, with the input voltage at 5 V and $V_{BE} = V_{BE(sat)} = 0.8$ V.

Now with the step change of input voltage at t_3, the saturation mode charge-control equation becomes

$$i_{BR} - \frac{I_{C(EOS)}}{\beta_F} = \frac{Q_S}{\tau_S} + \frac{dQ_S}{dt} \tag{6.3.4}$$

where i_{BR} is now due to the input voltage of 0 V; but with the transistor still in saturation, $V_{BE} = V_{BE(sat)} = 0.8$ V.

The solution of Eq. (6.3.4) is of the form

$$Q_S(t) = A + Be^{-t/\tau_S}$$

That is,

$$Q_S(t) = \tau_S \left(i_{BR} - \frac{I_{C(EOS)}}{\beta_F} \right) + Ke^{-t/\tau_S} \tag{6.3.5}$$

To solve for the constant K, we make use of the initial conditions from Eq. (6.3.3), since

$$Q_S(t_3) = \tau_S \left(i_{BF} - \frac{I_{C(EOS)}}{\beta_F} \right) = \tau_S \left(i_{BR} - \frac{I_{C(EOS)}}{\beta_F} \right) + K \tag{6.3.6}$$

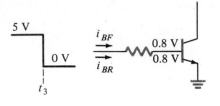

FIGURE 6.18
Definition of input base current as input voltage changes from 5 to 0 V at t_3.

therefore, $$K = \tau_S(i_{BF} - i_{BR})$$

Substituting for K in Eq. (6.3.5), we have

$$Q_S(t) = \tau_S\left[\left(i_{BR} - \frac{I_{C(EOS)}}{\beta_F}\right) + (i_{BF} - i_{BR})e^{-t/\tau_S}\right] \tag{6.3.7}$$

Now at time t_4, all the overdrive charge Q_S has been reduced to zero, and the transistor is again at the edge of saturation. Consequently, setting Q_S to zero in Eq. (6.3.7), we can solve for the saturation time $t = t_s$. That is,

$$t_s = t_4 - t_3 = \tau_S \ln \frac{i_{BF} - i_{BR}}{(I_{C(EOS)}/\beta_F) - i_{BR}} \tag{6.3.8}$$

where $$i_{BF} = \frac{5 - 0.8}{10 \text{ k}\Omega} = 0.42 \text{ mA}$$

and $$i_{BR} = \frac{0 - 0.8}{10 \text{ k}\Omega} = -0.08 \text{ mA}$$

Therefore $$t_s = 20 \ln \frac{0.42 + 0.08}{(4.9/70) + 0.08} = 24 \text{ ns}$$

ACTIVE REGION ($t_4 \rightarrow t_5$). The situation in the active region during turn-off is analogous to that during turn-on. That is, from Eq. (6.3.2),

$$\int_{t_4}^{t_5} i_B(t) \, dt = \frac{1}{\tau_{BF}} \int_{t_4}^{t_5} Q_F(t) \, dt + \Delta Q_F + \Delta Q_E + \Delta Q_C \tag{6.3.9}$$

During turn-off, the total ΔQ terms will have the same magnitude as for the turn-on, but with opposite sign.

The average base current during turn-off is obtained from

$$i_B(t_4) = \frac{0 - 0.8}{10 \text{ k}\Omega} = -0.08 \text{ mA}$$

$$i_B(t_5) = \frac{0 - 0.7}{10 \text{ k}\Omega} = -0.07 \text{ mA}$$

Therefore $$i_B = -0.075 \text{ mA}$$

That we might note the effect of recombination, we again initially neglect the recombination term in Eq. (6.3.9). Then

$$\int_{t_4}^{t_5} -0.075 \text{ mA } dt = -1.64 \text{ pC}$$

Therefore, the rise time

$$t_r = t_5 - t_4 = \frac{-1.64 \text{ pC}}{-0.075 \text{ mA}} = 22 \text{ ns}$$

Including the recombination term, and assuming a linear decrease of Q_F with time,

$$t_r = t_5 - t_4 = \frac{-1.64 \text{ pC}}{-(0.075 + 0.035) \text{ mA}} = 15 \text{ ns}$$

Now we note that recombination can be a substantial aid in the removal of charge from the base of the transistor.

RECOVERY REGION ($t_5 \rightarrow t_6$). With the transistor turned off there is still a final recovery time for the depletion region charges to reach their final state. Again, the situation is similar to the initial delay time prior to turn-on.

To obtain the average value for the current i_B during this period, we note that

$$i_B(t_5) = -0.07 \text{ mA} \qquad \text{and} \qquad i_B(t_6) = 0$$

therefore, $$i_B = -0.035 \text{ mA}$$

The change in the depletion region charge will be, from the cutoff region,

$$\Delta Q_V = -0.338 \text{ pC}$$

Hence, the final recovery time will be

$$t_{fr} = t_6 - t_5 = \frac{-0.338 \text{ pC}}{-0.035 \text{ mA}} = 9.7 \text{ ns}$$

6.3.1 Propagation Delay Time

The switching times as calculated above are of very much interest to the digital IC designer. However, of more interest to the digital IC user are the propagation delay times. These times are measured between two reference levels on the input and output voltage waveforms, as illustrated in Fig. 6.19.

For the inverter circuit of Fig. 6.16, the turn-on delay time t_{PHL} is measured as the output is changing from a high voltage level to a low voltage level. That is,

$$t_{PHL} = t_d + \frac{t_f}{2} = 0.73 + \frac{4.2}{2} = 2.8 \text{ ns}$$

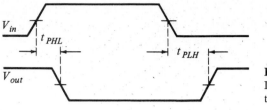

FIGURE 6.19
Input and output voltage waveforms, illustrating propagation delay times.

For the turn-off delay time t_{PLH}, the output is changing from a low level to a high level. Then

$$t_{PLH} = t_s + \frac{t_r}{2} = 24 + \frac{15}{2} = 31 \text{ ns}$$

The average propagation delay time is defined as

$$t_p = \frac{t_{PHL} + t_{PLH}}{2} = \frac{2.8 + 31}{2} = 17 \text{ ns}$$

6.4 SCHOTTKY-CLAMPED INVERTER

When the saturation time is a sufficiently large fraction of the total turn-off time of the inverter circuit, the delay time can be improved with a *Schottky clamp*. A Schottky-barrier diode (SBD) is connected between the base and the collector of the inverter transistor, as in Fig. 6.20. By extending the base metal to contact the collector region, this extra component is included by a simple addition to the fabrication procedure.

The Schottky clamp effectively prevents the transistor from saturating; hence there is no saturation storage time for the inverter circuit. When the transistor is on, because of a high voltage level at the input, the overdrive base current is diverted from the base into the Schottky diode and then into the collector of the transistor. The forward voltage drop of the Schottky diode is less than the forward

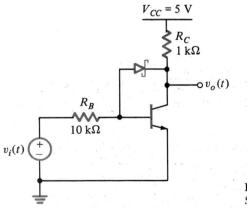

FIGURE 6.20
Schottky-clamped inverter circuit.

voltage drop of the base-collector junction diode of the transistor. The forward voltage drop of an aluminum n-type silicon Schottky diode is about 0.4 V. Thus for a clamped inverter circuit, and a transistor with $V_{BE(on)} = 0.7$ V, with the SBD conducting the voltage between collector and emitter, $V_{CE} = 0.7 - 0.4 = 0.3$ V. This is to be compared with $V_{CE(sat)} = 0.1$ V for the transistor. In the saturation mode, both junctions of the transistor are forward-biased. Hence, with $V_{BE(sat)} = 0.8$ V and $V_{CE(sat)} = 0.1$ V, $V_{BC(sat)} = 0.7$ V. The Schottky diode effectively clamps the base-collector voltage to 0.4 V; therefore, although the base-collector junction is forward-biased (by 0.4 V), it is conducting little current. The Schottky diode is truly forward-biased, but as noted in Sec. 4.9 there is no minority carrier charge storage in an SBD. The only recovery time is due to charge stored in the depletion region, which takes time to change.

6.4.1 Static Characteristics

While the inclusion of the Schottky diode has improved the transient character-istics of the bipolar transistor inverter, unfortunately it has had an opposite effect on the static characteristics. In particular, the low noise margin NM_L has been made worse. From Eq. (6.1.8), NM_L is defined as

$$NM_L = V_{IL} - V_{OL}$$

but V_{OL} has been increased, in the example, from 0.1 to 0.3 V. Hence, NM_L has been decreased by the same amount.

The Schottky diode has a minimal effect on the high noise margin NM_H. A small improvement is owing to the increase in V_{OL} leading to a decrease in V_{IH}, since from Eq. (6.1.7),

$$NM_H = V_{OH} - V_{IH}$$

6.4.2 Transient Characteristics

As noted above, the Schottky diode exhibits no minority carrier charge storage time, but there is a depletion-layer or junction capacitance that must be charged or discharged as the voltage across the diode increases or decreases. Therefore, compared to the saturating inverter circuit, the nonsaturating Schottky-clamped circuit does have longer delay, fall, and rise times. This is not always a disad-vantage, since the slower transition times do result in less ringing on the signal lines, as well as a reduction in the effects of capacitive coupling between sig-nal lines. However, since for an IC Schottky-clamped inverter the metal-silicon junction capacitance is typically less than one-third the base-collector junction capacitance, the increase in the transition times is minimal.

The junction capacitance of a Schottky-barrier diode is similar to that of a pn abrupt junction. It can therefore be included in the charge-control models of Figs. 6.11 and 6.12b, across the collector-base diode, as another voltage-dependent charge storage element. The charge stored can be included in the calculations as for Q_C.

Exercise 6.5. For the Schottky-clamped inverter of Fig. 6.20 calculate:
(a) The noise margins NM_H and NM_L.
(b) The propagation delay times t_{PHL} and t_{PLH}.
Use the transistor data from Fig. 6.16a, and for the Schottky diode,

$$V_{SBD(on)} = 0.4 \text{ V} \qquad C_{j0} = 0.05 \text{ pF} \qquad \phi_0 = 0.7 \text{ V} \qquad m = 0.5$$

6.5 COMPARISON WITH SPICE

It is both interesting and informative to compare the hand calculations of the switching times for the BJT inverter with the results obtained by computer-aided simulation programs such as SPICE.

TRANSISTOR PARAMETERS. The transistor parameters used in the SPICE runs are listed in Table 6.1. Comparing the data in Table 6.1 with that given in Fig. 6.16a, we note from the table that with $\beta_F = 70$ and $\tau_F = 0.2$ ns, $\tau_{BF} = \beta_F \tau_F = 14$ ns, as given in Fig. 6.16a.

For the transistor saturation time constant τ_S, from Eq. (6.2.77),

$$\tau_S = \frac{\tau_{BF}(\beta_R + 1) + \tau_{BR}(\beta_F)}{\beta_F + \beta_R + 1}$$

By using the data from the table we obtain

$$\tau_S = \frac{(14 \text{ ns})(2) + (20 \text{ ns})(70)}{70 + 1 + 1} = 20 \text{ ns}$$

Again, this is as given in Fig. 6.16a.

The turn-on voltage for the transistor, given as $V_{BE(on)} = 0.7$ V in Fig. 6.16a, follows from the arbitrary assumption that the transistor turns on when the collector current is 1% of its final value. Since from the diode equation [Eq. (4.4.1)], V_{BE} changes 60 mV for each factor-of-10 change in I_C, with $V_{BE(sat)} = 0.8$ V we assume $V_{BE(on)} = 0.68$ V ≈ 0.7 V.

The transistor saturation current I_s in Table 6.1 is obtained from the calculations in Sec. 6.3, where $I_{C(EOS)} = 4.9$ mA and $V_{BE(sat)} = 0.8$ V. Then from the diode equation, with heavy forward bias,

$$I_C = \alpha_F I_E = I_s e^{V_{BE}/V_T}$$

TABLE 6.1

$\beta_F = 70$	$\tau_F = 0.2$ ns
$\beta_R = 1$	$\tau_R = 20$ ns
$I_S = 2.1 \times 10^{-16}$ A	$C_{je0} = 0.3$ pF
$r_b = 0$	$\phi_e = 0.9$ V
$r_c = 5.5 \ \Omega$	$m_e = 0.5$
$r_e = 0$	$C_{jc0} = 0.15$ pF
	$\phi_c = 0.7$ V
	$m_c = 0.33$

then
$$I_s = \frac{4.9 \text{ mA}}{e^{800/26}} = 2.1 \times 10^{-16} \text{ A}$$

The ohmic bulk collector resistance for the transistor r_c is due to the difference in $V_{CE(sat)}$ given in Fig. 6.16a and that calculated from the tabulated data. From Eq. (5.3.22),

$$V_{CE(sat)} = V_T \ln \left[\frac{1/\alpha_R + (I_C/I_B)(1/\beta_R)}{1 - (I_C/I_B)(1/\beta_F)} \right]$$

Hence using the data from the table and using $I_{C(sat)} = 4.9$ mA and $I_{B(sat)} = 0.42$ mA from the calculations in Sec. 6.3, we calculate that

$$V_{CE(sat)} = 26 \text{ mV} \ln \left[\frac{1/0.5 + (4.9/0.42)(1/1)}{1 - (4.9/0.42)1/70} \right] = 73 \text{ mV}$$

But in Fig. 6.16a, $V_{CE(sat)}$ is given as 100 mV. Therefore, with $I_{C(sat)} = 4.9$ mA, the transistor collector resistance is given as

$$r_c = \frac{(100 - 73) \text{ mV}}{4.9 \text{ mA}} - 5.5 \ \Omega$$

The ohmic bulk base and emitter resistances r_b and r_e are assumed to be very small and are therefore made equal to zero. The junction capacitance data in Fig. 6.16a is taken directly from Table 6.1.

6.5.1 Saturated Inverter Results

The static voltage transfer characteristic obtained from the SPICE simulation is shown in Fig. 6.21a. Note that $V_{OH} = 5$ V and $V_{OL} = 0.1$ V, $V_{IL} = 0.67$ V and $V_{IH} = 1.5$ V. Except for a little rounding at the breakpoints, the curve is almost an exact replica of the voltage transfer characteristic shown in Fig. 6.2.

The switching times data were obtained directly from the SPICE printout of $v_{out}(t)$ shown in Fig. 6.22. With $V_{BE(on)} = 0.7$ V and $V_{BE(EOS)} = 0.8$ V, $\Delta V_{BE} = 100$ mV causes the transistor to pass through the active region. Now since $I_C = I_s e^{V_{BE}/V_T}$, the corresponding change in collector current is a factor of 47.5 $(= e^{\Delta V_{BE}/V_T})$. Hence with $I_{C(sat)} = 4.9$ mA, the time marks were taken when $I_{C(on)} = 0.1$ mA and $I_{C(EOS)} = 4.8$ mA, for a ratio of 48. In the collector circuit, $R_C = 1$ kΩ so that the resulting points on the output voltage are 100 mV from the initial or final value. That is, from Fig. 6.16b,

$$v_{out}(t_1) = 4.9 \text{ V} \qquad v_{out}(t_4) = 0.2 \text{ V}$$

$$v_{out}(t_2) = 0.2 \text{ V} \qquad v_{out}(t_5) = 4.9 \text{ V}$$

The switching times are tabulated and compared with the hand calculations in Table 6.2. Again, the excellent agreement is evident. The small peak in the output voltage, observable in Fig. 6.22 slightly after t_0, is because of the feedthrough (via the collector junction capacitance) of the input pulse while the transistor is still off.

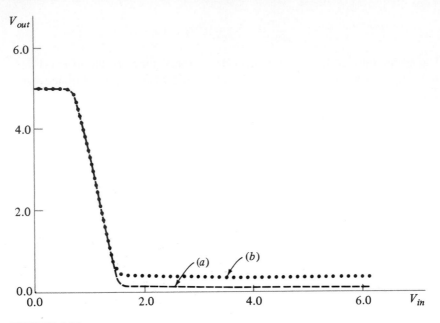

FIGURE 6.21
SPICE simulation of voltage transfer characteristics (*a*) for the saturating inverter circuit of Fig. 6.1*a* (shown dashed) and (*b*) for the nonsaturating Schottky-clamped inverter circuit of Fig. 6.20 (shown dotted).

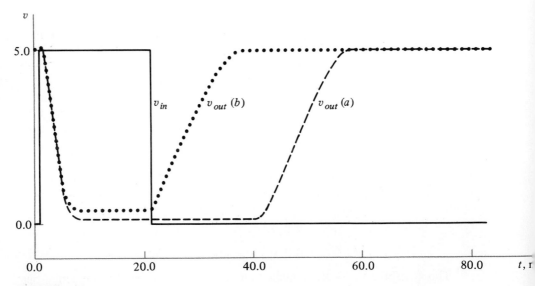

FIGURE 6.22
SPICE simulation of the transient response (*a*) for the saturating inverter circuit of Fig. 6.16*a* (shown dashed) and (*b*) for the nonsaturating Schottky-clamped inverter circuit of Fig. 6.20 (shown dotted).

TABLE 6.2

		Hand analysis, ns	SPICE, ns
Delay time	t_d	0.7	0.7
Fall time	t_f	4.2	4.5
Saturation time	t_s	24	22
Rise time	t_r	15	15

6.5.2 Schottky-Clamped Inverter Results

For the SPICE simulations of the Schottky-clamped circuit, use was made of the data given in Exercise 6.5. The transistor data were taken from Table 6.1. For the Schottky diode the depletion-layer capacitance data were taken directly from Exercise 6.5. From the problem it can be calculated that with 5 V at the input of the inverter, the Schottky diode current is 0.35 mA. Hence with $V_{SBD(on)} = 0.4$ V, the diode saturation current I_s can be computed from the diode equation. The result is $I_s = 7.3 \times 10^{-11}$ A.

From the computer-aided simulation printout of the voltage transfer characteristic, the breakpoints are determined to be $V_{OH} = 5.0$ V and $V_{OL} = 0.4$ V, $V_{IL} = 0.67$ V and $V_{IH} = 1.46$ V. This compares with the hand calculations of $V_{OH} = 5.0$ V and $V_{OL} = 0.4$ V, $V_{IL} = 0.7$ V and $V_{IH} = 1.5$ V. The values are almost identical.

To determine the switching times, the output voltage levels used were again 100 mV below or above the initial or final values, that is, 4.9 and 0.4 V. The computer results are compared with the hand calculations in Table 6.3. The agreement is excellent.

The computer-aided voltage transfer characteristic for the nonsaturating Schottky-clamped inverter is compared with the similar characteristic for the saturating circuit in Fig. 6.21b. Likewise, the switching times are compared in Fig. 6.22b.

6.6 SUMMARY

This chapter provides the analysis techniques for the static and dynamic characteristics of a saturating and nonsaturating (Schottky-clamped) BJT inverter circuit.

TABLE 6.3

		Hand analysis, ns	SPICE, ns
Delay time	t_d	0.8	0.8
Fall time	t_f	4.4	4.5
Saturation time	t_s	0.0	0.5*
Rise time	t_r	15	15

*This is the time difference $t_4 - t_3$. It is the time from the initiating of turn-off for the output voltage to change from 0.4 to 0.5 V.

The principal static characteristics are:

- Voltage transfer characteristic (VTC)
- Noise margins NM_H and NM_L
- Fan-out N

The dynamic characteristics are described in terms of the *charge-control* model of the BJT, which leads to a determination of the:

- Propagation delay times (t_{PHL} and t_{PLH})

The results of the hand calculations for an unclamped and Schottky-clamped inverter are compared with the SPICE results with very good correlation.

REFERENCES

1. P. E. Gray and C. L. Searle, *Electronic Principles, Physics, Models and Circuits*. Wiley, New York, 1969.
2. A. B. Glaser and G. E. Subak-Sharpe, *Integrated Circuit Engineering*, Addison-Wesley, Reading, Mass., 1977.

DEMONSTRATIONS

D6.1   BJT Inverter:
Voltage Transfer Characteristic

The objective of this experiment is to examine the voltage transfer characteristic of a simple BJT inverter.

1. (a) Connect the circuit shown in Fig. D6.1, but with no load at V_{O1}.* Start with $V_{I1} = 0$ V, and measure V_{O1} at each, 0.1-V increment of V_{I1}. Plot V_{O1} versus V_{I1}, and determine V_{OH}, V_{OL}, V_{IH}, V_{IL}, NM_H, and NM_L.
   (b) Repeat part (a) at temperatures of 0 and 100°C. Control the temperature with an ice bath and boiling water as in Demonstration D2.2.
2. (a) Use the measured parameters from Demonstration D5.2 to calculate the VTC, and compare with the observed result at $T_A = 25°C$.
   (b) Note the shift in the transition width of the VTC as the temperature is varied. What is the major cause of this?
   (c) Use the measured parameters to calculate the fan-out N_{max} with $NM_H = NM_L$. Simulate this load at V_{O1} as in Fig. D6.1, and repeat part 1(a) to confirm $NM_H = NM_L$.

---

*Use the same transistor as in Demonstration D5.1.

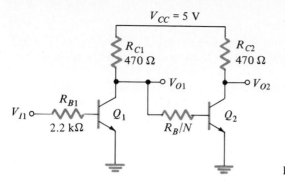

FIGURE D6.1

D6.2 BJT Inverter: Switching Times

In this experiment we will measure the switching times of a BJT inverter and compare them with the results obtained from SPICE.

1. (a) Connect the circuit as shown in Fig. D6.1, but with $R_B/N = 2.2$ kΩ.
   (b) With a rectangular voltage pulse at V_{I1}, measure the voltages and times as indicated in Fig. D6.2 and sketch the output voltage waveforms.

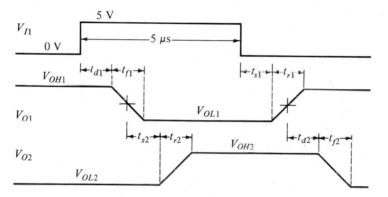

FIGURE D6.2

2. Use the measured parameters from Demonstrations D5.2 and D5.3 in a SPICE simulation to compute the voltages and times as shown in Fig. D6.2. Compare with the measured results and comment on any differences.
3. Use a Schottky-barrier diode* to clamp Q_1 and Q_2 out of saturation, and repeat part 1.
4. Repeat part 2, using "good estimates" for the SBD parameters.

---

*The HPA 2800, or a germanium switching diode such as the 1N34A.

PROBLEMS

P6.1. Sketch the voltage transfer curve for the circuit in Fig. 6.1 with $I_{ES} = 10^{-16}$ A and $I_{CS} = 2 \times 10^{-16}$ A, $\alpha_F = 0.98$ and $\alpha_R = 0.49$, and determine the noise margins NM_H and NM_L.

P6.2. For the circuit in Fig. 6.1 if the resistors can vary $\pm 20\%$ and V_{CC} varies $\pm 10\%$:

(a) Calculate $V_{IL(max)}$, the most positive value of low-level input that guarantees the transistor is off, and $V_{IH(min)}$, the least positive value of high-level input that guarantees the transistor is saturated.

(b) Repeat (a) if $0.95 \le \alpha_F \le 0.99$ and $0.3 \le \alpha_R \le 0.7$.

P6.3. In Fig. 6.1 a 20-kΩ pulldown resistor is added from the base of Q_1 to -2 V. Using the data in Fig. 6.1, sketch the VTC and calculate the noise margins.

P6.4. For the inverter of P6.3:

(a) Calculate the maximum number of load gates.

(b) Calculate the fan-out with $NM_H = NM_L$.

P6.5. In the circuit of Fig. 6.7a assume $V = 5$ V, $V_{BE(on)} = 0.7$ V, and $R = 10$ kΩ. The final value of the collector current is 43 mA. When $C = 10$ pF the collector current reaches its final value essentially instantaneously.

(a) Determine the value of the charge-control parameters τ_F, τ_{BF}, and β_F.

(b) Assume C is fixed at 10 pF but R is decreased to 5 kΩ. Sketch and dimension $i_C(t)$.

(c) Assume R is fixed at 10 kΩ but C is increased to 20 pF. Sketch and dimension $i_C(t)$.

P6.6. In the circuit of Fig. 6.16a, a 20-kΩ pulldown resistor is added from the base of the transistor to -2 V. Using the data in Fig. 6.16:

(a) Calculate t_d, t_f, t_s, t_r.

(b) Verify with SPICE.

P6.7. A transistor has the following charge-control parameters (assumed to be independent of bias currents): $\tau_F = 0.2$ ns, $\tau_R = 20$ ns, $\beta_F = 100, \beta_R = 1$.

(a) Evaluate the forward base charge Q_F if the collector current $I_C = 2$ mA and the transistor operates just at the edge of saturation with $V_{BC} = 0$ V.

(b) Determine the base charge components Q_F and Q_R if the base current is now made 0.2 mA with $I_C = 2$ mA. Compare the total base charge in parts (a) and (b).

P6.8. The transistor of P6.7 is operated with $I_{BF} = +0.2$ mA, $I_C = 2$ mA, and $I_{BR} = -0.2$ mA. Neglect charge storage in the depletion regions and calculate the fall time t_f, the saturation time t_s, and the rise time t_r.

P6.9. For the transistor of P6.7 with $I_C = 2$ mA and $I_B = 0.2$ mA:

(a) Calculate $V_{CE(sat)}$.

(b) Show that with $|I_B| \gg |I_C|$, $V_{CE(sat)} = V_T \ln (1/\alpha_R)$. Evaluate and compare for the conditions of part (a).

P6.10. An npn transistor has a collector junction capacitance of 0.5 pF with $V_{CB} = 5$ V. How much charge must be removed from the collector depletion region when V_{CB} changes from -0.1 to 10 V? Assume $\phi_0 = 0.7$ V.

P6.11. In the circuit of Fig. 6.16a, the input voltage changes at t_0 from -5 to $+5$ V, and at t_3 from $+5$ to -5 V. Using the data in Fig. 6.16:

(a) Calculate t_d, t_f, t_s, t_r.

(b) Verify with SPICE.

P6.12. Repeat P6.11 using the following data: $I_s = 2 \times 10^{-16}$ A, $r_b = 100$ Ω, $r_c = 20$ Ω, $\alpha_F = 0.96$, $\alpha_R = 0.40$, $\tau_F = 0.2$ ns, $\tau_R = 10$ ns, $C_{je0} = 1.2$ pF, $\phi_e = 0.9$ V, $m_e = 0.33$, $C_{jc0} = 0.6$ pF, $\phi_c = 0.8$ V, and $m_c = 0.5$.

P6.13. Two inverters such as those in Fig. 6.16a are cascaded together.
  (a) Calculate the average propagation delay time t_p of each inverter and of the combination.
  (b) Verify with SPICE.

P6.14. The circuit in Fig. P6.14 simulates two cascaded RTL inverters. Calculate and sketch $V_{ce}(t)$ and $V_d(t)$ for a 5-V 1-μs-wide pulse at V_{in}. Use the transistor data given in P6.12. For the diode, $V_D = 0.8$ V and $\tau_s = 5$ ns.

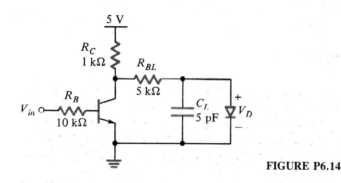

FIGURE P6.14

P6.15. For the circuit in Fig. P6.15, calculate and sketch $V_{be}(t)$ and $V_{out}(t)$. Use the transistor data given in P6.12.

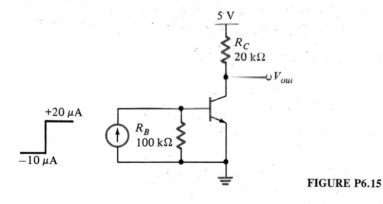

FIGURE P6.15

P6.16. For the circuit in Fig. 6.20, choose I_0 for the Schottky diode such that $V_{CE(on)} = 0.3$ V. Use the transistor data given in P6.12. $V_{in} = 5$ V.

P6.17. For the circuit of Fig. 6.20, calculate $V_{out}(t)$ for a 5-V 1-μs-wide pulse at V_{in}. For the Schottky diode, $I_0 = 10^{-10}$ A, $C_{j0} = 0.1$ pF, $\phi_0 = 0.6$ V, and $m = 0.5$. Use the transistor data given in P6.12. Verify with SPICE.

P6.18. Shown in Fig. P6.18 is an alternative method of preventing the inverter transistor from saturating. This arrangement is known as the *Baker clamp*, after its originator. The diodes are base-collector shorted transistors. Use the transistor data given in Table 6.1.

    (*a*) Sketch the VTC and determine the noise margins.
    (*b*) Compute the fan-out with $NM_H = NM_L$.

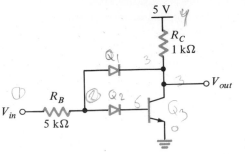

FIGURE P6.18

P6.19. For the circuit in Fig. P6.18, use the transistor data given in Table 6.1.
    (*a*) Calculate t_{PHL} and t_{PLH}.
    (*b*) Verify with SPICE.

P6.20. Use SPICE to investigate the effect of different diode-connected transistors on the average propagation delay time of the circuit in Fig. P6.18. Use the transistor data given in Table 6.1. What connection gives the best response time?

P6.21. Use the transistor parameters given in P6.12 to design an RTL inverter circuit such that with $NM_H = NM_L$ the fan-out N is 5 and $t_{PHL} = t_{PLH} = 5$ ns. Let $V_{CC} = 5$ V, and let $R_C = 1$ kΩ and V_{in} vary from 0 to 5 V. Verify your design with SPICE.

P6.22. Draw a layout of the BJT inverter shown in Fig. 6.1 if for the transistor $C_{je0} = 0.3$ pF, $C_{jc0} = 0.15$ pF, and $C_{js0} = 1.0$ pF. Assume $N_E = 10^{20}$, $N_B = 10^{18}$, $N_{EPI} = 10^{16}$, $N_{BL} = 10^{19}$, and $N_{sub} = 10^{17}$ cm^{-3}. The area of the buried layer is one-third the total substrate area. For the resistors, $R_S = 200$ Ω per square. Use the simplified design rules described in Sec. 5.5.1 and a minimum feature size of 3 μm.

CHAPTER
7

BIPOLAR
DIGITAL
GATE
CIRCUITS

7.0 INTRODUCTION

We come now to describe the use of the bipolar junction transistor (BJT) in digital circuits. In particular, we are concerned with digital integrated circuits (digital ICs).

The first form of digital ICs to receive general usage when they were introduced in 1962 was a simple connection of bipolar transistor inverter circuits (as described in Chap. 6) to yield a NOR gate. Since the circuit consisted of only resistors and transistors, it was named *resistor-transistor logic (RTL)*. An IC development a short time later consisted of a diode AND circuit followed by a bipolar transistor inverter. This was called *diode-transistor logic (DTL)*. Both RTL and DTL were IC versions of logic circuits made with discrete diodes, transistors, and resistors that had long been popular with digital circuit designers. The advantage of the integrated form was that the batch processing of the integrated circuits resulted in a product of very small size and high reliability, eventually at a very low price.

The first "new" digital design made possible by the IC fabrication process was *transistor-transistor logic (TTL)*. In these circuits the diode AND function of DTL was performed by a multiemitter transistor, a bipolar transistor with as many as eight emitters. The output of the TTL circuit was from a transistor inverter.

233

Hence, all these early digital ICs made use of the saturating transistor inverter. As such, the propagation delay time included some saturation delay time. Now, as noted in Sec. 6.4, Schottky diodes can usefully be employed to clamp the transistor out of saturation and hence decrease the propagation delay time. This is done in present-day TTL circuits, which are a variation of the basic TTL circuit (which is now nearly 20 years old).

Another way of avoiding saturation in a digital circuit is found in the *emitter-coupled logic (ECL)* circuits. These circuits exhibit the shortest propagation delay time of any of the commercially available digital ICs. They also require the most power to operate.

At the present time, the most popular of the bipolar digital IC families are the low-power Schottky-clamped TTL circuits (74LS and 74ALS) for general-purpose use and the high-speed TTL circuits (74F and 74AS) and ECL circuits (10K and 100K) for very high speed applications. The old forms of RTL, DTL, and TTL are no longer being produced commercially, except on special order for replacement purposes.

A family of bipolar logic circuits that challenges the MOSFET circuits in large-scale integration (LSI) is *integrated injection logic (I^2L)*. This digital IC is designed around a multicollector inverting transistor that operates in the reverse, or upside-down, mode. Two later developments in bipolar digital LSI approach the density of I^2L but use the inverting transistor operating in the normal forward mode. They are known as *Schottky transistor logic (STL)* and *integrated Schottky logic (ISL)*.

The engineer's choice of logic family for use in the design of a specific digital system is influenced by many factors. Among them are the need for low power, high speed, availability of more complex digital functions (MSI), compatibility with other parts of the system, and, of course, cost. Many systems mix logic families, e.g., TTL and NMOS or CMOS.

Since the development of bipolar digital ICs has been evolutionary rather than revolutionary, in this chapter we will first briefly review the older forms of digital ICs and discuss the principles of their operation. We will then make an in-depth study of the two most popular forms of bipolar digital ICs (74LS and 10K). The impact of LSI on bipolar digital gate circuit design will be included in a section describing the I^2L circuit and its variations. We conclude by presenting some solutions to the problems that arise when we have to mix the logic families for a complete digital system.

7.1   RESISTOR-TRANSISTOR LOGIC

The first digital IC to receive wide acceptance in the commercial market was *resistor-transistor logic (RTL)*. The circuit is a simple connection of two or more saturating transistor inverters sharing the same collector resistor. A 2-input digital gate is illustrated in Fig. 7.1*a*. Since the output F is an inverted (*negated*) form of either input A *or* B, this is an example of a 2-input NOR gate; using Boolean notation, $F = \overline{A + B}$. The logic symbol for this circuit is shown in Fig. 7.1*b*.

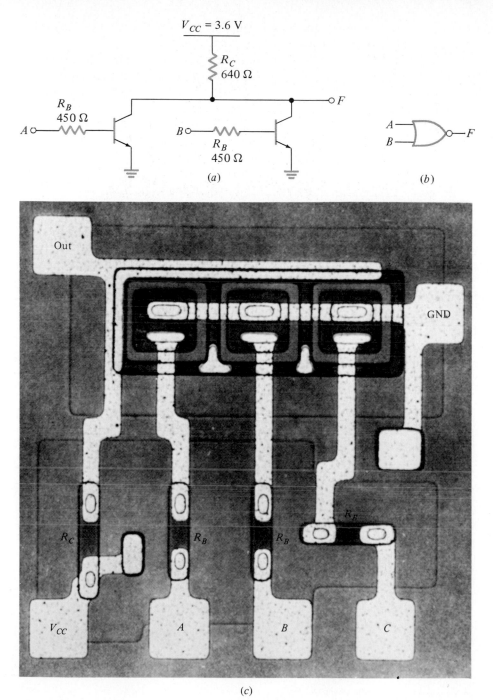

FIGURE 7.1
(a) A 2-input RTL NOR gate circuit. (b) Logic symbol for a 2-input NOR gate. (c) Photomicrograph of a 3-input RTL NOR gate.

Figure 7.1c is a photomicrograph of a 3-input RTL NOR gate that illustrates the layout of this simple digital IC.

Shown in Table 7.1 are some of the more important electrical characteristics for an RTL NOR gate. The voltage transfer characteristic can be derived from the static characteristics, as described in Sec. 6.1, and the noise margins and logic swing evaluated.

The power dissipation given in the table is an average value. That is,

$$P_{D(av)} = \frac{V_{CC}(I_{CCH} + I_{CCL})}{2} \tag{7.1.1}$$

where I_{CCH} is the supply current with the output high and I_{CCL} is the supply current with the output low. Similarly, the propagation delay time is given as an average:

$$t_P = \frac{t_{PHL} + t_{PLH}}{2} \tag{7.1.2}$$

It is sometimes possible to decrease the propagation delay time of a digital gate by increasing the operating currents in the circuit. However, this does increase the power dissipation of the gate. The product of the average propagation delay time and the average power dissipation is defined in Sec. 3.4.2 as a figure of merit for logic circuits. Normally a minimum value of the product is desired. When average values of power and delay are taken, the power-delay product is defined as

$$PDP = t_P P_{D(av)} \tag{7.1.3}$$

With the propagation delay time given in nanoseconds (ns) and the power dissipation in milliwatts (mW), the units of the product are picojoules (pJ). From Table 7.1, the figure of merit is

$$PDP = (12 \text{ ns})(16 \text{ mW}) = 192 \text{ pJ}$$

Both combinational (e.g., logic gates) and sequential (e.g., flip-flops) circuits were produced in the RTL family of digital ICs. The combinational circuits ranged from simple inverters to 4-input NOR gates. The sequential circuits ranged from single JK flip-flops to a 4-bit shift register. None of these circuits contained more than 10 gates; they therefore come under the generic title of *small-scale integrated* (SSI) circuits.

TABLE 7.1
Resistor-transistor logic: electrical characteristics

min V_{OH}/max V_{OL}	1.2 V/0.2 V	Fan-out	5
min V_{IH}/max V_{IL}	0.8 V/0.7 V	Supply volts	+3.6 V
min NM_H/min NM_L	0.4 V/0.5 V	Power dissipation per gate	16 mW
min logic swing	1.0 V	Propagation delay time	12 ns

In summary, the RTL circuits had the advantage of

1. Being simple
2. Using low power

But they had the disadvantages of

1. A small logic swing
2. A low noise margin

Primarily because of these disadvantages, the RTL family of circuits was overtaken within a couple of years by the next development in digital ICs, the diode-transistor logic (DTL) family of circuits.

Exercise 7.1. For the RTL circuit shown in Fig 7.1, assume $\beta_F = 30$, $V_{BE(on)} = V_{BE(sat)} = 0.65$ V, and $V_{CE(sat)} = 0.15$ V.
(a) Calculate the logic swing and noise margins with fan-out = 5.
(b) Compute the maximum fan-out if we require $NM_H = NM_L$.

7.2 DIODE-TRANSISTOR LOGIC

A schematic of the type of *diode-transistor logic (DTL)* that first appeared on the commercial market in 1962 is shown in Fig. 7.2a. In common with RTL, this digital IC was familiar to digital circuit designers who had been using such circuits for years, but with discrete diodes, transistors, and resistors. The circuit shown in Fig. 7.2a is basically a 2-input diode AND circuit followed by a transistor inverter. Hence this is a 2-input NAND gate, where $F = \overline{AB}$. The logic symbol for this circuit is shown in Fig. 7.2b.

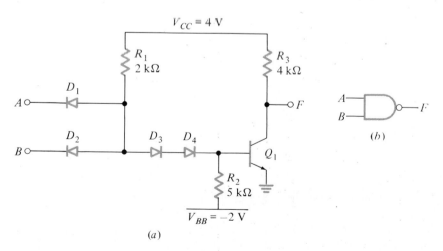

(a)

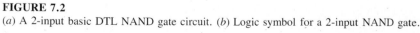

FIGURE 7.2
(a) A 2-input basic DTL NAND gate circuit. (b) Logic symbol for a 2-input NAND gate.

To determine the voltage transfer characteristic of this DTL circuit, note with the output transistor cut off that

$$V_F = V_{OH} = V_{CC}$$

$$= 4.0 \text{ V}$$

Unlike RTL circuits, in DTL circuits V_{OH} is independent of the number of load gates connected to the output F. This is because with $V_F = 4.0$ V the input diodes of any load gates are reverse-biased, and only reverse (or leakage) currents flow in the collector resistor R_3 of the driving gate. These currents are generally very small, and cause only a minute voltage drop across resistor R_3.

For the steady-state condition, the output transistor is assumed to be saturated when conducting. Then, typically,

$$V_F = V_{OL} = V_{CE(sat)}$$

$$= 0.1 \text{ V}$$

The inverting transistor will be at the *edge of conduction* ($I_C = I_{C(EOC)}$) when $V_{BE} = V_{BE(on)} = 0.7$ V. Assuming that $V_{D(on)} = 0.7$ V, the input voltage

$$V_{IL} = V_{BE(on)} + 2V_{D(on)} - V_{D(on)}$$

$$= 0.7 \text{ V} + 1.4 \text{ V} - 0.7 \text{ V} = 1.4 \text{ V}$$

The inverting transistor will be at the *edge of saturation* ($I_C = I_{C(EOS)}$) when $V_{BE} = V_{BE(sat)} = 0.8$ V. Again assuming $V_{D(on)} = 0.7$ V,

$$V_{IH} = V_{BE(sat)} + 2V_{D(on)} - V_{D(on)}$$

$$= 0.8 \text{ V} + 1.4 \text{ V} - 0.7 \text{ V} = 1.5 \text{ V}$$

Hence for DTL, the logic swing is given as

$$LS = V_{OH} - V_{OL}$$

$$= 4.0 - 0.1 = 3.9 \text{ V}$$

For the noise margins,

$$NM_H = V_{OH} - V_{IH}$$

$$= 4.0 - 1.5 = 2.5 \text{ V}$$

$$NM_L = V_{IL} - V_{OL}$$

$$= 1.4 - 0.1 = 1.3 \text{ V}$$

Note that this basic DTL circuit exhibits a much larger logic swing than the RTL circuit. Also, the noise margins are much larger. A disadvantage of this DTL circuit is the extra voltage supply V_{BB} that is required, as well as the inconvenience of having to assign one of the pins of the IC package to this extra supply. The pull-down resistor R_2 and supply V_{BB} were a carryover from the discrete component design, where it was thought necessary that the two series diodes D_3 and D_4 be conducting at all times—to improve (square up) the voltage transfer characteristic and decrease the turn-off time of the output transistor.

7.2.1 Modified DTL

A modified form of DTL, known as the 930 Series, appeared on the scene in 1964 and quickly became the standard digital IC in most digital systems of that era. Indeed it remained popular for almost 10 years. The schematic of this modified DTL is shown in Fig. 7.3.

In comparing the two DTL circuits, note that they are very similar; but in Fig. 7.3, one of the series diodes has been replaced by a transistor Q_1 that when on is self-biased to operate in the forward active mode. Thus, assuming $V_{BE(on)} = V_{D(on)} = 0.7$ V and $V_{BE(sat)} = 0.8$ V, for the modified DTL circuit,

$$V_{IL} = 2V_{BE(on)} + V_{D(on)} - V_{D(on)}$$

$$= 1.4 \text{ V}$$

$$V_{IH} = V_{BE(sat)} + V_{D(on)} + V_{BE(on)} - V_{D(on)}$$

$$= 1.5 \text{ V}$$

Hence V_{IL} and V_{IH} typically have the same value in each of the DTL circuits. Similarly, with the output transistor cut off in the high state, and saturated in the low state, the values of V_{OH} and V_{OL} are V_{CC} and $V_{CE(sat)}$, respectively, for each circuit.

The advantage of using transistor Q_1 in Fig. 7.3 in place of diode D_3 in

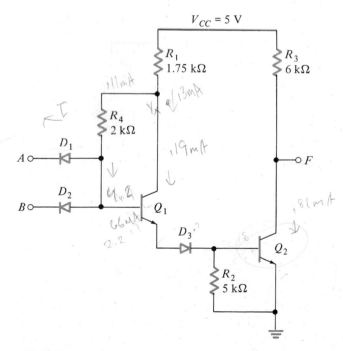

FIGURE 7.3
Modified DTL circuit.

Fig. 7.2a is that with the output transistor Q_2 conducting, transistor Q_1 is acting as an emitter-follower, supplying current to the base of Q_2. Hence, there is a current gain of $\beta_F + 1$ from the base of Q_1 to the base of Q_2. Since the fan-out of these DTL circuits is dependent on $I_{C(EOS)}$ of Q_2, the modified circuit yields a higher fan-out than the basic circuit.

Some of the more important electrical characteristics for the modified DTL circuit are given in Table 7.2.

With regard to the fan-out, a particular difference should be noted between the DTL and the RTL circuits. In DTL circuits, with the output of the driver gate in the high state, the input diodes of each load gate are reverse-biased and there is essentially an open circuit between the driver and the load. However, when the output of the driver gate is in the low state, the input diodes of each load gate are forward-biased. Hence, the load current I_L for each load gate is from V_{CC} through resistor R_1 in Fig. 7.2a and the forward-biased input diodes and into the collector of the output transistor of the driver gate. That is, in DTL circuits the direction of the load current is from the load gates to the driver gate. As such, the driver gate acts as a *current sink* for the load gates.

In RTL circuits, with the output of the driver gate in the low state, each of the load transistors is cut off. But when the output of the driver gate is in the high state, the driver is supplying base current to the load transistors. That is, in RTL circuits the direction of the load current is from the driver gate to the load gates. The driver gate acts as a *current source* for the load gates.

The power dissipation listed in Table 7.2 is about the same as that for RTL. But the propagation delay time for the modified DTL circuit is about twice that of the RTL circuit. This is obviously a disadvantage of the modified DTL circuit.

Another disadvantage of DTL circuits is the surface area of the silicon chip covered by the input diodes, especially in a multi-input logic gate. As the demand grew for more logic functions on a single chip, DTL was doomed, to be replaced by the more area-efficient and faster transistor-transistor logic (TTL).

Exercise 7.2. For the basic DTL gate circuit shown in Fig. 7.2, assume

$$V_{BE(on)} = 0.7 \text{ V} \qquad V_{BE(sat)} = 0.8 \text{ V}$$

$$V_{D(on)} = 0.7 \text{ V} \qquad V_{CE(sat)} = 0.1 \text{ V}$$

$$\beta_F = 30$$

(a) Sketch the voltage transfer characteristic.

(b) Calculate the fan-out.

TABLE 7.2
Modified diode-transistor logic: typical electrical characteristics at $T_A = 25°C$

V_{OH}/V_{OL}	4.8 V/0.2 V	Fan-out	8
V_{IH}/V_{IL}	1.5 V/1.2 V	Supply volts	+5.0 V
NH_H/NM_L	3.3 V/1.0 V	Power dissipation per gate	10 mW
Logic swing	4.6 V	Propagation delay time	30 ns

7.3 TRANSISTOR-TRANSISTOR LOGIC

The basic *transistor-transistor logic (TTL)* circuit is compared to the basic DTL circuit in Fig. 7.4. Note in the DTL circuit that resistor R_1 connects to the anode (a *p* region) of three diodes. This common *p* region becomes the base of an *npn* transistor Q_1 at the input of the TTL circuit. The input terminals of the TTL gate are connected to separate emitters of transistor Q_1, so that the input AND diodes of the DTL circuit become multiple emitter-base junctions in the TTL circuit. Furthermore, one of the level-shifting diodes in the DTL circuit becomes the base-collector junction for transistor Q_1 in the TTL circuit. The use of a multiemitter transistor in place of the diodes not only makes more efficient use of the silicon surface area in the input stage but, as we shall see later, it also serves to decrease the propagation delay time of the TTL gate.

7.3.1 Standard TTL

The circuit diagram of a TTL gate that was long the standard of digital circuit designers is shown in Fig. 7.5a. This is the Series 54/74,* which is exemplified here as a 2-input NAND gate, where $F = \overline{AB}$. Note that the level-shifting diode D_1 in Fig. 7.4b has been replaced by a transistor Q_2 in Fig. 7.5, which provides more base drive to the output transistor Q_3. This allows the output transistor to

---

*The Series 54 is a premium version for operation over the temperature range of −55 to +125°C. The Series 74 is similar but operates over the more limited temperature range of 0 to 70°C.

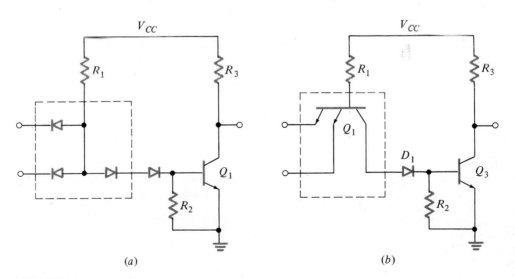

(a)   (b)

FIGURE 7.4
Circuit diagram of (a) basic DTL gate and (b) basic TTL gate.

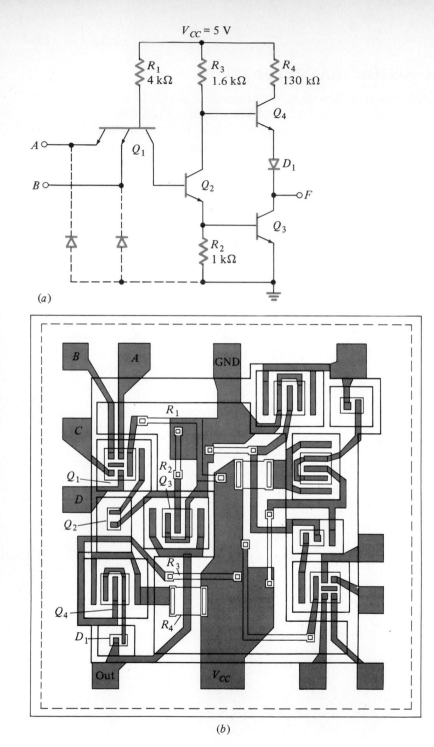

FIGURE 7.5
(a) Standard 2-input TTL NAND gate circuit. (b) Layout of a dual 4-input TTL NAND gate.

sink more current. The output stage also features an *active pull-up* (transistor Q_4). Hence this circuit is able to *source* more current than the *passive pull-up* (simply, a collector resistor) found in RTL and DTL circuits. The active pull-down and pull-up circuits, familiarly known as a *totem-pole* output circuit, provide more current to discharge and charge any parasitic capacitance associated with the load, thus decreasing the transition times at the turn-on and turn-off of these digital gates. Figure 7.5*b* illustrates the layout of a dual 4-input TTL NAND gate.

VOLTAGE TRANSFER CHARACTERISTIC. The voltage transfer characteristic of this standard TTL gate is displayed in Fig. 7.6. In describing the characteristic we assume the usual typical values, that is,

$$V_{BE(on)} = 0.7 \text{ V}$$

$$V_{BE(sat)} = 0.8 \text{ V}$$

$$V_{CE(sat)} = 0.1 \text{ V}$$

First note, with either input (A or B) low (that is, $V_{in} - 0.1$ V), transistor Q_1 is operating in the saturated mode. This is because with the base-emitter junction forward-biased, the base current of Q_1 will be approximately 1 mA. But the collector current of Q_1 is limited to the reverse, or leakage, current across the collector-base junction of transistor Q_2, and this will typically be approximately 1 nA. With $I_{C1} \ll \beta_F I_{B1}$, transistor Q_1 must be saturated. As a consequence,

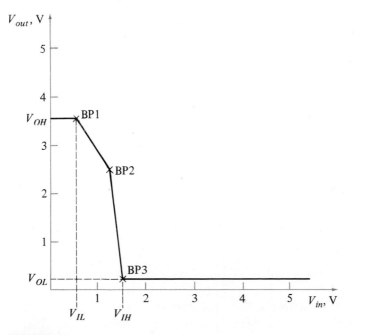

FIGURE 7.6
Voltage transfer characteristic of the standard TTL circuit.

$V_{CE1} = V_{CE(sat)} = 0.1$ V, and so $V_{C1} = 0.2$ V. But $V_{C1} = V_{B2}$; both transistors Q_2 and Q_3 must therefore be cut off. Hence, with either input low, the output is high. Now to provide even leakage current to the output node, transistor Q_4 must be on. Therefore, with $V_{BE4(on)} = V_{D1(on)} = 0.7$ V, we have that $V_{OH} = V_{CC} - 2V_{BE(on)} = 3.6$ V.

The first breakpoint in the characteristic occurs when transistor Q_2 turns on. That is, with $V_{C1} = V_{B2} = 0.7$ V and $V_{E2} = 0$ V, $V_{BE2} = 0.7$ V but $V_{BE3} = 0$ V. Base current to Q_2 is provided by the forward-biased base-collector junction of Q_1. Recall that in saturation, both junctions of the transistor are forward-biased. Hence with $V_{C1} = 0.7$ V and $V_{CE1(sat)} = 0.1$ V, $V_{E1} = V_{in} = 0.6$ V. Thus at the first breakpoint, $V_{IL} = 0.6$ V and $V_{OH} = 3.6$ V.

The second breakpoint occurs when transistor Q_3 turns on. But $V_{BE3} = V_{R2}$, and therefore $I_{E2} = V_{R2}/R_2$. Consequently, for $V_{BE3} = V_{BE(on)} = 0.7$ V, $I_{C2} \approx I_{E2} = 0.7$ V/1 kΩ = 0.7 mA. The voltage at the collector of Q_2 is $V_{C2} = V_{CC} - I_{C2}R_3 = 5 - (0.7)(1.6) = 3.9$ V. Note that with $V_{CE2} = V_{C2} - V_{E2} = 3.9 - 0.7 = 3.2$ V, Q_2 is operating in the forward active mode. The gate output voltage is $2V_{BE(on)}$ below V_{C2}; therefore $V_{out} = 2.5$ V. To determine the input voltage, transistor Q_1 is still in the saturation mode, and $V_{C1} = V_{BE2} + V_{BE3}$; that is, $2V_{BE(on)} = 1.4$ V. Hence, $V_{in} = 1.3$ V. The coordinates of the second breakpoint are $V_{in} = 1.3$ V and $V_{out} = 2.5$ V.

The third, and final, breakpoint occurs when Q_3 saturates. The gate output voltage is then $V_{CE(sat)} = 0.1$ V. With transistor Q_2 also saturated, $V_{C1} = 2V_{BE(sat)} = 1.6$ V. Transistor Q_1 is still saturated, so that $V_{in} = 1.6 - 0.1 = 1.5$ V. The coordinates are $V_{in} = 1.5$ V and $V_{out} = 0.1$ V.

With 1.5V $< V_{in} < 2.3$ V, transistor Q_1 operates in the reverse saturation mode. This is because the base of Q_1 is clamped at

$$V_{BC1(on)} + V_{BE2(sat)} + V_{BE3(sat)} = 0.7 + 0.8 + 0.8 = 2.3 \text{ V}$$

Both the base-collector and base-emitter junctions of Q_1 are forward-biased, but the gate input current is now *into* the emitter of Q_1.

With $V_{in} \geq 2.3$ V, transistor Q_1 operates in the reverse active mode, with the base-collector junction forward-biased but the base-emitter junction reverse-biased.

FAN-OUT. A consequence of the reverse active mode of Q_1 is that the input high current I_{IH} to the standard TTL circuit is much greater than that in the DTL circuit. In DTL circuits, I_{IH} is due simply to the leakage current of the input diodes. In standard TTL circuits, with Q_1 in the reverse active mode, all of the inputs are high, and

$$I_{IH} = \frac{\beta_R I_{B1}}{M} \tag{7.3.1}$$

where M is the number of emitters to the input transistor. To minimize I_{IH}, the reverse beta of the input transistor is purposely kept small, typically ≤ 0.2. Then, in the example of Fig. 7.5,

$$I_{IH} = \frac{0.2(5 - 2.3)}{(4 \text{ k}\Omega)(2)} = 67 \ \mu A$$

This value is much greater than even the worst-case maximum value of 5 μA for the modified DTL circuit of Fig. 7.3.

With only one input low, the input transistor of the TTL circuit operates in the saturation mode. The input low current I_{IL} is the emitter current of transistor Q_1 in Fig. 7.5, which is the sum of the base and collector currents. But since the collector current is essentially zero, $I_{IL} = I_{B1}$. That is,

$$I_{IL} = I_{B1} = \frac{V_{CC} - V_{BE1(sat)} - V_{CE3(sat)}}{R_1} \quad (7.3.2)$$

$$= \frac{5 - 0.8 - 0.1}{4 \text{ k}\Omega} = 1.0 \text{ mA}$$

To determine the fan-out N for the standard TTL gate, it is necessary that the output transistor Q_3 of the driver gate be able to sink NI_{IL}. That is,

$$N \le \frac{I_{C3(EOS)}}{I_{IL}} \quad (7.3.3)$$

PROPAGATION DELAY TIME. Mention was made at the beginning of this section that the use of a transistor at the input of the TTL circuit improves the propagation delay time, in particular t_{PLH}. Refer to Fig. 7.5. For the output to go high, it is necessary to turn off base current to transistor Q_3. Therefore, it is required to quickly turn off transistor Q_2. Prior to the transition the input is at a high level and both transistors Q_2 and Q_3 are saturated, and $V_{C1} = 1.6$ V. With a high-to-low transition at the input, $V_{E1} = 0.1$ V. Therefore initially, $V_{CE1} = 1.5$ V and transistor Q_1 operates in the *forward active mode*. That is, until Q_1 saturates, the collector current is $I_{C1} = \beta_F I_{B1}$. This current, which can be appreciable, is the turn-off base current for transistor Q_2. Note that it is not until *after* Q_2 turns off that Q_1 saturates. Hence, because of the forward current gain β_F of transistor Q_1, transistor Q_2 quickly turns off. With Q_2 off, transistor Q_4 becomes a source of current to the output, causing the gate output to go high, provided that Q_3 is off. The turn-off base current for Q_3 is through R_2. The turn-off time for Q_3 is usually longer than that of Q_2. But transistor Q_1 does serve to decrease the propagation delay time t_{PLH}.

OUTPUT STAGE. Two components in the output stage of this TTL circuit deserve comment. They are diode D_1 and resistor R_4. The necessity for the diode is as follows. With the output of the gate in the low state, both transistors Q_2 and Q_3 are saturated. Thus typically $V_{C2} = V_{BE3(sat)} + V_{CE2(sat)} = 0.9$ V, and $V_{C3} = V_{CE3(sat)} = 0.1$ V. Hence the voltage difference, $V_{C2} - V_{C3} = 0.8$ V. Without the diode D_1 in the circuit, this voltage is sufficient to turn on transistor Q_4, which would source current into the current sink of Q_3. The result would be an indeterminate output voltage. Including the diode D_1, the voltage difference ($V_{C2} - V_{C3}$) is insufficient to turn on both the diode and the transistor Q_4. Therefore, for this steady-state condition, with the output in the low state, transistor Q_3 is on and transistor Q_4 is off. With the output in the high state, it is vice versa. However, during the transition of the output from low to high, as transistor Q_3 turns off and transistor Q_4 turns on, there is the possibility of both transistors conducting at the same time. This could cause large currents to flow for a short time in the collector circuit of Q_4, resulting in large "current spikes" on the V_{CC} line. The inclusion of resistor R_4 allows transistor Q_4 to saturate, and the collector current is limited to a safe value. Resistor R_4 also serves to limit the gate output current should the output be shorted to ground when it is in the high state.

CLAMP DIODES. Finally, in the description of the TTL circuit of Fig. 7.5, the clamp diode from each input to ground should be noted. For the static conditions of a high or low level at the input, they are reverse-biased and play no part in the circuit. However as the output changes state, the switching transition times of these circuits are very short. Any inductance associated with the load causes high-frequency oscillations ("ringing") to appear on the output lines. Hence the input to a gate can be greater than 5 V on the positive swing and less than 0 V on the negative swing. The positive swing reverse-biases the emitter-base junction of the input transistor, but the base resistor R_1 limits the current and no harm is done. On the negative swing, with transistor Q_1 saturated, the voltage at the collector will only be 0.1 V more positive than the emitter voltage. The result is that the collector-substrate isolation diode of Q_1 can be forward-biased. This can lead to undesired voltage spikes at other nodes in the circuit or possible fatal damage to the IC. The diodes at the input "clamp" the input voltage so that it cannot swing more negative than about -0.7 V.

A listing of the electrical characteristics for this standard TTL gate circuit is given in Table 7.3. The supply and logic level voltages are compatible with the modified DTL characteristics listed in Table 7.2, although the noise margins and logic swing of the DTL circuits are superior. The power dissipation of each gate is about the same, but the propagation delay time of the TTL circuit is only one-third that of the DTL circuit. Hence the power-delay product of the TTL circuit is a factor of 3 better. The fan-out is also better.

This standard TTL circuit first appeared on the commercial market in 1965, but it was about 5 years before it overtook DTL in popularity. About that time (1970), a new form of TTL appeared that had improved dc and transient characteristics. This was the Schottky-clamped TTL circuit [TTL(S)].

Exercise 7.3. For the standard TTL gate circuit shown in Fig. 7.5 assume that

$$V_{BE(on)} = 0.7 \qquad V_{BE(sat)} = 0.8$$

$$V_{D(on)} = 0.7 \qquad V_{CE(sat)} = 0.1$$

$$\beta_F = 30 \qquad \beta_R = 0.3$$

(a) Calculate the fan-out for both output states.
(b) What is the average power dissipation?

TABLE 7.3
Standard transistor-transistor logic (54/74 TTL):
typical electrical characteristics at $T_A = 25°C$

V_{OH}/V_{OL}	3.5 V/0.2 V	Fan-out	10
V_{IH}/V_{IL}	1.5 V/0.5 V	Supply volts	+5.0 V
NM_H/NM_L	2.0 V/0.3 V	Power dissipation per gate	10 mW
Logic swing	3.3 V	Propagation delay time	10 ns

7.3.2 Schottky-Clamped TTL

A Schottky-barrier diode can readily be incorporated into the layout of a bipolar junction transistor by a simple extension of the base contact metal to include contact with the collector region. A simple illustration of this is given in Fig. 7.7a. Using aluminum (or platinum) as the metal, an Al (or Pt) Si (n-type) diode is formed at the metal-collector region interface. The forward voltage of the Schottky-barrier diode is less than the nominal $V_{BC(on)}$ of the transistor; hence, as noted in Sec. 6.4, the transistor is prevented from saturating when turned on. Consequently there is no saturation delay time when the transistor is turned off. The circuit symbol used to indicate a Schottky-clamped transistor is shown in Fig. 7.7b. In TTL circuits, the Schottky-barrier diodes commonly have platinum as the metal; typically $V_{SBD(on)} \approx 0.5$ V.

An example of a *Schottky-clamped* TTL [TTL(S)] gate is shown in Fig. 7.8. Compared to the standard TTL circuit of Fig. 7.5, there are other circuit improvements besides the use of Schottky-clamped transistors. In particular, the

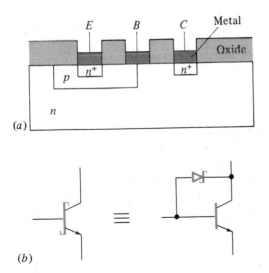

(a)

(b)

FIGURE 7.7
(a) Cross section of Schottky-clamped transistor. (b) Circuit symbol for Schottky-clamped transistor.

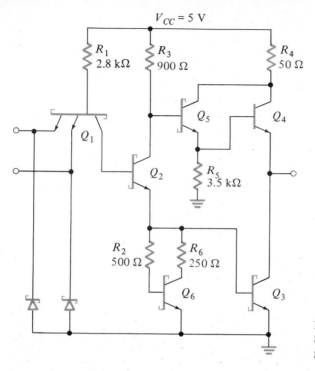

FIGURE 7.8
Schottky-clamped TTL NAND gate (54S/74S).

diode D_1 in the standard circuit is replaced by transistor Q_5 in this circuit. There is still $2V_{BE(on)}$ between the collector of Q_2 and the output of the gate, but Q_5 combines with Q_4 to form a cascaded emitter-follower, which is able to drive more source current into the load when the output of the gate goes to the high state. Thus the transition time t_{PLH} is reduced. Note that with the collectors of transistors Q_4 and Q_5 connected together and with both transistors conducting, the value of V_{CE4} is given as $V_{CE4} = V_{BE4} + V_{CE5}$. As a result, transistor Q_4 can never saturate, and it is unnecessary to provide a Schottky clamp for this transistor.

Transistor Q_6 in Fig. 7.8 is known as a *squaring circuit*, since it takes out the "knee" which appears between BP1 and BP2 in the voltage transfer characteristic (Fig. 7.6) of the standard TTL circuit. The cause of the knee is found in Fig. 7.5, where with V_{in} increasing from 0 V, transistor Q_2 is able to turn on before Q_3. Then, with Q_2 in the forward active mode, the gate output voltage steadily decreases until the current through R_2 is sufficient to turn on and saturate Q_3. In the new circuit, resistor R_2 is replaced by transistor Q_6. For current to flow in Q_2, transistors Q_3 and Q_6 must also be on. Hence Q_2 and Q_3 turn on at approximately the same gate input voltage, resulting in a voltage transfer characteristic with a narrower transition width and improved noise margins.

Resistors R_2 and R_6 ensure that most of the emitter current of Q_2 goes to the base of Q_3. The ratio of the base currents, I_{B3}/I_{B6}, is primarily determined by the value of R_2. Resistor R_6 is required, since without it, V_{BE3} would be clamped to $V_{BE6(on)} - V_{SBD(on)} = 0.2$ V.

Transistor Q_6 also serves to quickly turn off Q_3 and hence decrease the propagation delay time t_{PLH}. With a high-to-low transition at the input of the gate, transistor Q_2 quickly turns off because Q_1 is in the forward active mode. The turn-off base current of Q_3 is due to the collector current of Q_6, which is also in the forward active mode. With $V_{BE3} > V_{BE6}$, the base of Q_3 supplies current to the base of Q_6. But $I_{C6} = \beta_F I_{B6}$, resulting in a rapid turn-off of Q_3.

The resistor values in the Schottky-clamped circuit, namely, R_1 and R_3, are about one-half those for the standard circuit. The supply voltage for both circuits is 5 V. Consequently the power dissipation of the TTL(S) circuit is about twice that of the standard TTL, but the average propagation delay time of the Schottky-clamp circuit is a mere 3 ns. Compared to the 10 ns for the standard circuit, this is very fast. Also note that the power-delay product has been improved from 100 to 60 pJ.

The speed of the TTL(S) circuit is more than fast enough for most digital systems. Indeed, a propagation delay time of 10 ns is generally adequate. However, the need for more complex, and larger, circuits on a single chip appears to be a continuing one. But in the popular dual in-line package, the power dissipation of the chip is limited to about 500 mW. Hence the next development in TTL circuits, which was introduced in 1975, was a low-power version of the Schottky-clamped circuit (TTL(LS)).

7.3.3 Low-Power Schottky-Clamped TTL

The *low-power Schottky-clamped* TTL [TTL(LS)] gate circuit has a typical propagation delay time of 10 ns, which is the same as that for the standard TTL circuit. But the power dissipation is only 2 mW, which is only one-fifth of that for the standard circuit. To obtain this low power it is necessary to increase the resistance values by about a factor of 5 compared to those in the standard circuit.

The schematic of a TTL(LS) NAND gate circuit is shown in Fig. 7.9. The resistance values of R_1 and R_3 can be compared with those in the standard gate circuit of Fig. 7.5. The low-power Schottky circuit is similar to the regular Schottky circuit of Fig. 7.8, but the latest design has returned to a simple diode (although a Schottky-barrier diode) AND circuit at the input to the gate. Being Schottky-clamped, transistor Q_1 does not saturate, so a transistor is not required at the input to extract the excess overdrive charge from the base of Q_1. Also 10 years has passed since the introduction of the standard TTL circuit. During this time the IC fabrication technology has steadily improved and the minimum allowed dimension has decreased from about 12 to about 6 μm. As a result, the silicon surface area of the input diodes in Fig. 7.9 is only about one-third of the input transistor in Fig. 7.5. Consequently, the parasitic capacitances are likewise reduced.

In the low-power Schottky circuit, the Schottky-clamped transistors are at

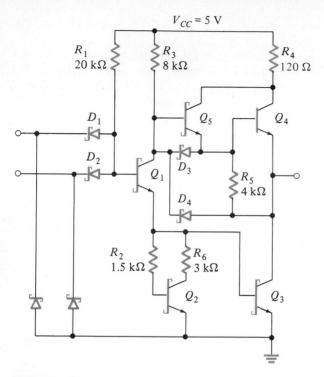

FIGURE 7.9
Low-power Schottky-clamped TTL NAND gate (54LS/74LS).

the edge of conduction with $V_{BE(EOC)} = 0.7$ V and are fully conducting with $V_{BE(on)} = 0.8$ V. Because of the Schottky clamp,

$$V_{CE(on)} = V_{BE(on)} - V_{SBD(on)} = 0.8 - 0.5 = 0.3 \text{ V}$$

A typical voltage transfer characteristic for the TTL(LS) circuit is given in Fig. 7.10. The first breakpoint is where transistors Q_1 and Q_3 turn on. Then $V_{B1} = 2V_{BE(EOC)} = 1.4$ V, and $V_{in} = V_{B1} - V_{SBD(on)} = 0.9$ V. With no load, there is no current in R_5, but due to Q_5 the output voltage is $V_{CC} - V_{BE(EOC)} = 4.3$ V. Hence at BP 1, $V_{IL} = 0.9$ V and $V_{OH} = 4.3$ V.

At the second breakpoint, Q_3 is "hard on," and $V_{out} = V_{OL} = V_{CE3(on)} = 0.3$ V. At the input, $V_{in} = V_{IH} = 2V_{BE(on)} - V_{SBD(on)} = 1.1$ V. Note that the inclusion of transistor Q_2 has eliminated the knee in the transfer characteristic.

The turn-off of transistor Q_4 is speeded up by including diode D_3. The turn-off base current for Q_4 is through D_3 and into the collector of the forward active mode transistor Q_1. This also aids in turning on Q_3. Since the base current of Q_3 is from the emitter of Q_1, its collector current is a component of the emitter current. Diode D_4 also helps speed up the transition of the output voltage from high to low.

For comparison purposes, listed in Table 7.4 are the principle performance

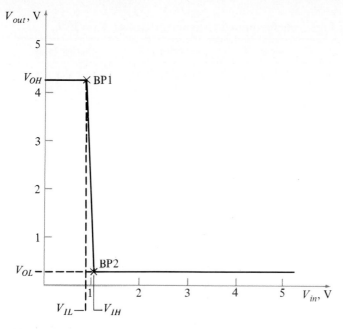

FIGURE 7.10
Voltage transfer characteristic of low-power Schottky-clamped TTL circuit.

characteristics of the three TTL circuits that have been described. The minimum and maximum values of the input and output voltages for both the high and low logic state are given. From these values the worst-case logic swing and transition width may be determined. Also given are the minimum and maximum values of the input and output currents for each logic state. From these, a safe value for the fan-out may be reckoned. The negative sign on I_{OH} and I_{IL} indicates that the current is *out* of the gate.

Exercise 7.4. For the 54LS/74LS gate circuit shown in Fig. 7.9 with

$$V_{BE(EOC)} = 0.7 \text{ V} \qquad V_{SBD(on)} = 0.5 \text{ V}$$
$$V_{BE(on)} = 0.8 \text{ V} \qquad \beta_F = 60$$

and for Q_2,

$$V_{BE(on)} = 0.7 \text{ V}$$

(a) Calculate the maximum fan-out.
(b) Find the average power dissipation.

7.3.4   Advanced Schottky-Clamped TTL

Chronologically in the past a new form of TTL has been introduced about every 5 years, namely, the standard TTL in 1965, the Schottky-clamped circuit around

TABLE 7.4
Transistor-transistor logic: performance characteristics at $T_A = 25°C$

	Series 74	Series 74S	Series 74LS
min V_{OH}/max V_{OL}	2.4 V/0.4 V	2.7 V/0.5 V	2.7 V/0.5 V
min V_{IH}/max V_{IL}	2.0 V/0.8 V	2.0 V/0.8 V	2.0 V/0.8 V
min I_{OH}/min I_{OL}	−0.4 mA/16 mA	−1.0 mA/20 mA	−0.4 mA/8 mA
max I_{IH}/max I_{IL}	40 µA/−1.6 mA	50 µA/−2.0 mA	20 µA/−0.4 mA
Typical propagation delay time	10 ns	3 ns	10 ns
Typical power dissipation per gate	10 mW	20 mW	2 mW

1970, and the low-power Schottky-clamped circuit about 1975. Hence, it is not surprising that by 1980 there was another new development. But this time there were two new circuits. They were:

54F/74F (fast) Series

54AS/74AS (advanced Schottky)

54ALS/74ALS (advanced low-power Schottky) Series, both derivatives of the 54S/74S family

These circuits are similar to the Schottky-clamped circuits that have just been described. The basic AS and ALS circuits are illustrated in Fig. 7.11. Note that the AS circuit and component values are almost identical to the 54S/74S circuit of Fig. 7.8, but the multiemitter input transistor has been replaced by a Schottky-diode AND circuit as in the 54LS/74LS Series. In the ALS circuit, except for the inclusion of transistors Q_1 and Q_2, the circuit is very similar to the LS circuit of Fig. 7.9. Additionally, note that the component values in the ALS circuit are about twice the value of those in the LS circuit. By including the emitter-follower Q_3, the ALS circuit is able to operate internally at very high speed, but compared to the LS circuit, the input current I_{IL} is halved. The *pnp* emitter-follower Q_1 is required to compensate for the V_{BE} voltage level shift of the *npn* transistor Q_3. The Schottky diode D_2 provides for a rapid turn-off of Q_4 when the input goes low. As a result of these changes, the input and output voltages of the advanced Schottky circuits are essentially compatible with the earlier series of TTL circuits.

These new circuits make use of the latest developments in bipolar IC fabrication technology. A 3-µm minimum internal feature size is used, as well as oxide isolation with walled base and emitters for extremely small device size and reduced parasitic capacitances. Also, ion implantation is used to give well-controlled shallow junctions for the active devices.

The result of these advances is given in Table 7.5. Note that the AS series aims for the highest speed: the propagation delay time of 1.5 ns is one-half that of the earlier Schottky series, but the power dissipation is the same at 20 mW. The ALS series aims for the lowest power: the power dissipation is only 1 mW, one-half that of the LS series, but in addition, the propagation delay time has been reduced from 10 to 4 ns. The F series is intermediate between the AS and the

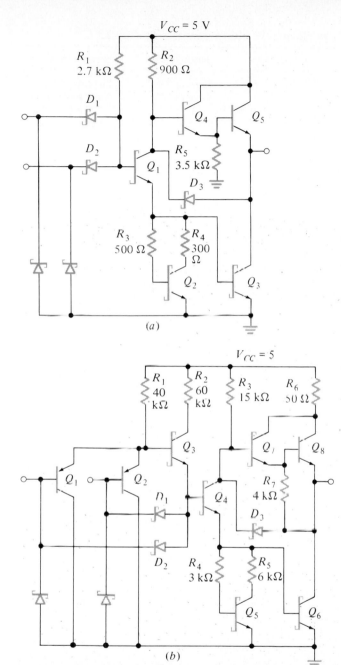

FIGURE 7.11
Advanced Schottky-clamped TTL circuits. (*a*) Advanced Schottky NAND gate (54AS/74AS).
(*b*) Advanced low-power Schottky NAND gate (54ALS/74ALS).

TABLE 7.5
Advanced Schottky TTL: performance charactersitics at $T_A = 25°C$

	Series 74F	Series 74AS	Series 74ALS
min V_{OH}/max V_{OL}	2.7 V/0.5 V	Same as 74S	Same as 74ALS
min V_{IH}/max V_{IL}	2.0 V/0.8 V	Same as 74S	Same as 74ALS
min I_{OH}/min I_{OL}	−1.0 mA/20 mA	−2.0 mA/20 mA	−0.4 mA/4.0 mA
max I_{IH}/max I_{IL}	20 μA/−0.6 mA	0.2 mA/−2.0 mA	20 μA/−0.2 mA
Typical propagation delay time	2.5 ns	1.5 ns	4 ns
Typical power dissipation per gate	4 mW	20 mW	1 mW

ALS, with a propagation delay time of 2.5 ns and a power dissipation of 4 mW. The improvement in the output capability of the advanced Schottky circuits is illustrated in Fig. 7.12, where the average propagation delay time is plotted as a function of the load capacitance.

In the chronological development of advanced digital ICs it should be noted that in 1985 a new CMOS series of SSI and MSI digital circuits was announced. This was the 54ACT/74ACT Series, a CMOS family of circuits that have similar voltage transfer characteristics and pinouts as the 54LS/74LS Series, but with improved speed parameters and much less power consumption. The new CMOS circuits are described in Sec. 3.7.5.

In summary of the TTL circuits it may be stated that at the time of writ-

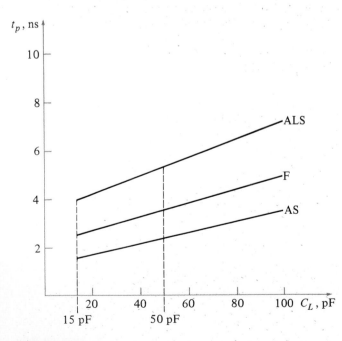

FIGURE 7.12
Average propagation delay time versus load capacitance for advanced Schottky TTL.

ing (1986), the 74LS series is still the most popular in the design of commercial general-purpose digital systems. Though it is anticipated that the CMOS 74AC/ACT Series, with improved characteristics and with much less average power dissipation, will eventually replace the 74LS Series. However, with the ongoing improvements in the bipolar technology, such as decreasing design rules leading to reduced area of the devices and oxide-walled base and emitter regions to reduce sidewall capacitance, such developments as the 74F, 74AS, and 74ALS Series will continue and will always find a niche in the higher-speed applications.

7.3.5 TTL Gate Circuits

Each of the TTL circuits that has been described has been exemplified by a 2-input NAND gate. However, in addition, all the basic combinational logic forms of AND, OR, NAND, and NOR as well as the INVERT function are available in each of the TTL circuits. The number of inputs to these gates is from one to as many as eight. Furthermore, the supply of SSI (small-scale integration) in combinational logic also includes more complex circuits such as exclusive-OR and AND-OR-INVERT gates. As an example, shown in Fig. 7.13 is the schematic for a two-wide, 2-input AND-OR-INVERT (2-2 input AOI) gate in TTL(LS). In

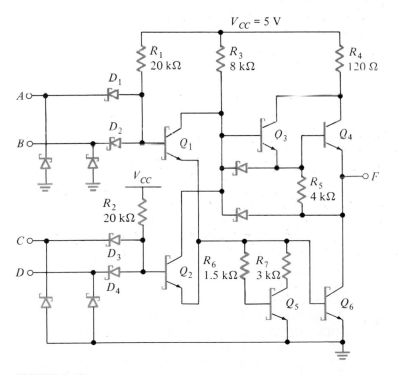

FIGURE 7.13
Two-wide, 2-input AND-OR-INVERT gate [TTL(LS)].

this circuit, diodes D_1, D_2 and D_3, D_4 constitute the two 2-input AND gates. The OR logic is performed at the common connection for the emitters of transistors Q_1 and Q_2. Transistor Q_6 provides the INVERT function. With the AND logic at the input true, either Q_1 or Q_2 will turn on Q_6, resulting in a low state at the output. The characteristics of the input circuit and the totem-pole output circuit are similar in all TTL(LS) circuits. This is true of all the TTL circuits; the input and output characteristics are particular to a given TTL circuit type.

Another popular logic function is the wired-AND, wherein the outputs of two (or more) gates are simply wired together. The need of a separate AND gate is then eliminated. In effect, the collectors of the output current sink transistors are connected together. The output, the wired-AND, is high only with every gate output high. With just one output low, the wired-AND function will be low. This connection is not feasible with the totem-pole output circuit, since with some outputs high and the others low, the simple wire connection provides for a current source (the active high output) driving directly into a current sink (the active low output). The result is that the voltage level of the wired-AND is indeterminate. To avoid this problem, some of the basic logic gates are also available with *open-collector* outputs. As an example, in Fig. 7.9, transistors Q_4 and Q_5 are left out of the circuit. The output is simply taken from the collector of Q_3. The circuit then has an active pull-down, but a resistor, connected from V_{CC} to the output, has to be provided to give a passive pull-up. The value of the resistor cannot be too low, since the output low current I_{OL} for each pull-down transistor has to include the current through the pull-up resistor as well as the total input low current I_{IL} for the load gates. Also, the value of the resistor cannot be too high, since the total input high current I_{IH} for the load gates flows from V_{CC} through the pull-up resistor, lowering the output high voltage V_{OH} and consequently reducing the noise margin NM_H.

The problem of using a passive pull-up resistor in wired-AND logic is avoided with gate circuits using *three-state outputs*. These gates have a totem-pole output circuit to give an active high output and an active low output, but they also have a third state in which both output transistors are off, to yield a high-impedance output. As illustrated in Fig. 7.14 for TTL(LS), the three-state circuit is similar to the schematic in Fig. 7.9, with the addition of the two diodes D_3 and D_4 and the CONTROL input. Normally, with the logic inputs A and B in the high state, transistor Q_1 turns on due to base current from V_{CC} through resistor R_1. But with CONTROL $=$ LOW, the base current to Q_1 is diverted through diode D_4, and Q_1 is off, as are transistors Q_2 and Q_3. Also, with CONTROL $=$ LOW, the base current for Q_5 is to diode D_3, so that Q_5 is off as well as Q_4. Hence with CONTROL $=$ LOW, both Q_3 and Q_4 are off and the output of the gate is in its third, or high-impedance, state. With CONTROL $=$ HIGH, both D_3 and D_4 are reverse-biased and the output is able to follow the voltage levels at A and B in the normal way. In particular, rapid charge and discharge of any load capacitance is possible because of the active pull-up and pull-down circuits. The three-state circuits are very useful in data bus applications, where data from various locations are gated onto a common data bus.

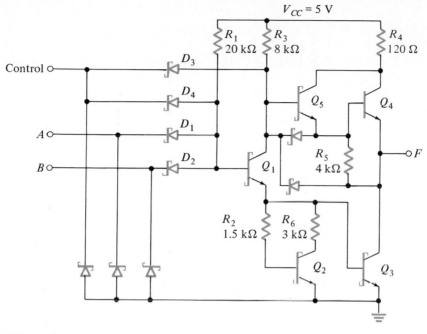

FIGURE 7.14
Three-state output NAND gate [TTL(LS)].

7.4 EMITTER-COUPLED LOGIC

Another series of digital ICs that has been commercially available for as long as the TTL circuits is the *emitter-coupled logic (ECL)* circuit. This is another one of the circuits that had its origins long before the invention of integrated circuits, or even the transistor. But the IC process has permitted the development of this circuit such that it presently is the fastest commercially available form of digital IC, with typical propagation delay times of less than 1 ns and clock rates approaching 1 GHz.

The basis of all ECL circuits is the nonsaturating *current switch*, shown in Fig. 7.15. As can be seen from the figure, another name for this circuit is the *emitter-coupled pair*. By a judicious choice of values for R_C and I_{EE}, the circuit can be designed so that the transistors in the current switch do not saturate. This is one reason for the short propagation delay time typical of ECL circuits. As presented in Fig. 7.15, the reference voltage V_R at the base of Q_2 is at ground, and V_{in} is applied to the base of Q_1. The current source I_{EE} is 2 mA, and $R_{C1} = R_{C2} = 1$ kΩ. The supply voltage is +5 V.

With $V_{in} = -1$ V, the voltage at the base of transistor Q_2 is more positive than that at the base of Q_1; therefore Q_2 is on and Q_1 is off. The forward bias at the base-emitter junction of Q_2 causes the base-emitter junction of Q_1 to be reverse-biased—in this case by about 0.3 V. Consequently,

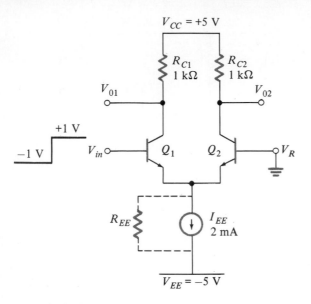

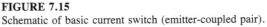

FIGURE 7.15
Schematic of basic current switch (emitter-coupled pair).

$$V_{01} \approx V_{CC} = +5 \text{ V}$$

$$V_{02} = V_{CC} - I_{C2}R_{C2} = +3 \text{ V}$$

With $V_{in} = +1$ V, the situation is reversed with Q_1 on and Q_2 off. A reverse bias of about 0.3 V is now at the base-emitter junction of Q_2. Therefore,

$$V_{01} = V_{CC} - I_{C1}R_{C1} = +3 \text{ V}$$

$$V_{02} \approx V_{CC} = +5 \text{ V}$$

The current I_{EE} is switched from one transistor to the other, as the input voltage is either less than, or greater than, the reference voltage. With $V_{in} = V_R$, then $I_{C1} = I_{C2}$. But from the ideal diode equation, $V_{BE} = V_T \ln (I_E/I_{ES})$. That is, at $T_A = 25°C$ only a 60-mV change in V_{BE} is required to cause a decade change in I_E. Hence in Fig. 7.15, with $V_{in} = +120$ mV, $I_{C1} = 100I_{C2}$. With I_{C2} only 1% of I_{C1}, transistor Q_2 is essentially off and Q_1 is conducting almost all the current I_{EE}. With $V_{in} = -120$ mV the situation is reversed.

Note that the transition from one logic state to the other is centered on the reference voltage V_R and is, to a first-order, independent of the transistor parameters. The transition width is approximately 200 mV. That is,

$$V_{IL} \approx V_R - 100 \text{ mV}$$

$$V_{IH} \approx V_R + 100 \text{ mV}$$

An elementary current source for the current switch may be obtained by simply connecting a resistor R_{EE} from the coupled emitters to a negative voltage

V_{EE}. Also, since the transistors do not saturate and with $\beta_F = 100$,

$$I_C \approx I_E = \frac{V_R - V_{BE} - V_{EE}}{R_{EE}}$$

With $|V_{EE}| \gg |V_{BE}|$, the V_{BE} term may be neglected. The output voltage is therefore quantized for the two logic states as

$$V_{OH} - V_{CC}$$

$$V_{OL} = V_{CC} - I_C R_C = V_{CC} - \frac{R_C}{R_{EE}}(V_R - V_{EE})$$

Note that V_{OL} is dependent on the *ratio* of the resistors R_C and R_{EE}. In the IC fabrication process it is difficult to maintain close control of the absolute value of resistors ($\pm 20\%$ is typical), but the *ratio* of resistor values can be well-controlled (typically, $\pm 2\%$).

Also note in Fig. 7.15, that the 2-V logic swing at the output of the current switch ($V_{OH} - V_{OL} = 2$ V) is equal to the logic swing applied to V_{in}. But for compatibility of output to input a voltage-level-shifting stage is necessary. In this case, the output voltage level needs to be shifted down by 4 V.

7.4.1 ECL 10K Series

The most popular form of ECL is the 10K Series, which is produced by many manufacturers. The basic schematic is shown in Fig. 7.16. In this circuit, three

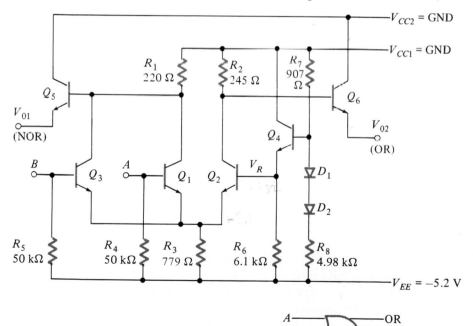

FIGURE 7.16
Schematic of ECL 10K 2-input OR/NOR gate.

transistors (Q_1, Q_2, and Q_3) form the current switch. Transistor Q_2 is the reference transistor; the reference voltage at the base of Q_2 is obtained from a low-impedance voltage source due to the emitter-follower Q_4. Transistors Q_1 and Q_3 are the input transistors and share the same collector resistor, so the current switch is described as having a *fan-in* of 2. The emitters of all three current switch transistors are connected together. The current source I_{EE} is due to R_3 and V_{EE}. The output of the current switch is through the two emitter-followers Q_5 and Q_6. These transistors serve as voltage level shifters but also as low impedance output voltage drivers. Note there are no resistors shown in the emitter circuit of Q_5 and Q_6. They are provided by the load gates, similar to R_4 and R_5 at the base of Q_1 and Q_3. Hence the output circuits in the 10K series of ECL have active pull-ups but passive pull-downs. The pull-down resistors are pinch resistors on the chip, but their 50-kΩ value limits the pull-down current. For high-speed operation these resistors must be paralleled by much smaller discrete resistors, generally 2 kΩ to -5.2 V or 50 Ω to -2 V.

Also note in the circuit of Fig. 7.16 that two V_{CC} connections are provided. One is to the basic current switch and reference voltage circuit, the other is only to the collectors of the output emitter-followers. Because of the fast switching times of these circuits and the inevitable parasitic capacitance associated with the load gates, large steplike changes of current flow in the output emitter-follower circuits. As a result, large voltage transients ("spikes") appear in the collector circuit of transistors Q_5 and Q_6. On the other hand, the supply current to the current switch is almost constant; as the current in one side of the switch increases, the current in the other side decreases. By keeping the "dirty" $V_{CC}(=V_{CC2})$ separate from the "clean" $V_{CC}(=V_{CC1})$, the effects of the voltage transients in other parts of the system are minimized. The two supplies come together only at the system power source. The V_{CC} is usually at ground potential in ECL circuits, since the output voltage level for both logic states is directly related to V_{CC1} and ground can be considered the best available fixed potential. This requires that V_{EE} be at a negative potential, and the standard value in ECL circuits is -5.2 V.

Since outputs are taken from both sides of the current switch in Fig. 7.16, this circuit, in common with most ECL circuits, provides *complementary outputs*. That is, with either input *A or B* = high, transistor Q_1 or Q_3 is on and Q_2 is off. As a result the output from Q_2, that is, V_{02}, is high, and from Q_1, that is, V_{01}, is low. Hence,

$$V_{02} = A + B \qquad \text{the OR output}$$

$$V_{01} = \overline{A + B} \qquad \text{the NOR output}$$

The logic function of the circuit is described as a 2-input OR/NOR gate. The logic symbol is included in Fig. 7.16. The provision of complementary outputs from one digital gate is a very useful feature of the ECL circuits.

Another useful feature is the active pull-up and passive pull-down output circuit that allows many outputs to be simply wired together, readily providing a wired-OR circuit. Then, of course, only one pull-down resistor is required for the wired-OR.

As previously mentioned, the output emitter-followers Q_5 and Q_6 also serve as low output impedance voltage drivers. In the 10K Series, the output is capable of providing the full logic swing across a load resistance even as small as 50 Ω to -2 V. This is very useful in high-speed digital circuits where the interconnection between systems is often done with transmission lines that have a characteristic impedance of 50 to 100 Ω.

Finally, in describing the schematic of Fig. 7.16, it should be noticed that the on-chip 50-kΩ resistors at the base of the input transistors provide that with no connection applied the input is automatically in the logic low state.

VOLTAGE TRANSFER CHARACTERISTIC. In the analysis to determine the voltage transfer characteristic, assume that when on, all transistors operate in the forward active mode, with $V_{BE(on)} = 0.75$ V and β_F sufficiently high that the effects of base current may be neglected. First, calculate the reference voltage V_R.

$$V_R = V_{B4} - V_{BE(on)}$$

where

$$V_{B4} = V_{CC} - \frac{R_7}{R_7 + R_8}(V_{CC} - 2V_{BE(on)} - V_{EE})$$

$$= 0 - \frac{0.907 \text{ k}\Omega}{(0.907 + 4.98) \text{ k}\Omega}(0 - 1.5 + 5.2) = -0.57 \text{ V}$$

therefore

$$V_R = -0.57 - 0.75 = -1.32 \text{ V}$$

Next, for the current switch assume that Q_2 is on and Q_1 and Q_3 off. Then

$$V_{E2} = V_R - V_{BE(on)}$$

$$= -1.32 - 0.75 = -2.07 \text{ V}$$

and

$$I_{E2} = \frac{V_{E2} - V_{EE}}{R_3}$$

$$= \frac{-2.07 + 5.2}{0.779 \text{ k}\Omega} = 4.02 \text{ mA}$$

With $I_{C2} = I_{E2}$,

$$V_{C2} = V_{CC} - I_{C2}R_2$$

$$= 0 - (4.02)(0.245) = -0.984 \text{ V}$$

The assumption that Q_2 is in the forward active mode is correct, since

$$V_{CE2} = -0.984 + 2.07 = +1.09 \text{ V}$$

At the output,

$$V_{02} = V_{C2} - V_{BE(on)}$$

$$= -0.98 - 0.75 = -1.73 \text{ V}$$

With Q_1 and Q_3 off,

$$V_{01} = V_{C2} - V_{BE(on)}$$

$$= 0 - 0.75 = -0.75 \text{ V}$$

In practice, the typical output voltage levels for the ECL 10K gate are $V_{02} = -1.74$ V and $V_{01} = -0.90$ V. The typical value for V_{01} is less than the calculated value because of the finite base current that flows in R_1 when a load is applied to V_{01}. Note that the typical output voltage levels are symmetrical about the reference voltage ($V_R = -1.32$ V). The outputs may therefore be directly connected to the inputs of similar load gates. The current switch in the load gates will operate correctly.

With V_A and $V_B = -1.74$ V, both Q_1 and Q_3 are practically cut off. Since with

$$V_{E1} = V_{E3} = V_{E2} = -2.07 \text{ V}$$

$$V_{BE1} = V_{BE3} = -1.74 + 2.07 = +0.33 \text{ V}$$

The base-emitter junctions are forward-biased, but at this voltage the collector current will be extremely small (less than 1 nA).

With V_A or $V_B = -0.90$ V, the current in the switch is to either Q_1 or Q_3 and Q_2 is cut off. With Q_1 conducting,

$$I_{E1} = \frac{V_{B1} - V_{BE(on)} - V_{EE}}{R_3}$$

$$= \frac{-0.90 - 0.75 + 5.2}{0.779 \text{ k}\Omega} = 4.56 \text{ mA}$$

With $I_{C1} = I_{E1}$,

$$V_{C1} = V_{CC} - I_{C1}R_1$$

$$= 0 - (4.56)(0.220) = -1.00 \text{ V}$$

As a result,

$$V_{01} = V_{C1} - V_{BE(on)}$$

$$= -1.00 - 0.75 = -1.75 \text{ V}$$

With Q_2 off, but including the effects of loading,

$$V_{02} = -0.90 \text{ V}$$

Hence the output voltage levels from each side of the current switch are indeed complementary, both in the high and the low logic state.

The voltage transfer characteristic for this circuit is shown in Fig. 7.17. The symmetry is complete, except for the low state at the NOR output. The explanation is as follows. As the input voltage at A is steadily increased above the reference voltage, the collector current of Q_1 also steadily increases. Consequently the voltage V_{CE1} will decrease. With sufficient input voltage, Q_1 will saturate. Assuming $V_{CE(sat)} = 0$ V, saturation is reached for Q_1 with

$$V_{E1} = V_{C1} = V_{CC} - \frac{R_1}{R_1 + R_3}(V_{CC} - V_{CE(sat)} - V_{EE})$$

$$= 0 - \frac{0.22 \text{ k}\Omega}{(0.22 + 0.779) \text{ k}\Omega}(0 + 5.2) = -1.15 \text{ V}$$

Therefore, at the edge of saturation,

$$V_{in} = V_{E1} + V_{BE(on)}$$

$$= -1.15 + 0.75 = -0.4 \text{ V}$$

and

$$V_{01} = V_{C1} - V_{BE(on)}$$

$$= -1.15 - 0.75 = -1.9 \text{ V}$$

This point of saturation is indicated on the transfer characteristic in Fig. 7.17. With Q_1 saturated, the base-collector junction is forward-biased, and with $V_{CE(sat)} = 0$ V, $V_{BC(on)} = V_{BE(on)}$. Therefore with Q_1 saturated,

$$V_{01} = V_{in} \quad V_{BC1(on)} - V_{BE5(on)}$$

$$= V_{in} - 2V_{BE(on)}$$

Of course, a similar explanation can be made for Q_3 saturating. However, in the normal mode of operation none of the transistors in the ECL circuit saturate, because the input voltages do not rise higher than -0.7 V.

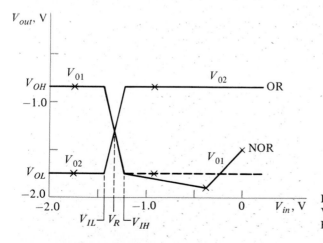

FIGURE 7.17
Voltage transfer characteristic of ECL 10K circuit.

A listing of typical electrical characteristics for the ECL 10K Series is given in Table 7.6. Especially note the symmetrical noise margins for the high and low states and also the small logic swing.

FAN-OUT. Since the output current of the gate is from emitter-followers and the input current to the load gates is the base current of the nonsaturating current switch transistors, there is the potential of a very high fan-out. At a low clock rate, the fan-out is of the order of I_E/I_B, that is, the current gain β_F, which is of the order of 100. However, associated with each load gate is a finite load capacitance, which must be charged and discharged as the driving gate changes state. Hence changing the voltage at the input of a load gate takes time. In most applications, this limits the number of load gates, or fan-out, to about 10.

PROPAGATION DELAY TIME. The small logic swing and the fact that in the normal mode of operation none of the transistors saturates lead to very short propagation delay times in ECL circuits. In the 10K Series it is about 2 ns.

POWER DISSIPATION. The power dissipation of the gate is calculated as follows. For the current switch,

$$\text{Average emitter current} = \frac{4.02 + 4.56}{2} = 4.28 \text{ mA}$$

$$\text{Power consumption} = (5.2 \text{ V})(4.28 \text{ mA}) = 22.3 \text{ mW}$$

For the reference voltage supply,

$$\text{Current in } Q_4 = \frac{V_{R6}}{R_6} = \frac{-1.32 + 5.2}{6.1 \text{ k}\Omega} = 0.64 \text{ mA}$$

$$\text{Current in bias supply} = \frac{V_{R7}}{R_7} = \frac{0 - 0.57}{0.907 \text{ k}\Omega} = 0.63 \text{ mA}$$

$$\text{Power consumption} = (5.2 \text{ V})(0.64 + 0.63) \text{ mA} = 6.6 \text{ mW}$$

In many ECL circuits it is possible for the reference voltage supply to serve more than one current switch. In a quad 2-input OR/NOR gate, the average power dissipation per gate is

$$22.3 + \frac{6.6}{4} = 24 \text{ mW}$$

TABLE 7.6
Emitter-coupled logic (10K): typical electrical characteristics at $T_A = 25°C$

V_{OH}/V_{OL}	−0.9 V/−1.7 V	Fan-out	10
V_{IH}/V_{IL}	−1.2 V/−1.4 V	Supply volts	−5.2 V
NM_H/NM_L	0.3 V/0.3 V	Power dissipation per gate	24 mW
Logic swing	0.8 V	Propagation delay time	2 ns

The power-delay product of the 10K Series is therefore approximately

$$(2 \text{ ns})(24 \text{ mW}) = 48 \text{ pJ}$$

As a figure of merit, this figure is a little better than the 60 pJ of the regular Schottky TTL but not as good as the 30 pJ of the advanced Schottky TTL. However, such a comparison is not really valid, since the latter figure includes the effects of an advanced processing technology. A better comparison would be with an ECL circuit that uses similar design rules and technology, such as the 100K Series that is described in the next section.

Exercise 7.5. The circuit of an earlier form of ECL is similar to Fig. 7.16, except that

$$R_1 = 290 \ \Omega \qquad R_4 = 50 \text{ k}\Omega \qquad R_7 = 300 \ \Omega$$
$$R_2 = 300 \ \Omega \qquad R_5 = 50 \text{ k}\Omega \qquad R_8 = 2.3 \text{ k}\Omega$$
$$R_3 = 1.18 \text{ k}\Omega \qquad R_6 = 2 \text{ k}\Omega$$

Also, emitter resistors are included for Q_5 and Q_6, from the output nodes to V_{EE}:

$$R_{E5} = R_{E6} = 1.5 \text{ k}\Omega$$
$$V_{CC1} = V_{CC2} = 0 \text{ V} \qquad V_{EE} = -5.22 \text{ V}$$

Assume

$$V_{BE(on)} = 0.75 \text{ V} \qquad \beta_F = 100$$

Calculate:
(a) V_R, V_{OH}, and V_{OL}.
(b) The fan-out with $NM_{II} = NM_L$.
(c) Noise margins with fan-out = 10.

7.4.2 ECL 100K Series

In conventional ECL circuits, as demonstrated in the 10K Series, the reference voltage supply is designed so that V_R is midway between V_{OH} and V_{OL}. This ensures equal noise margins for the logic high and logic low state. Further, by including the two diodes D_1 and D_2 in the reference voltage supply, compensation is provided for V_{BE} variations with temperature, so that V_R is centered with respect to V_{OH} and V_{OL} even with changes of temperature. However, the reference voltage and output voltage levels do change with temperature, increasing as temperature increases. Moreover, changes in the V_{EE} supply also cause changes in V_R and V_{OL}.

This shift in the voltage transfer characteristic can be troublesome in a large digital system incorporating many smaller units, each with its own supply voltage and ambient temperature. In a later development, ingenious circuit design is used to make the transfer characteristic almost independent of supply voltage and temperature variations. This is the ECL 100K Series.

A schematic of a 2-input OR/NOR gate of the 100K Series is shown in Fig. 7.18a. The circuit is similar to that for the 10K Series, but a transistor current

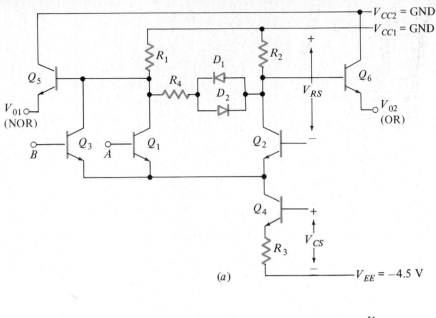

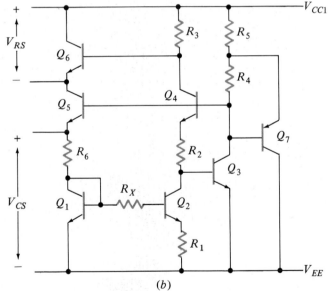

FIGURE 7.18
Schematic of ECL 100K 2-input OR/NOR gate. (*a*) Current switch including input and output transistors. (*b*) Bias network for V_{RS} and V_{CS}.

source Q_4 is used in the emitter-coupled current switch, and back-to-back diodes D_1 and D_2 with resistor R_4 are connected between the complementary outputs of the current switch. The reference voltages V_{RS} and V_{CS} are both invariant to supply and temperature changes. Also note that the V_{EE} supply is changed to -4.5 V in order to reduce the power dissipation of the circuit.

The bias network for V_{RS} and V_{CS} is given in Fig. 7.18b. Notice that V_{RS} is referenced to V_{CC1} and

$$V_{RS} = V_{BE6} + V_{R3}$$

Also, V_{CS} is referenced to V_{EE} and

$$V_{CS} = V_{BE3} + V_{R2} \qquad (V_{BE4} \text{ is canceled by } V_{BE5})$$

Because of the shunt regulator Q_7 the collector current of Q_3 is held constant, even as V_{EE} is changed with respect to V_{CC1}. Should I_{C3} tend to increase, due say to V_{EE} decreasing (made more negative), the voltage drop across R_4 would increase, causing Q_7 to conduct harder and the extra current to be shunted away from Q_3 into Q_7. As a consequence, changes in the supply voltage have no effect on the collector currents I_{C3}, I_{C2}, and I_{C1}. With no change in I_{C3}, there is no change in V_{BE3}; and with no change in I_{C2}, there is no change in V_{R2} or V_{R3}. No change in I_{C1} means V_{BE6} does not change. The result is that both V_{RS} and V_{CS} are insensitive to changes in the supply voltage.

In the bias network, transistor Q_1 is operated at a higher current density than is Q_2, with the result that $V_{R1} = V_{BE1} - V_{BE2}$. Furthermore, the difference of the two V_{BE}'s yields a positive temperature coefficient for V_{R1}. The voltage V_{R1} is amplified by the resistor ratio R_3/R_1 to produce V_{R3}, which is part of V_{RS}, and the positive temperature coefficient of V_{R3} is used to compensate for the negative temperature coefficient of V_{BE6}. Similarly, V_{R2} is amplified from V_{R1} by the ratio R_2/R_1, and the temperature coefficient of V_{R2} compensates for the temperature coefficient of V_{BE3}. As a result, both V_{RS} and V_{CS} are insensitive to changes in temperature. The resistor R_X compensates for process variations in β_F and V_{BE}.

With V_{RS} and V_{CS} invariant to changes in the supply voltage, V_{OH} and V_{OL} are also unaffected by changes in V_{EE}. However a change in temperature does cause the current in the current source transistor (Q_4 in Fig. 7.18a) to change. In particular, as temperature increases, V_{BE4} changes approximately -1.5 mV/°C. This results in an increase in the voltage across R_3, causing the current in the current switch to increase. With Q_2 off and Q_1 conducting, most of the collector current I_{C1} is through R_1, but D_1 is also conducting and a small portion of I_{C1} is through R_2 and R_4. As a consequence, the voltage drop across R_1, which increases approximately 1.5 mV/°C, serves to compensate for the temperature dependence of the base-emitter junction of Q_5, which decreases about 1.5 mV/°C. Also D_1 exhibits a temperature coefficient of -1.5 mV/°C. The net result is a 3 mV/°C increase of voltage across the series combination of R_4 and R_2. But $R_4 = R_2$, so the voltage at the base of Q_6 decreases about 1.5 mV/°C. This compensates for the temperature dependence of the base-emitter junction of Q_6. With Q_1 off and Q_2 conducting, diode D_2 is forward-biased and a similar compensation occurs for V_{BE5}. Hence V_{OH} and V_{OL} tend to be unaffected by variations in temperature.

The voltage transfer characteristic of the fully compensated 100K Series and the conventional ECL circuit are compared in Fig. 7.19. Due to supply voltage variations, the conventional circuit shows a typical change of 150 mV/V for V_R and 250 mV/V for V_{OL}, whereas the change in the fully compensated circuit is an order of magnitude improved at 10 mV/V and 15 mV/V for the same parameters.

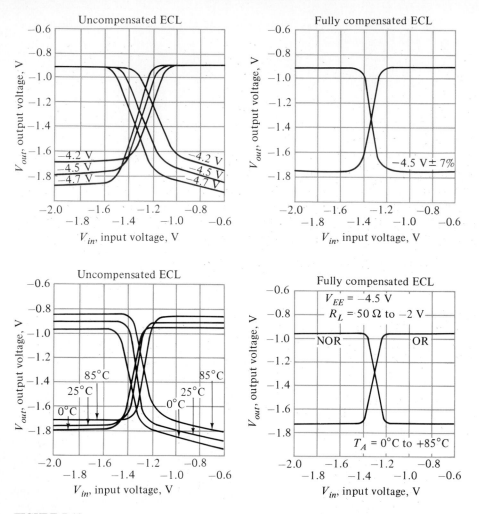

FIGURE 7.19
Voltage transfer characteristics comparing fully compensated ECL with uncompensated ECL.
(*Fairchild Semiconductor.*)

The temperature coefficients for the 10K Series are 1.1 mV/°C for V_R, 0.6 mV/°C for V_{OL}, and 1.5 mV/°C for V_{OH}. For the 100K Series the temperature coefficients are less than 0.1 mV/°C for each of these parameters.

As well as the improved voltage transfer characteristic, the transistors used in the 100K Series are of an improved design. Similar to the transistors used in the advanced Schottky TTL gates, the transistors used in the 100K Series make full use of an advanced fabrication technology, including oxide isolation and walled base and emitter regions for minimum-sized devices and extremely small parasitic capacitances. The layout and cross section of these latest developments in transistor fabrication are compared with the preceding design in Fig. 7.20. The

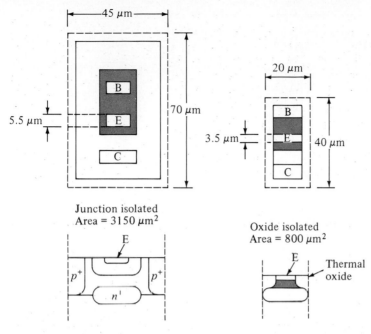

FIGURE 7.20
Layout and cross section of transistors used in 100K ECL compared with the junction-isolated transistors used in 10K ECL. (*Fairchild Semiconductor.*)

result is that the typical propagation delay time of a 100K Series gate is 0.75 ns, with a power dissipation of 40 mW, resulting in a power-delay product of 30 pJ. Among the advanced bipolar digital gate circuits, the 100K Series dissipates the most power but exhibits the shortest propagation delay time.

7.5   INTEGRATED INJECTION LOGIC

The circuits that have so far been described in this chapter readily lend themselves for use in small-scale (SSI) and medium-scale (MSI) integrated circuits, but their use in large-scale (LSI) integrated circuits is generally prohibited because of their relatively large area and power dissipation. Even a modest 500-gate array, designed with TTL(LS), would dissipate 1 W of power and cover a surface area of 25 mm². The surface area is about maximum for a high-yield IC, but the power dissipation is excessive.

The bipolar transistor answer to the challenge of MOSFETs in LSI is *integrated injection logic (I²L)*, also known as *merged transistor logic (MTL)*. A simple I²L logic circuit is shown in Fig. 7.21. The circuit consists of only three multicollector *npn* transistor inverters, with collectors connected together to form wired-AND logic. The multicollectors allow intermediate logic functions to be utilized. So the circuit of Fig. 7.21 provides the inverted functions $\overline{A}$ and $\overline{B}$, the

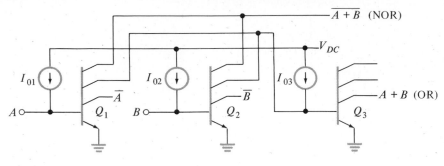

FIGURE 7.21
Simple I^2L logic circuit.

NOR function $(\overline{A\,B} = \overline{A} + \overline{B})$, and the OR function $(A + B)$. Unused outputs can be left as open circuits.

At the input of each transistor inverter there is a current source I_0, known as the *injector current*. With, for example, input A in the logic high state, the current I_{01} is into the base of transistor Q_1, so that Q_1 saturates and each output of Q_1 is in the logic low state. The current in each collector of Q_1 is from a similar current source as I_{01}, located at the base of a load transistor—for example, I_{03} at the base of Q_3. With $V_{CE(sat)} < V_{BE(EOC)}$, there is no current to the base of Q_3, and Q_3 is off. Hence with an output low, the current I_0 is the collector current of the driving transistor; and with the output high, the current I_0 is the base current of the load transistor.

7.5.1   Standard I^2L

Three ways of drawing a basic I^2L circuit are shown in Fig. 7.22, where it can be seen that the injector current is from the collector of a *pnp* transistor that is connected to the base of an *npn* transistor. Since the collector of the *pnp* and the base of the *npn* are both *p* regions, they can be formed together as a common *p* region. Similarly, the base of the *pnp*, an *n* region that is connected to ground, can be formed with the emitter of the *npn*, which is also an *n* region that is connected to ground. This *integrating*, or *merging*, of the two transistors is illustrated in the cross section of the gate shown in Fig. 7.22*d*.

The fabrication of the I^2L gate starts with a heavily doped (n^+) silicon substrate on which is grown an *n*-type epitaxial layer. The *p* and n^+ regions are then diffused as in the normal processing of an *npn* transistor. The n^+ substrate, which is connected to ground, serves as the emitter of the *npn* and the base contact to the *pnp* transistor. Note that the I^2L gate structure comprises a *lateral pnp* and a *vertical npn* transistor. With the multiple n^+ regions at the surface of the silicon, the *npn* transistor is similar to the multiemitter transistor at the input of the standard TTL circuit. However, in the I^2L circuit, the collector and emitter connections of the *npn* inverter are reversed. That is, the transistor is

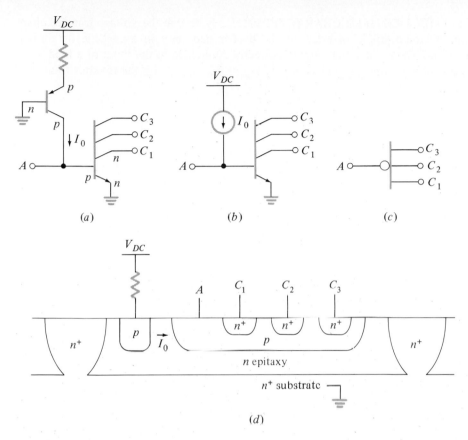

FIGURE 7.22
I^2L digital gate. (*a*) Circuit diagram; (*b*) equivalent circuit; (*c*) logic symbol; (*d*) cross section.

operated upside-down. The advantage here is that with the emitter connected to the substrate, the substrate becomes the common ground of the circuit. Hence, the area of the silicon surface required for emitter connections is considerably reduced, allowing for denser packing of the inverter circuits.

For a bipolar transistor operated in the forward mode, the main flow of carriers is from the emitter to the collector. Hence the forward current gain of the normally connected transistor may be described as β_D (beta down), since the flow of carriers is *down* into the silicon. But for the *npn* transistor in the I^2L circuit, the forward current gain is β_U (beta up), since the flow of carriers is now *up* to the surface of the silicon. Compared to the normal connection, β_D corresponds to beta forward (β_F), which is generally very large, while β_U is the same as beta reverse (β_R), which is usually rather small. That is,

$$\beta_D \equiv \beta_F \approx 100$$

$$\beta_U \equiv \beta_R \approx 5$$

VOLTAGE TRANSFER CHARACTERISTIC. Note that the voltage transfer characteristic of a single inverter can be determined only in a cascade of similar inverters. Then, since the output is directly connected to the input of a load transistor, with the driving transistor off, $V_{out} = V_{BE(sat)}$. With the inverter transistor on, $V_{out} = V_{CE(sat)}$. Hence,

$$V_{OH} = V_{BE(sat)} \qquad V_{IH} = V_{BE(EOS)}$$

$$V_{OL} = V_{CE(sat)} \qquad V_{IL} = V_{BE(EOC)}$$

The transition width is from the transistor at the edge of conduction ($V_{BE(EOC)}$) to the transistor at the edge of saturation ($V_{BE(EOS)}$). Using typical values,

$$V_{OH} = 0.8 \text{ V} \qquad V_{IH} = 0.7 \text{ V}$$

$$V_{OL} = 0.1 \text{ V} \qquad V_{IL} = 0.6 \text{ V}$$

From this data, note the small logic swing (LS = 0.7 V) and noise margins (NM_H = 0.1 V, NM_L = 0.5 V). Operation with these low values is possible only because a complete digital system can be designed for a single chip. Translators and buffers are used at the input and output of the I^2L circuit, typically giving TTL voltage levels and noise margins off the chip.

FAN-OUT. To determine the fan-out of the circuit, with an input in the high state, each collector of the inverting transistor must be capable of sinking the injector current I_0. Therefore the total collector current I_{CT} is simply I_0 multiplied by the number N of collectors for each transistor, or

$$I_{CT} = NI_0$$

But the base current for the transistor is the injector current I_0:

$$I_B = I_0$$

Therefore the current gain of the transistor is

$$\beta_U \geq \frac{I_{CT}}{I_B} = \frac{NI_0}{I_0} = N$$

Hence for this circuit configuration, the current gain β_U restricts the number of collectors N_{max} of the multicollector transistor to be $\leq$ 5. However the low fan-out is not a serious limitation. Most I^2L circuits are designed with from two to five collectors.

Any number of collectors may be connected together to form the wired-AND logic function, but the connection can be to only one input. Connecting two inputs together causes the total collector current to double, because two current sources are wired in parallel.

POWER-DELAY PRODUCT. For the I^2L gate the power dissipation is simply the product of the injector current I_0 and the supply voltage V_{DC}. The supply voltage is in the range of 0.7 to 1 V and is generally current-limited by a series resistor to the emitter of the current source.

A feature of the circuit is that the injector current can be changed by simply varying the supply voltage or series resistance. Increasing I_0 allows for faster switching of the inverting transistor, but there is an increase in the power dissipation. However the power-delay product in I²L does remain essentially constant over a very large range of injector currents. This is illustrated in the graph of Fig. 7.23, where the average propagation delay time is plotted against the average power dissipation. The dashed lines are for a constant power-delay product of 0.1 and 1.0 pJ. The solid lines represent the power-delay product of a production microprocessor unit using 8-μm design rules and a development laboratory ring oscillator using 5-μm design rules. At the larger injection currents, charge storage and bulk series resistance effects in the neutral base region of the switching transistor limit the minimum value of the propagation delay time.

Exercise 7.6. For the I²L gate shown in Fig 7.22, assume one output only and replace the current source I_0 with a 10-kΩ resistor. $V_{DC} = 1.5$ V. Use the transistor data shown in Table 6.1, but with $\alpha_F = 0.98$ and $\alpha_R = 0.8$.
(a) Sketch the VTC and determine the noise margins.
(b) Verify with SPICE.

INTEGRATED INJECTION LOGIC IN LSI. A prime requirement for LSI is a large packing density. That is, the number of gates per unit surface area of the silicon should be high. In the I²L gate this is a result of:

1. The simple circuit configuration
2. The merging of the injector (*pnp*) and inverter (*npn*) transistors
3. The multiple collectors at the surface of the silicon

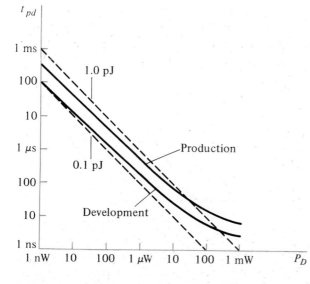

FIGURE 7.23
Power-delay product curves for production I²L and development I²L.

Furthermore, for an even larger packing density, a common *injector rail* can serve several *gate bars*, as is illustrated in the layout of a portion of a gate array in Fig. 7.24. In the array, metal lines running vertically in the figure are used to connect the desired collector regions. Second-layer metal, or polysilicon, running horizontally makes connection to the required base regions.

Using 5-μm design rules a packing density of 200 gates/mm^2 may be obtained for the I^{2}L circuit. Hence approximately 5000 gates may be contained within a chip area of 25 mm^2. This is 10 times larger than the 500 gates calculated for a TTL(LS) chip of the same area. Also, with a power-delay product of 1 pJ, an average propagation delay time of 20 ns is possible with only 50 μW of power dissipation per gate. This is equivalent to an average power dissipation of only 250 mW for the complete 5000-gate array. A further comparison of parameters for I^{2}L and TTL(LS) is given in Table 7.7.

7.5.2 Schottky I^{2}L

We have seen in Table 7.7 that with similar design rules the packing density of I^{2}L is about 10 times that of TTL(LS). However, the speed of I^{2}L is somewhat less than that of TTL(LS). In an effort to reduce the minimum propagation delay time but still have a small-area circuit which is useful in LSI, many variations of the basic I^{2}L circuit have been reported. One of these is *Schottky I^{2}L*.

The schematic of a basic Schottky I^{2}L gate is shown in Fig. 7.25. In this circuit the Schottky diodes are used solely to decrease the logic swing; thus, for any given injector current the propagation delay time is reduced, typically to one-half or one-third of that for a standard I^{2}L gate. The delay time is decreased, since the smaller voltage change results in a more rapid charge and discharge of the parasitic capacitances at the input and output of the gate.

For the Schottky-barrier diode to be effective a reasonably large value is desired for the barrier potential of the diode. Thus a low conductivity region

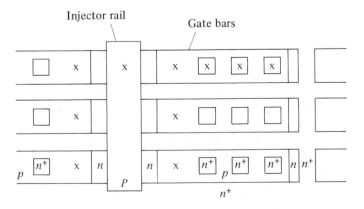

FIGURE 7.24
Layout of I^{2}L gate array.

TABLE 7.7
**Comparison of parameters for I²L and TTL(LS) made
with the same process technology and design rules (5 μm)**

	I²L	TTL(LS)
Packing density, gates/mm²	100–200	10–20
Power-delay product, pJ	1–2	20
Propagation delay time, ns	10–20 min	5–10 min
Supply volts	≈1.0 V	5 V
Logic swing	≈0.7 V	≈4 V
Current range per gate	1 nA to 1 mA	0.2 to 1 mA

is required at the collector of the *npn* transistor. By diffusion, it is difficult to control the impurity concentration of a lightly doped *n*-region into a more heavily doped *p*-region. Hence the collector region is made by ion implantation. The metal barrier is either platinum (Pt) or palladium (Pd). With Pt-Si (*n*-type), a typical forward voltage $V_{SBD(on)} = 0.4$ V.

With the input to the gate in the high state the inverting transistor still saturates, and for $V_{CE(sat)} = 0.1$ V the output voltage $V_{OL} = 0.5$ V. A typical voltage transfer characteristic for a Schottky I²L gate is shown in Fig. 7.26, where

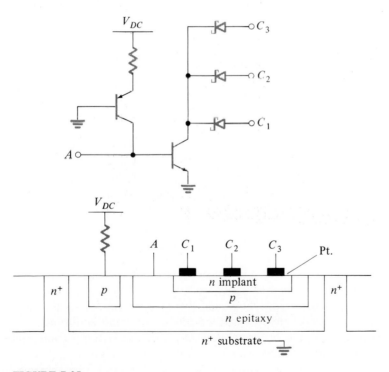

FIGURE 7.25
Basic Schottky I²L gate circuit.

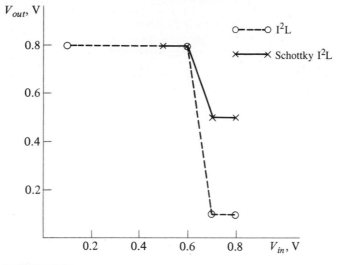

FIGURE 7.26
Voltage transfer characteristic for I^2L and Schottky I^2L.

$$V_{OH} = 0.8 \text{ V} \qquad V_{IH} = 0.7 \text{ V}$$
$$V_{OL} = 0.5 \text{ V} \qquad V_{IL} = 0.6 \text{ V}$$

Note the very small logic swing (LS = 0.3 V) and noise margins ($NM_H = NM_L = 0.1$ V).

It should be observed that the Schottky diodes also effectively isolate the outputs, one from another, at the collector of the *npn* transistor. Hence, separate collector regions are not required as they are in standard I^2L. This permits a closer spacing of the collector contacts and therefore an increased packing density for Schottky I^2L.

7.5.3   Schottky Transistor Logic

The power-delay product of Schottky I^2L could be further improved if the *npn* transistor were clamped to prevent saturation when on, as in other Schottky-clamped inverters that have previously been described. The clamp diode would be connected between the base and collector and could be made by extending the *npn* base metal to cover a portion of the collector region. However, this requires a careful choice of metals for the Schottky-barrier diodes, since a simple analysis of the VTC will show that the logic swing of such a gate is due to the difference between the on voltage of the clamp diode and the isolation diode. As a result, the fabrication process is more complex and the production yield is diminished.

However, such a circuit has been reported. Described as *Schottky transistor logic (STL)*, the circuit uses aluminum (Al) for the clamp diode ($V_{SBD(on)} = 0.5$ V) and titanium (Ti) for the isolation diode ($V_{SBD(on)} = 0.3$ V). Using 5-μm design rules, the power-delay product is given as 0.2 pJ and the minimum average propagation delay time is 2.5 ns. The packing density is estimated to be approximately 250 gates/mm^2.

7.6 INTEGRATED SCHOTTKY LOGIC

Another circuit that shows promise of high speed at a low operating power with a high packing density is *integrated Schottky logic (ISL)*. This circuit, shown in Fig. 7.27, is similar to Schottky transistor logic but with two major differences.

One difference is that in ISL the *npn* transistor is operated in the normal forward mode. That is, there is a topside emitter contact that must be connected to ground. As a result of the forward (or down) operation of the transistor, the charge stored in the base and collector regions is reduced. Compared to I^{2}L, there is a significant improvement in the transient response of the inverter. Like STL, multi-output logic is performed with multiple Schottky diodes at the surface of the *n*-type collector of the transistor.

The other difference is that heavy saturation of the *npn* transistor is avoided by a *pnp* clamp transistor. As shown in Fig. 7.27, when the base-collector junction of the *npn* becomes forward-biased, the emitter-base junction of the *pnp* is also forward-biased. The *pnp* operates in the forward active mode, since its collector-base junction is reverse-biased by $V_{CE(sat)}$ of the *npn* transistor.

From the cross-sectional view of Fig. 7.28a, the clamp transistor can be seen as a composite of two substrate *pnp* transistors that are merged with the *npn* transistor. The *p* substrate is the collector region common to a lateral *pnp* transistor and a vertical *pnp* transistor. The extra *p* implant that overlaps the *p*$^+$ isolation ring reduces the base width of the lateral *pnp* and reduces the charge stored in the base region of the clamp transistor.

Because the emitter contact is on top in ISL, the current source I_0 cannot be merged with the *npn* transistor. A simple resistor ($R \approx 5$ kΩ) which is best formed by ion implantation is used. Also, because of the need to route the ground connection in metal (rather than through the *n*$^+$ substrate as in I^{2}L), the packing

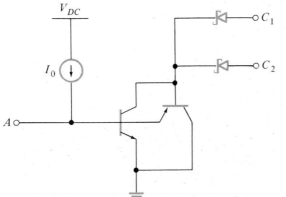

FIGURE 7.27
Basic ISL gate circuit.

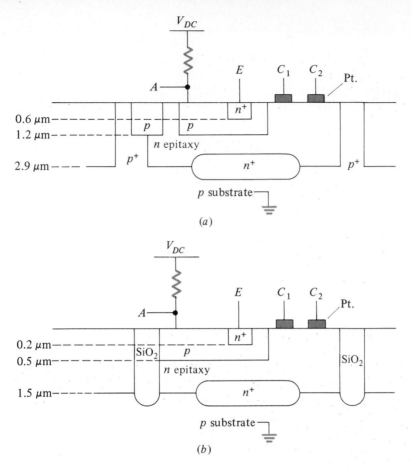

FIGURE 7.28
Cross section of ISL circuit: (*a*) Junction-isolated; (*b*) oxide-isolated.

density of ISL is lower than that for I^{2}L, but the average propagation delay time of ISL is about a factor of 5 better than for I^{2}L.

Using 5-μm design rules, the following data have been reported for the junction-isolated ISL circuit illustrated in Fig. 7.28*a*. With an injector current of 200 μA, the minimum propagation delay time was 2.7 ns, yielding a power-delay product of 0.5 pJ/V.

A later development in ISL is the oxide-isolated version shown in Fig. 7.28*b*. Due to the thin epitaxial layer and shallow implants, the base width of the vertical *pnp* clamp transistor is only approximately 1 μm thick, and a lateral *pnp* is no longer necessary. Also, by using oxide isolation, the sidewall junction capacitance to the substrate is eliminated. As a result, with 5-μm design rules the following data have been reported. With an injector current of 65 μA, the minimum propagation delay time was 2.3 ns, resulting in a power-delay product of 0.1 pJ/V.

In summary, I²L has the highest packing density, but it is relatively slow. Both STL and ISL are about 5 times faster than I²L, but the packing density is only about two-thirds that of I²L.

7.7 INTERFACING

Generally a digital system is constructed using only a single logic family, such as TTL, ECL, CMOS, or NMOS. However, sometimes this is impossible or not advantageous. Sometimes the desired components such as gates, registers, memories, and microprocessors are not available in a single logic family. Also, it may be only the "front end" logic that requires the high speed of ECL; for the slower logic the lower-power TTL(LS), or even CMOS, may be adequate. Then it becomes necessary to interconnect components from different logic families. In this section we consider this *interconnection*, or *interfacing*, problem.

At present (1987), TTL, in particular TTL(LS), is still the most widely used form of digital IC, especially in the area of SSI and MSI. Hence we will primarily be concerned with interfacing between TTL(LS) and the other logic families.

It is required that the input characteristics of the interface be similar to those of the driving gate, while the output characteristics of the interface match the output characteristics of the receiver gate. Then it will appear as if the driving gate is driving a gate of the same logic family as the driver, and the receiver gate is being driven from a gate of the same logic family as the receiver.

INTERFACING ECL AND TTL. When going from ECL to TTL it is necessary to voltage-level-shift the output from the ECL circuit to be compatible with the input of the TTL circuit. The loading on the ECL output should be that typical of an ECL gate, but the driving capability of the interface should be similar to that of a TTL gate. A circuit that has these features is shown in Fig. 7.29. This is a commercial ECL-to-TTL *translator*.

The input of the translator is a typical 2-input ECL circuit, consisting of the current switch transistors Q_1 (or Q_3) and Q_2. From the voltage divider circuit at the base of Q_5, the reference voltage V_R is found to be -1.3 V. With both inputs A and B low ($= -1.7$ V), the current ($I_0 = 6.7$ mA) from the current-source transistor Q_4 is switched to transistor Q_2, causing a voltage drop across R_7. This results in the base of Q_9 being at about 0 V. That is, Q_9 is off. Since both Q_1 and Q_3 are off, with no load at the output of the translator the voltage at the base of Q_{10} is about 5 V. Consequently, the output voltage of the translator is at a high level, and $V_{OH} = 3.6$ V. Note that the output circuit of Q_9, Q_{10}, and Q_{11} is very similar to the output circuit of the TTL(S) circuit in Fig. 7.8.

With either A or B in the high state ($= -0.9$ V), the current I_0 is switched to either Q_1 or Q_3. This causes a voltage drop across R_3, which results in the base of Q_{10} being at about 0 V. Now Q_{10} is off, as is Q_{11}. With Q_2 also off, the emitter-follower Q_8 is now able to feed current to the base of Q_9. Consequently Q_9 can be hard on, but being a Schottky-barrier transistor, it does not saturate.

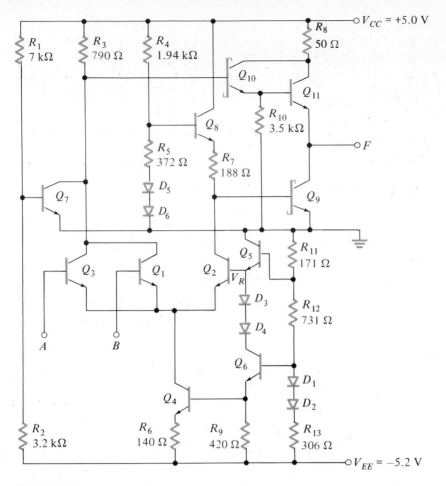

FIGURE 7.29
ECL-to-TTL translator.

The result is the output of the translator is at a low level, and $V_{OL} = 0.5$ V with $I_{OL} \leq 20$ mA.

Shown in Fig. 7.30 is a TTL-to-ECL translator. The input circuit is a 2-input diode AND circuit formed by Schottky diodes D_1 and D_2. Voltage level shifting is accomplished by the current-source transistor Q_2 and resistor R_3, as well as the base-emitter junction of Q_1 and diodes D_3 and D_4. The transistors Q_4 and Q_5 form the current switch with current-source transistor Q_6. The reference voltage V_R is nominally -1.5 V. Complementary ECL outputs are obtained from the emitter-followers Q_9 and Q_{10}, in a similar manner as in the ECL 10K OR/NOR gate of Fig. 7.16.

Both these translator circuits combine the nonsaturating current switch of ECL and the Schottky diodes of TTL(S) to yield a fast average propagation delay time of, typically, 5 ns.

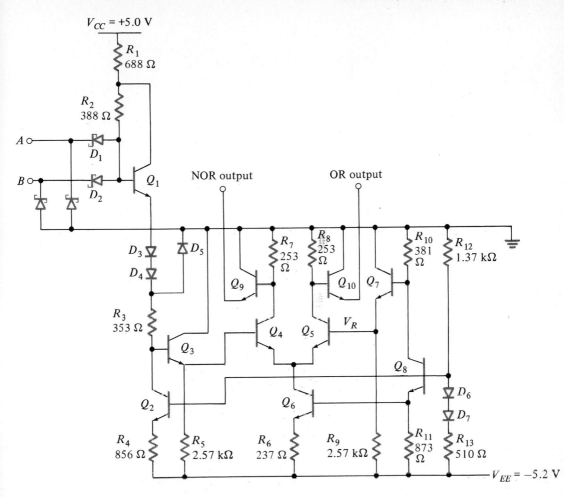

FIGURE 7.30
TTL-to-ECL translator.

INTERFACING I²L AND TTL. The logic family with which I²L is most generally interfaced is TTL(LS). Hence we initially list the particular characteristics of interest for the TTL(LS) circuit. They are, from Table 7.4:

min V_{OH}/max V_{OL}: 2.7 V/0.5 V
min I_{OH}/min I_{OL}: −0.4 mA/8 mA
max I_{IH}/max I_{IL}: 20 μA/−0.4 mA

Interfacing I²L and TTL(LS) is very simple, as is illustrated in Fig. 7.31a. A pull-up resistor R_A is connected from V_{CC} of the LS circuit, or some other 5-V supply, to the collector of the output transistor Q_A in the I²L driver. Resistor R_A can be on or off the chip. The value of R_A is made as large as possible,

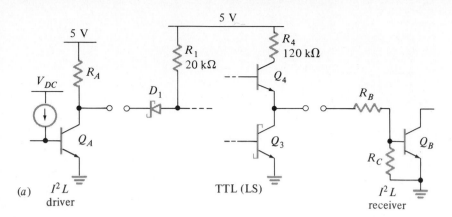

(a) I^2L driver   TTL (LS) I^2L receiver

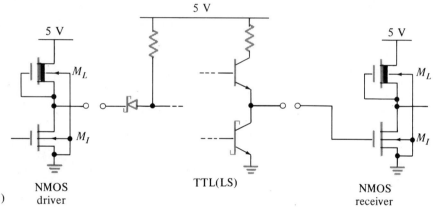

(b)   NMOS driver   TTL(LS)   NMOS receiver

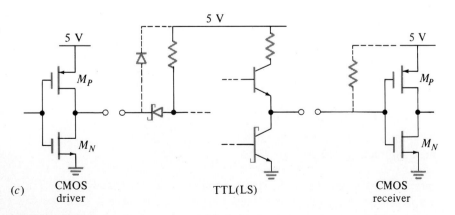

(c)   CMOS driver   TTL(LS)   CMOS receiver

FIGURE 7.31
Interfacing.

provided that the static and dynamic characteristics of the interface are met. The static requirement is that in the high state, V_{in} to the LS circuit must be 2.7 V with $I_{IH} \leq 20$ μA. For the dynamic requirement we consider the transition time t_{LH}, which is primarily due to resistor R_A and the output load capacitance C_L. In the low state, the output transistor Q_A must be able to sink the receiver input current, but this is generally no problem since for TTL(LS), $I_{IL} \leq 0.4$ mA with $V_{in} = 0.4$ V.

Driving an I²L gate from TTL(LS) also provides no problem. Generally all that is required is a voltage divider, as is shown in Fig. 7.31a. With the LS output in the high state (≥ 2.7 V), the Thévenin equivalent resistance of R_B and R_C must provide adequate drive to the I²L receiver input transistor Q_B. With the LS output in the low state, the voltage divider ensures that Q_B is off. The values of R_B and R_C are as large as possible considering the speed of operation of the circuit.

INTERFACING NMOS AND TTL. Interfacing between an NMOS depletion load inverter and TTL(LS) is illustrated in Fig. 7.31b. Again, provided that $V_{DD} = 5$ V for the NMOS circuit, there is little problem.

With the output of the NMOS driver in the high state, the transistor M_I is off but the W/L of transistor M_L must be sufficient to source I_{IH} to the LS circuit with V_{out} of the driver $\geq V_{OH}$. But with output of the driver in the low state, transistor M_I must sink I_{IL} from the LS circuit, as well as the current from M_L, with $V_{out} \leq V_{OL}$.

Driving an NMOS receiver from TTL(LS) is simply done by a direct connection, as is shown in Fig. 7.31b. Since the NMOS input provides only a capacitive load, there are no dc currents to or from the TTL circuit. Consequently, in the high state the output will be very close to V_{CC}, while in the low state the output will be very close to ground.

INTERFACING CMOS AND TTL. Interfacing CMOS and TTL(LS) is the simplest of all, especially if V_{DD} for the CMOS is 5 V. As shown in Fig. 7.31c, either coming from or going to CMOS, a simple direct connection is used.

For the CMOS driver, the W/L of the p-channel transistor should be sufficient to source I_{IH} to the LS circuit with $V_{in} = 2.7$ V, while the W/L of the n-channel transistor should permit sinking I_{IL} with $V_{in} = 0.5$ V. Normally, this will require a much larger area for M_N than for M_P. As a result, for the CMOS driver, the propagation delay time t_{PHL} will be much shorter than t_{PLH}. For more equal delay times the W/L of M_P would have to be considerably increased.

In connecting TTL(LS) to CMOS, a pull-up resistor is advisable to ensure that M_P is off with $V_{out} = V_{OH}$. With the pull-up resistor (≈ 10 kΩ) the only static load, there is generally no problem in M_N being off with $V_{out} = V_{OL}$.

With V_{DD} for the CMOS > 5 V, a clamp diode is necessary at the LS receiver input to prevent the input going much above 5 V. In addition, the pull-up resistor at the CMOS receiver input will require special attention to the output of the LS circuit. An open-collector transistor output may be needed, one with a higher collector voltage rating than the normal 7 V.

Exercise 7.7. Design a CMOS inverter for driving 10 TTL(LS) loads. Calculate the W/L for the p-channel and n-channel transistors and sketch your circuit. The voltage levels into the inverter are 0 and 5 V. $V_{DD} = 5$ V. CMOS data:

$$V_{TP} = -2 \text{ V} \qquad k'_p = 10 \ \mu\text{A/V}^2$$
$$V_{TN} = +2 \text{ V} \qquad k'_n = 20 \ \mu\text{A/V}^2$$

TTL(LS) data: See Table 7.4.

7.8 SUMMARY

In this chapter the development of bipolar gate circuits is described and their principal static and dynamic characteristics are presented. Specifically described in detail are:

- Resistor-transistor logic (RTL)
- Diode-transistor logic (DTL)
- Transistor-transistor logic (TTL), including:
  - Standard TTL
  - Schottky-clamped (S)
  - Low-power Schottky (LS)
  - Advanced Schottky (AS)
  - Advanced low-power Schottky (ALS)
- Emitter-coupled logic (ECL), including the 10K and 100K Series
- Integrated injection logic (I^2L)
- Schottky-transistor logic (STL)
- Integrated Schottky logic (ISL)

In addition, the problem of interconnecting or interfacing between the circuits of the various logic families has been discussed and particular solutions have been presented.

REFERENCES

1. V. H. Grinich and H. G. Jackson, *Introduction to Integrated Circuits*, McGraw-Hill, New York, 1975.
2. D. J. Hamilton and W. G. Howard, *Basic Integrated Circuit Engineering*, McGraw-Hill, New York, 1975.
3. H. Taub and D. Schilling, *Digital Integrated Electronics*, McGraw-Hill, New York, 1977.
4. H. H. Muller, W. K. Owens, and P. W. J. Verhofstadt, "Fully Compensated Emitter-Coupled Logic: Eliminating the Drawbacks of Conventional ECL," *IEEE Journal of Solid-State Circuits*, vol. SC-8, October 1973, pp. 362–367.
5. J. E. Smith (ed.), *Integrated Injection Logic*, IEEE Press, New York, 1980.
6. J. Lohstroh, "Devices and Circuits for Bipolar (V)LSI," *Proceedings of the IEEE*, vol. 69, July 1981, pp. 812–826.

DEMONSTRATIONS

D7.1   TTL Gates: Voltage Transfer Characteristic

The objective of this experiment is to examine the voltage transfer characteristic of a TTL(LS) gate, as represented by the 74LS00—a quad 2-input NAND gate.

1. The voltage transfer characteristic of a 74LS00 may be directly observed with the circuit shown in Fig. D7.1. (Calibrate the oscilloscope for equal vertical and horizontal scales.)

      Carefully sketch the VTC, labeling all asymptotic voltages and breakpoints. Do this for:

   (a) A fan-out of 1, as in Fig D7.1a.
   (b) A fan-out of 10, simulated by the load circuit of Fig. D7.1b.

2. What are the noise margins NM_H and NM_L for the two cases of fan-out?

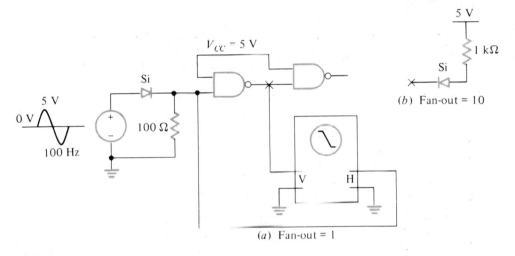

(a) Fan-out = 1

(b) Fan-out = 10

FIGURE D7.1

D7.2   TTL Gates: Input/Output Characteristics

In this experiment we will examine the V-I characteristics at the input and output terminals of the 74LS00.

1. (a) Connect the circuit shown in Fig. D7.2a, and plot I_{in} versus V_{in} as V_{in} is varied from -0.5 to $+5.0$ V.
   (b) Connect the circuit shown in Fig. D7.2b, and plot V_{OL} versus I_{OL} as V_{CS} is varied from -5.0 to $+15$ V.

2. Comment on the reason for the slope of these curves, especially as the input and output voltages go negative.

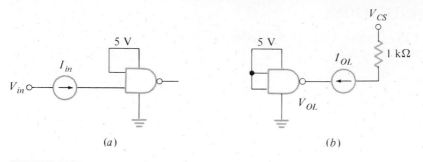

(a) (b)

FIGURE D7.2

D7.3 TTL Gates: Propagation Delay Time

In this experiment we will measure the propagation delay time of the 74LS00.

1. (a) Use the ring oscillator circuit shown in Fig. D7.3 to measure the average propagation delay time. For further comments on the ring oscillator circuit, see Demonstration D3.2.
   (b) With the connection as in Fig. D7.3 measure the V_{CC} supply current for the complete package and determine the power-delay product per gate for the 74LS00.
   (c) The effective capacitance for a TTL(LS) input is 5 pF. Simulate a fan-out of 10 by adding 50 pF at each node in Fig. D7.3 and repeat parts (a) and (b).

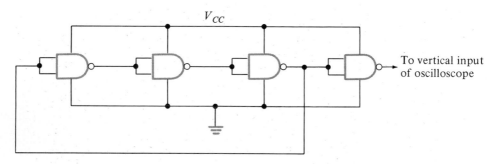

FIGURE D7.3

PROBLEMS

For all problems use the following data unless otherwise specified.

$$I_{ES} = 10^{-16} \text{ A} \qquad I_{CS} = 4 \times 10^{-16} \text{ A}$$
$$\alpha_F = 0.98 \qquad \alpha_R = 0.25$$
$$C_{je0} = 1 \text{ pF} \qquad C_{jc0} = 0.5 \text{ pF}$$
$$\phi_e = 0.9 \text{ V} \qquad \phi_c = 0.8 \text{ V}$$
$$m_e = 0.5 \qquad m_c = 0.33$$
$$\tau_F = 0.2 \text{ ns} \qquad \tau_R = 10 \text{ ns}$$

$$C_{js0} = 3.0 \text{ pF} \qquad r_b = 50 \ \Omega$$
$$\phi_s = 0.7 \text{ V} \qquad r_c = 20 \ \Omega$$
$$m_s = 0.33 \qquad r_e = 1 \ \Omega$$

Assume a base-collector shorted transistor for all pn junction diodes. For the Schottky-barrier diodes:

$$I_0 = 10^{-12} \text{ A} \qquad C_{j0} = 0.2 \text{ pF} \qquad \phi_0 = 0.7 \text{ V} \qquad m = 0.5$$

P7.1. A low-power version of RTL is similar to that shown in Fig. 7.1, but with $R_B = 1.5 \text{ k}\Omega$ and $R_C = 3.6 \text{ k}\Omega$. Use the standard transistor data and calculate:
(a) The noise margins NM_H and NM_L with a fan-out of 5.
(b) The fan-out N with $NM_H = NM_L$.

P7.2. For the RTL circuit shown in Fig. 7.1 assume

$$0.6 < V_{BE(on)} < 0.8 \qquad 20 < \beta_F < 200$$

$$0.7 < V_{BE(sat)} < 0.9 \qquad R's \pm 20\%$$

$$0 < V_{CE(sat)} < 0.2 \qquad V_{CC} \pm 10\%$$

Calculate:
(a) Worst-case noise margins.
(b) Worst-case fan-out.

P7.3. For the modified DTL circuit shown in Fig. 7.3:
(a) Solve for the noise margins.
(b) Find the fan-out.

P7.4. Repeat P7.2 for the modified DTL circuit shown in Fig. 7.3.

P7.5. Shown in Fig. P7.5 is a simple DTL NAND gate used in an LSI circuit. Note the reduced value for V_{CC}. Use the standard transistor data, but for the Schottky-barrier input diodes use $I_0 = 10^{-14}$ A.
(a) Sketch the VTC showing values of all the breakpoints.
(b) Calculate the fan-out.

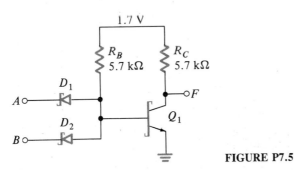

FIGURE P7.5

P7.6. Compare the basic DTL and TTL gates shown in Fig. 7.4. $R_1 = 4 \text{ k}\Omega$, $R_2 = 5 \text{ k}\Omega$, $R_3 = 1 \text{ k}\Omega$, and $V_{CC} = 5$ V. For each gate calculate:
(a) The noise margins.
(b) The fan-out. (For the TTL gate let $NM_H = NM_L$.)

P7.7. Repeat P7.6:

(a) In terms of the average propagation delay time t_p and power dissipation P_D. Hence determine the power-delay product (PDP).

(b) Verify with SPICE.

P7.8. A low-power version of TTL is similar to that shown in Fig. 7.5, but with $R_1 = 40$ kΩ, $R_2 = 12$ kΩ, $R_3 = 20$ kΩ, and $R_4 = 500$ Ω.

(a) Sketch the VTC showing values of all the breakpoints.

(b) Determine the noise margins.

P7.9. For the low-power TTL circuit of P7.8:

(a) Calculate the fan-out for a low- and high-output level ($NM_H = NM_L$).

(b) Calculate the V_{CC} supply currents I_{CCH} and I_{CCL} and determine the average power dissipation.

P7.10. For the 54S/74S gate shown in Fig. 7.8:

(a) Sketch the VTC showing values of all the breakpoints.

(b) Determine the noise margins.

P7.11. For the 54S/74S gate shown in Fig. 7.8:

(a) Calculate the fan-out.

(b) Determine the average power dissipation.

P7.12. Shown in Fig. P7.12 is an "on-chip" TTL inverter circuit. Use the standard transistor data but with $\beta_F = 20$ and $\beta_R = 0.2$.

(a) Sketch the VTC showing values of all the breakpoints.

(b) Compute the fan-out with $NM_H = NM_L$.

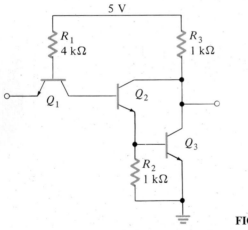

FIGURE P7.12

P7.13. Use SPICE to compare the propagation delay times t_{PLH} and t_{PHL} of the 54LS/74LS gate shown in Fig. 7.9.

(a) A fan-out $N = 1$ and load capacitance $C_L = 15$ pF.

(b) A fan-out $N = 10$ and load capacitance $C_L = 50$ pF.

P7.14. Shown in Fig. P7.14 is an early ECL circuit.

(a) Calculate V_{OH} and V_{OL} at X.

(b) Find the value of R_1 so that $V_Y = \overline{V_X}$.

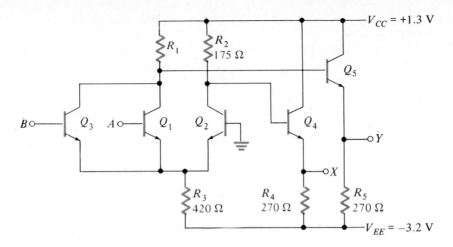

FIGURE P7.14

P7.15. For the ECL circuit shown in Fig. P7.15, $V_{CC} = 0$ V, $V_{EE} = -4.5$ V, $V_{BE(on)} = 0.8$ V, $V_R = -1.2$ V, logic swing $= 0.8$ V symmetrical about the reference voltage, and $I_{E(max)} = 5$ mA for all transistors. Assume that the logic voltage levels V_{in} and V_{out} are compatible and neglect base currents.
(a) Calculate the value of resistors R_3 and R_4.
(b) Determine the value of resistors R_1, R_2, and R_E.

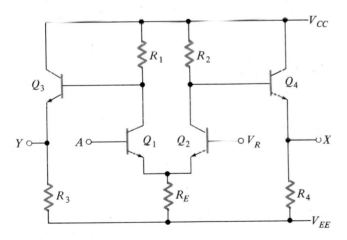

FIGURE P7.15

P7.16. From the results of P7.15:
(a) Sketch the VTC showing the values of all the breakpoints.
(b) Compute the noise margins.

P7.17. Use SPICE to determine the propagation delay times t_{PHL} and t_{PLH} of the ECL gate shown in Fig. 7.16. Assume the load is due to a parallel combination ($R_L = 500$ Ω,

$C_L = 10$ pF) connected from the OR output to V_{EE}. Also determine the power-delay product for this gate.

P7.18. Given the four variables A, B, C, and D, sketch an I^2L circuit that performs the AND-OR-INVERT function

$$Y = \overline{AB + CD}$$

P7.19. For the I^2L inverter shown in Fig. 7.22, $I_0 = 10~\mu A$. Use the standard transistor data, but $\alpha_R = 0.8$.
   (a) Assume only one collector C_1 is connected to a similar inverter and calculate V_{OH} and V_{OL} at C_1.
   (b) Now assume all three collectors are each connected to one load and compute V_{OH} and V_{OL} at C_1.

P7.20. Use SPICE to compare the VTC and propagation delay times of a basic STL and ISL inverter. Assume the "device under test" is being driven from, and is driving, one similar inverter. Let $I_0 = 10~\mu A$.
   STL data: Use the standard transistor data, but $\alpha_R = 0.8$.
      For the clamp diode,

$$I_0 = 10^{-14}~A \qquad C_{j0} = 0.2~pF \qquad \phi_0 = 0.8~V \qquad m = 0.5.$$

   For the isolation diode,

$$I_0 = 10^{-11}~A \qquad C_{j0} = 0.2~pF \qquad \phi_0 = 0.6~V \qquad m = 0.5.$$

   ISL data: Use standard transistor data for both *npn* and *pnp* transistor.
      For the isolation diode the data are as for STL.

P7.21. Sometimes it is necessary to interconnect the various TTL families.
   (a) Determine the maximum number of TTL(S) gates that a TTL(LS) circuit can drive.
   (b) Likewise, determine how many TTL(LS) gates can be applied to the output of a TTL(S) circuit.

P7.22. Design an NMOS E-D inverter to drive two TTL(LS) loads. Calculate the W/L for each of the NMOS devices and sketch your circuit. Voltage levels into the inverter are 5 and 0.5 V. $V_{DD} = 5$ V.

NMOS data: See device parameters at beginning of problems in Chap. 3.
TTL(LS) data: See Table 7.4.

CHAPTER 8

REGENERATIVE LOGIC CIRCUITS

8.0 INTRODUCTION

In all the logic circuits that have so far been described, at any point in time the output has been directly related to the input by some logic combination, except for some short propagation delay. Hence, as a class these circuits are known as *combinational logic* circuits. Common examples of this type of circuit are the simple digital gates NOR, NAND, etc., that have been described in the previous chapters. Combinational circuits, which lack intentional connections between outputs and inputs, are also known as *nonregenerative circuits*.

There is another class of circuits, known as *sequential logic circuits*, in which not only immediately previous input data affect the outputs. Outputs also are dependent on preceding values of input data. These circuits also find ready application in digital systems. Examples are clock counters and data registers as well as clock oscillators and time delay circuits. A characteristic of sequential circuits is that one or more output nodes are intentionally connected back to inputs to give *positive feedback*, or *regeneration*.

Basic to sequential circuits is the *bistable* circuit. Prominent examples of the bistable circuit found in digital ICs are:

1. Latches
2. Flip-flops

Since these circuits can become very complex, they will initially be described with logic diagrams. Following this, description will be given of these circuits implemented with:

1. Bipolar junction transistors (BJT)
2. MOS field-effect transistors (NMOS)
3. Complementary MOS (CMOS)

Another family of regenerative circuits that are especially useful in the generation of voltages that are a function of time are *multivibrator circuits*. Multivibrator circuits include:

1. The bistable circuit
2. The monostable, or one-shot, circuit
3. The astable, or oscillator, circuit

This chapter will include an analysis and description of the design of these circuits using the above IC technologies.

8.1   BASIC BISTABLE CIRCUIT

Shown in Fig. 8.1a are two logic inverters connected in cascade, along with a voltage transfer characteristic that is typical of such a circuit. Shown plotted is the VTC for Q_1, that is, V_{O1} versus V_{I1}, and since $V_{I2} = V_{O1}$ we also show V_{O2}

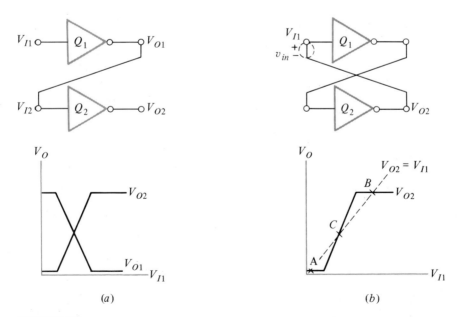

(a)   (b)

FIGURE 8.1
(a) Series of connection of two logic inverters with voltage characteristics. (b) Bistable circuit with voltage transfer characteristic.

versus V_{I1}. Notice that V_{O2} is in phase with V_{I1}, as should be expected from a cascade of two inverters. Hence V_{O2} may be connected to V_{I1}, with no change in the transfer characteristic. This connection has been made in Fig. 8.1b; also repeated is the VTC, V_{O2} versus V_{I1}, along with the linear constraint $V_{O2} = V_{I1}$. Now note that there are only three possible operating points for the circuit. It will be shown that A and B are *stable operating points*, and C is an *unstable point*.

The low voltage level at point A results in Q_1 being off, and a consequent high voltage level at V_{O1} causes Q_2 to be on, yielding a low level at V_{O2}. The high voltage level at point B causes Q_1 to be on and Q_2 to be off, resulting in A and B being stable points of operation. Alternatively, it may be seen that the voltage gain of the circuit is given by the slope of the characteristic, that is, dv_{o2}/dv_{i1}. At the stable points the slope, and hence the gain, is less than unity.

However at point C the slope is positive and greater than unity. For this case both Q_1 and Q_2 must be on. Point C is unstable, since any small voltage v_{in} (even noise) introduced into the circuit at, say, V_{I1} will be amplified and *regenerated* around the circuit loop, causing the operating point to move to one of the stable points. With V_{I1} made positive with respect to V_{O2}, as shown dashed in Fig. 8.1b, stability is reached at point B. With the polarity of the noise source reversed, stability is realized at point A.

Hence the cross-coupling of two inverter circuits results in a *bistable* circuit, that is, a circuit with two stable states. For stability it is necessary that the voltage gain of the complete circuit be less than 1. This is generally established in the circuit by having one of the inverters off.

For the bistable circuit to change state it is necessary that the voltage gain of the circuit be made greater than 1. This can be accomplished by introducing a *trigger voltage pulse* at v_{in}. With, say, Q_1 off, a positive trigger pulse is required at v_{in}. The voltage amplitude of the pulse should be sufficient to raise the voltage gain of the circuit to a little greater than unity; then due to the cross-coupled circuit and resulting positive feedback, the trigger pulse will be regenerated and a change of stable state will be effected. The width of the trigger pulse need be only a little greater than the total propagation delay time around the circuit loop, that is, twice the average propagation delay time of the logic inverters.

In summary, a bistable circuit is a circuit with two stable states. The circuit may be triggered from one stable state to the other; but in the absence of a trigger the circuit remains in that stable state indefinitely, provided, of course, that power remains applied to the circuit. Another common name for the bistable circuit is *flip-flop*. However, in digital ICs the name flip-flop is reserved for a more complex circuit than a simple cross-coupling of two inverters.

8.2   SR LATCH

The simplest form of the bistable circuit is the *latch*. The bistable "latches" on (remembers) the trigger pulse. Hence, we can say that the trigger pulse is stored in the latch. This introduces the concept of the latch as a *memory* circuit—an idea that will be further developed in the next chapter, on semiconductor memories.

8.2.1   SR Latch with NOR Gates

In Fig. 8.2a the SR latch is implemented with two 2-input NOR gates. One of the inputs of the NOR gate is used to cross-couple to the output of the other NOR gate, while the second input provides a means of triggering the latch from one stable state to the other. The logic symbol for the latch is shown in Fig. 8.2b. The two outputs Q and $\overline{Q}$ are complementary, and by definition the latch is in the *set state* with a logic 1 at the Q output and logic 0 at the $\overline{Q}$ output. Conversely, with a logic 0 at the Q output and logic 1 at the $\overline{Q}$ output, the latch is in the *reset state*.

From Fig. 8.2a it can be seen that a logic 1[*] at the set (S) input will cause a logic 0 at the $\overline{Q}$ output and, with the reset (R) input also a logic 0, this will result in a logic 1 at the Q output. That is, the latch is now set, and the S input can safely be returned to the 0 state. Alternatively, a logic 1 at the R input will cause a logic 0 at the Q output and with the S input now at a logic 0 the result is a logic 1 at the $\overline{Q}$ output. Since the latch responds to high voltage levels at the inputs, they are referred to as *active high inputs*.

These results are also given in the *characteristic table* for the latch, as

---

[*]Following general practice, in this text positive logic is assumed unless stated otherwise. That is, logic 1 = high (H) voltage level and logic 0 = low (L) voltage level.

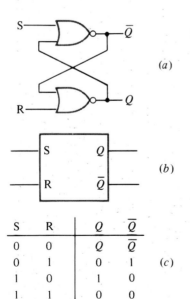

(a)

(b)

S	R	Q	$\overline{Q}$	
0	0	Q	$\overline{Q}$	
0	1	0	1	(c)
1	0	1	0	
1	1	0	0	

FIGURE 8.2
(a) Latch made with NOR gates. (b) Block diagram indicating active high input. (c) Characteristic table.

shown in Fig. 8.2c. The characteristic table is the truth table for the latch that lists the output states for all the possible combinations of the input states. The first entry in the table shows that with low voltage levels at both S and R inputs there is no change in the Q and $\overline{Q}$ outputs. With high levels at both S and R inputs the initial result is that both Q and $\overline{Q}$ are low. Since the outputs are not complementary, a logic 1 at both S and R inputs is considered to be a *forbidden*, or *not-allowed*, condition. Actually what happens is that when the input trigger pulses are returned to their quiescent levels, the state of the latch is due to whichever input is last to go low.

8.2.2 SR Latch with NAND Gates

The SR latch can also be designed with NAND gates, as in Fig. 8.3a. Similar to the NOR circuit, the reset condition of the latch is with a logic 0 at the Q output and logic 1 at the $\overline{Q}$ output, but with the NAND circuit the S and R inputs are normally at a logic 1. The latch is set with a logic 0 at the S input and reset with a logic 0 at the R input. Hence this latch responds to *active low inputs*, as is indicated by the small circles at the S and R inputs of the logic symbol shown in Fig. 8.3b. It is also indicated in the characteristic table of Fig. 8.3c. For the NAND configuration of the SR latch, the not-allowed condition is with both S and R inputs in the logic 0 state—since this leads to both Q and $\overline{Q}$ outputs being in the logic 1 state.

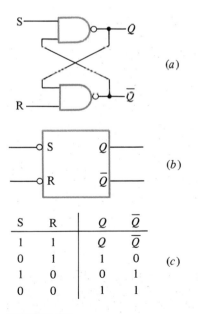

(a)

(b)

S	R	Q	$\overline{Q}$	
1	1	Q	$\overline{Q}$	
0	1	1	0	(c)
1	0	0	1	
0	0	1	1	

FIGURE 8.3
(a) Latch made with NAND gates. (b) Block diagram indicating active low input (c) Characteristic table.

8.3   JK FLIP-FLOP

By the addition of two feedback lines the ambiguity at the output, because of both S and R inputs being activated at the same time, can be overcome. The device is then known as a *JK flip-flop*. An all-NAND version of this circuit is shown in Fig. 8.4a. An important addition is a *clock* (CK) input, provided so that the change in the output logic states of the flip-flop can be synchronized with a system clock. Hence the J and K inputs are termed the *synchronous inputs*; the J is the *clocked-set* and the K the *clocked-reset*. Note from the logic symbol for this circuit given in Fig. 8.4b that all three inputs are activated by high voltage levels.

The JK flip-flop is set and reset in a similar manner to the SR latch, except for the synchronizing with CK by way of the two NAND gates at the input of the flip-flop. However, with both J and K inputs high, when CK goes high, entry into the flip-flop is only into the side that causes the flip-flop to change state, due to the J and K feedback lines. That is, with the flip-flop, say, reset, Q is low and $\overline{Q}$ is high, and entry is only by way of the J input, causing the flip-flop to set. The Q output is then high and $\overline{Q}$ is low. This now enables the K input; and if CK is still high, the flip-flop will again change state. Hence for this version of the JK flip-flop there is a very definite restriction on the pulse width of CK. It must be less than the propagation delay time of the flip-flop. This undesirable limitation can be eliminated with the master-slave principle, to be described in the next section.

The characteristic table of the JK flip-flop is given in Fig. 8.4c. Since the two outputs are always complementary, only the state of the output following the nth CK is given, indicated as Q_{n+}. With logic 0s at both J and K inputs there is

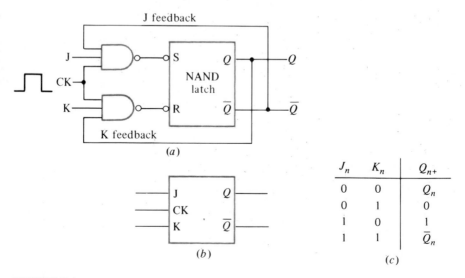

J_n	K_n	Q_{n+}
0	0	Q_n
0	1	0
1	0	1
1	1	$\overline{Q}_n$

(c)

FIGURE 8.4
JK flip-flop. (a) Logic diagram. (b) Logic symbol. (c) Characteristic table.

no change in the state of the flip-flop after clocking, indicated as Q_n. But with logic 1s at both J and K inputs the flip-flop changes state when clocked, indicated by $\overline{Q}_n$. A characteristic of all JK flip-flops is that with both J and K at a high level the unit will *toggle* when clocked, that is, change state irrespective of its original state. Operated exclusively in this latter mode, this circuit is then termed a *T flip-flop*.

8.3.1   JK Master-Slave Flip-Flop

As shown in the logic diagram of Fig. 8.5a, the JK *master-slave* flip-flop is simply a cascade of two JK flip-flops, with the master driving the slave. An important point to notice is that the master is activated by CK, but the slave by an inverted form of CK, namely, $\overline{CK}$.

   The operation of the master-slave principle is best explained with the aid of the timing diagram of the clock pulse shown in Fig. 8.5b. As CK rises, $\overline{CK}$ is going down, so that at time t_1, $\overline{CK}$ has fallen sufficiently to disable the input NAND gates of the slave. This isolates the slave from the master and freezes the state of the slave latch. At time t_2 CK has risen to enable the input NAND gates of the master. Hence depending on the state of the J and K feedback lines, the state of the J or K input can be entered into the master. Now on the back, or falling, edge of the clock pulse, CK goes down; and at time t_3, the input NAND gates to the master are disabled, freezing the state of the master latch. Finally at

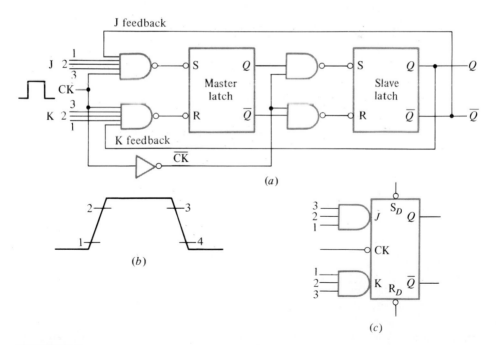

(a)

(b)

(c)

FIGURE 8.5
JK master-slave flip-flop. (*a*) Logic diagram. (*b*) Clock pulse waveform. (*c*) Logic symbol.

time t_4, the NAND gates to the slave are enabled and the state of the master is transferred to the slave. The outputs Q and $\overline{Q}$ are obtained from the slave latch. Due to this master-slave principle there is no limit to how wide CK can be, since when the output changes state at time t_4, entry into the master has been disabled. Only on the next clock pulse will the master input gates be enabled, and then not until a time corresponding to time t_2. Note that there is a minimum limit to the width of CK. It must be wider than the propagation delay time through the master flip-flop.

The logic symbol for a JK master-slave flip-flop is shown in Fig. 8.5c. As illustrated here, it is common for the flip-flop to have more than one J and K input. In this example there are three J inputs, all of which must form an AND for the flip-flop to be set by CK. Similarly, the three K inputs cause the flip-flop to be reset. The small circle on the CK line indicates that the flip-flop outputs Q and $\overline{Q}$ change state following the negative, or falling, edge of CK. Also included in the symbol (but not in the logic diagram) are the direct-set (S_D) and direct-reset (R_D) inputs. These are the *direct*, or *asynchronous*, inputs that operate independently of CK. As shown, the flip-flop is preset, that is, both master and slave latches are directly set following a negative transition at the S_D input. The flip-flop is directly reset with a similar transition at the R_D input.

A problem with the master-slave flip-flop is that it is subject to what is known as *ones catching*. That is, with CK high, and supposing the slave to be in the reset state, the J input gate is enabled. Thus any "spike" or "glitch" on the J input line will cause the master latch to be set. It is impossible to reset this latch, since entry into the K input gates is disabled by the K feedback line. Thus we say the J input has caught a 1 that will subsequently be transferred to the slave when CK goes down. One way to prevent ones catching is to keep the duration of the l state for CK as short as possible. Another way is to make use of JK *edge-triggered* flip-flops.

8.3.2   JK Edge-Triggered Flip-Flop

The logic diagram of a JK edge-triggered flip-flop is shown in Fig. 8.6a. With CK at a high level, entry into the input NAND gates is controlled by the JK feedback lines in a similar manner as for the master-slave flip-flop. But entry into the NAND latch is prevented until CK goes low. When CK does go low the input NAND gates are disabled; but provided that the transition time of the clock is not too long, a narrow negative-going pulse, whose width is approximately that of the gate delay time, will appear at either the S or R input dependent on the JK logic at the input of the flip-flop. The output of the flip-flop changes as the CK goes low, with the output state due to the state of the J and/or K inputs just prior to CK going low. This *set-up time* is generally less than 20 ns. The circuit is forgiving of any spikes or glitches on the J or K lines prior to this time. Some flip-flop designs require the state of the J and K input lines to remain stable for a time after clocking. This is known as the *hold time*. Usually the minimum hold time is 0 ns.

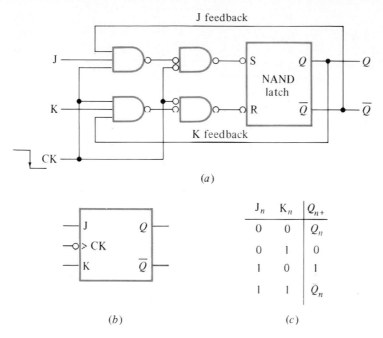

FIGURE 8.6
JK negative edge-triggered flip-flop. (*a*) Logic diagram. (*b*) Logic symbol. (*c*) Characteristic table.

As shown in Fig. 8.6*b*, the logic symbol for an edge-triggered flip-flop is differentiated from the master-slave by a small > sign at the CK input. The presence or absence of the inverting symbol is again used to indicate whether the outputs Q and $\overline{Q}$ change on the falling or rising edge of the clock pulse.

In summary, of the JK flip-flops, we have noted that the device may be of the master-slave or edge-triggered type. The outputs may change following or on the rising or falling edge of the clock pulse. There may be from one to as many as four J and K inputs ($\overline{J}$ and $\overline{K}$ inputs may also be used). The device may or may not have direct set and reset inputs.

8.4 D FLIP-FLOP

A very useful flip-flop, widely used in digital circuits and systems, for the temporary store of data is the *D flip-flop*. The logic diagram of one type of D flip-flop is shown in Fig. 8.7*a*. Note that the inverter at the D input ensures that the S and R inputs to the latch will always be complementary when clocked by CK. The characteristic table for the flip-flop is very simple. As shown in Fig. 8.7*c*, with clocking, the Q output simply follows the D input. The $\overline{Q}$ output is always complementary to the Q output.

The D flip-flop may be of two types. It may be *edge-triggered*, as in Fig. 8.7*a*. When similar to the edge-triggered JK with CK either high or low, the D input has no effect on the output. Data are transferred from the D input to the

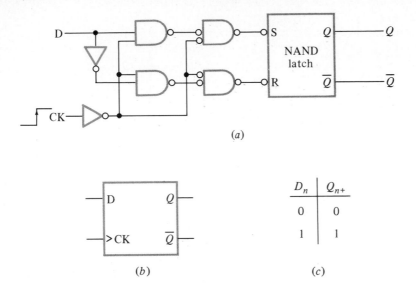

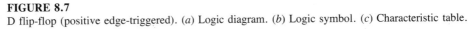

(a)

D_n	Q_{n+}
0	0
1	1

(b)   (c)

FIGURE 8.7
D flip-flop (positive edge-triggered). (a) Logic diagram. (b) Logic symbol. (c) Characteristic table.

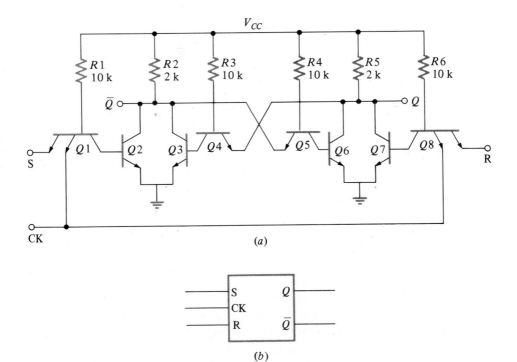

(a)

(b)

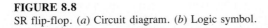

FIGURE 8.8
SR flip-flop. (a) Circuit diagram. (b) Logic symbol.

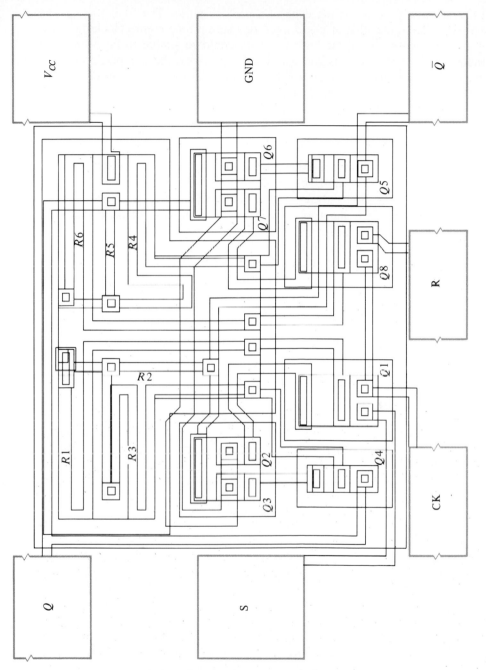

FIGURE 8.8 (*cont.*)
SR flip-flop. (*c*) Layout design.

NAND latch only on the rising edge of the clock pulse. The final state of the flip-flop is due to the state of the D input one set-up time before clocking. Or it may be *transparent*, when the logic diagram would be similar to Fig. 8.4a, but without the JK feedback lines and with the J input, now the D input, connected to the K input through an inverter. The CK input is then really an enable (E) input that when high allows the Q output to follow the D input. The state of the flip-flop is frozen only when E goes low.

8.5 TTL CIRCUITS

Since the basic TTL circuit is a diode AND gate followed by an inverter, that is, a NAND gate, most TTL flip-flops are configured with NAND gates. A simple example is the SR flip-flop shown in Fig. 8.8. An SR flip-flop is a basic SR latch with the addition of a CK input for synchronizing purposes. The logic symbol is included in the figure. Something of the complexity of the layout design for even this simple flip-flop is illustrated in Fig. 8.8c.

A more complicated circuit is the D-type positive edge-triggered flip-flop in Fig. 8.9. With the aid of the logic diagram, the basic TTL(LS) NAND gates that constitute the output latch can soon be identified in the circuit diagram. The input latch circuits are a little more complicated, but note that each NAND gate of the input latches is a 3-input TTL(LS) compatible circuit. The D input is equivalent to 1-unit load for TTL(LS), but the CK and S_D inputs are 2-unit loads, while the R_D input is a 3-unit load.

Similar to other edge-triggered circuits, only the state of the D input one

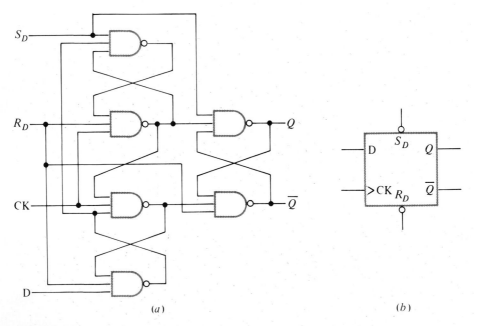

(a)   (b)

FIGURE 8.9
TTL(LS) D flip-flop (positive edge-triggered). (a) Logic diagram. (b) Logic symbol.

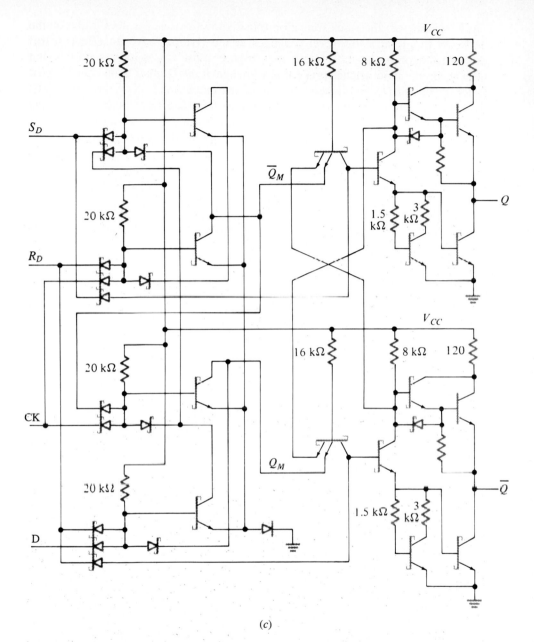

(c)

FIGURE 8.9 (*cont.*)
TTL(LS) D flip-flop (positive edge-triggered). (*c*) Circuit diagram.

set-up time before the clock transition (positive in this case) is held in the output latch. With CK at a low level, changes at the D input have no effect on the output latch since Q_M and $\overline{Q}_M$ are held high because CK is low. With CK at a high level, the input latches prevent any change at the D input from changing the state of Q_M and $\overline{Q}_M$ and hence the state of the output latch.

In TTL flip-flops the asynchronous direct set and reset inputs are usually active low inputs. The synchronous operation of the flip-flop can be on either the positive (rising) or the negative (falling) edge of the clock pulse, but edge-triggering is generally the preferred mode of operation.

The logic diagram of a JK negative edge-triggered flip-flop is shown in Fig. 8.10. It is left as an exercise for the reader to implement this circuit in TTL, but notice the dominant use of NAND gates.

Exercise 8.1. Implement the JK flip-flop shown in Fig. 8.10 in Schottky TTL, that is, TTL(S).

8.6   ECL CIRCUITS

The ECL circuit is a basic OR/NOR circuit so that a simple NOR-type SR latch can readily be implemented in ECL, as in Fig. 8.11. However, the use of ECL

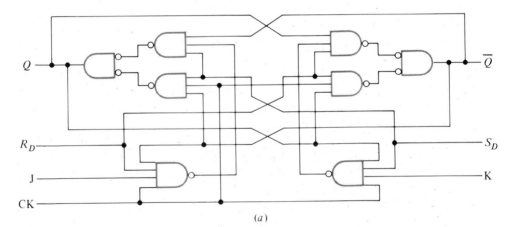

(a)

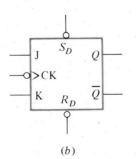

(b)

FIGURE 8.10
TTL JK flip-flop (negative edge-triggered). (a) Logic diagram. (b) Logic symbol.

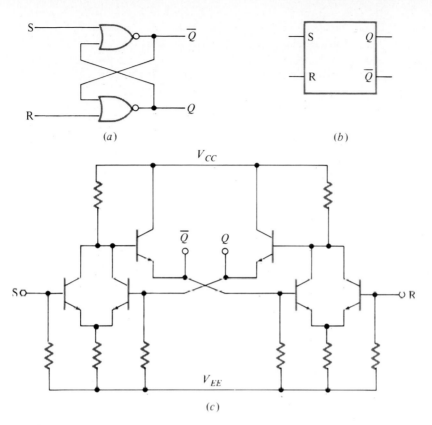

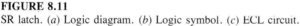

FIGURE 8.11
SR latch. (*a*) Logic diagram. (*b*) Logic symbol. (*c*) ECL circuit.

circuits is generally restricted to very high speed operation, so the flip-flops are more complex than a simple intraconnection of two SR latches.

An example of this complexity is shown in the circuit diagram for a D-type positive edge-triggered flip-flop of Fig. 8.12. However, with the aid of the logic diagram the cross-coupled NOR gates of the input and output latches can be recognized. Especially note that the AND function (of the clock pulses and data) at the input of both latches is performed by *series gating*. That is, the source of the emitter current for the ECL NOR circuits is another emitter-coupled pair. The current in the lower pair of transistors is controlled by the clock lines, which must be level-shifted downward to be compatible with the reference voltages V_{RM} and V_{RS}. Therefore current is steered to the NOR circuits so that edge-triggering is effected. For example, with CK at a low level (-1.7 V) the steering current is in the two reference transistors (Q_{RM} and Q_{RS}) and the output of the master latch (Q_M and $\overline{Q}_M$) can follow the D input. But the input transistors of the slave latch are disabled since the steering current is in Q_{RS}. Now when CK goes to a high level (-0.9 V) the input transistors of the slave latch become active and the output of the slave takes on the state of Q_M and $\overline{Q}_M$. But also the D input transistor has become disabled and the master latch has been isolated from the D input. The

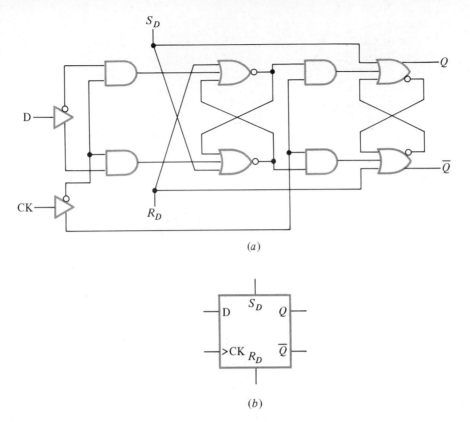

(a)

(b)

FIGURE 8.12
ECL D flip-flop (positive edge-triggered). (a) Logic diagram. (b) Logic symbol.

state of the D input one set-up time before CK went high is retained in the now-active cross-coupled NOR gates of the master latch. The biasing of V_{RS} is made a little more positive than V_{RM} so that when CK goes high the clocking of the slave latch is just a little later than that of the master. This ensures that the D input to the master latch is disabled before the input transistors of the slave are turned on.

Note also that the active high inputs of the direct set and reset override the clock pulses and are fed directly to both the master and slave latches.

Space prohibits a complete circuit schematic of an ECL JK flip-flop, but from the logic diagram of Fig. 8.13 the cross-coupled NOR gates of the input and output latches are readily apparent. Series gating is also used to perform the AND functions at the input of the flip-flop. The use of current steering with transistors that do not saturate along with the small voltage swings results in typical toggle frequencies of greater than 600 MHz for ECL flip-flops of the 100K Series.

Exercise 8.2. Draw a circuit diagram of the ECL JK flip-flop given in Fig. 8.13.

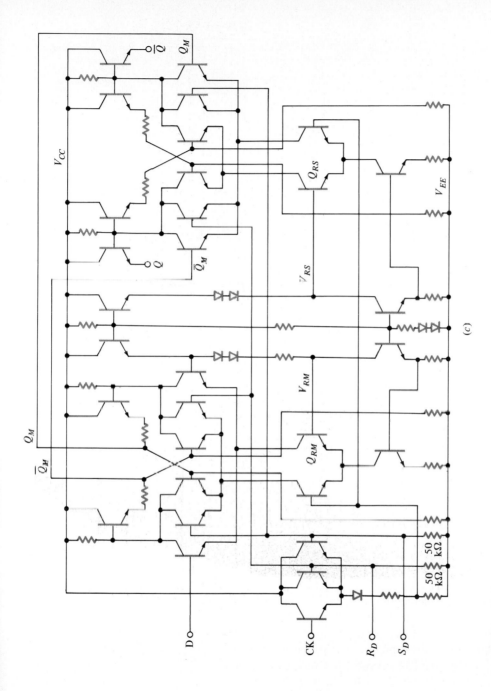

FIGURE 8.12 (cont.)
ECL D flip-flop (positive edge-triggered). (c) Circuit diagram.

307

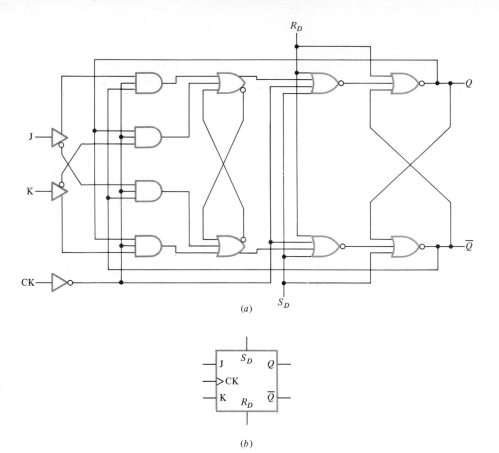

FIGURE 8.13
ECL JK flip-flop (positive edge-triggered). (*a*) Logic diagram. (*b*) Logic symbol.

8.7 I²L CIRCUITS

A simple SR latch implemented with I²L is shown in Fig. 8.14. Note in the circuit diagram the wire-AND at the base of the output inverters. Hence these devices are represented as 2-input NAND gates in the logic diagram. Two collectors are required for each output inverter. Recall that in I²L circuits the connection of two input bases in parallel is prohibited since this would lead to increased collector current for the driver, and possible fan-out problems. The multicollectors at the output avoid this problem. The extra inverters at the input, for S and R, provide active high inputs to control the state of the latch.

Included in Fig. 8.14*b* is a simpler symbol used to represent the merged *pnp* current source and *npn* inverter. This format allows for a much clearer presentation of the I²L inverter when used in LSI circuits.

In Fig. 8.15 the latch is merged with similar circuits in the design of an I²L D-type negative edge-triggered flip-flop. Representation of the I²L gate as a

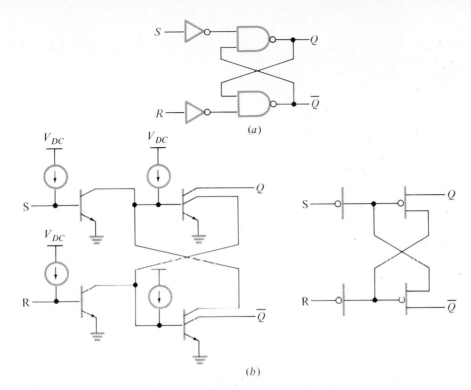

FIGURE 8.14
SR latch. (*a*) Logic diagram. (*b*) I^2L circuit diagram.

multi-input NAND gate makes for an easy transition from the logic to the circuit diagram. Indeed, this is the preferred method when designing with custom I^2L gate arrays. In the layout of this circuit the nine gate bars, one for each transistor, are ranged perpendicular to a common injector rail. The complete flip-flop is contained within a surface area of 91 × 130 μm.

Exercise 8.3. For the I^2L D flip-flop of Fig. 8.15, sketch a layout diagram similar to that shown in Fig. 7.24. Draw lines to form all the necessary logic connections without allowing any lines to cross, i.e., use single-layer metal intraconnects.

8.8 NMOS CIRCUITS

In most NMOS designs, the NOR gate is generally preferable to the NAND gate, since, as explained in Sec. 3.6, for a similar VTC the NOR gate takes less surface area than the NAND gate. Hence most NMOS flip-flops use NOR gates in preference to NAND gates whenever feasible. The analysis of these flip-flops follows that of the NMOS inverter and gates described in Chap. 3.

An example of a simple SR latch using enhancement-depletion (E-D) inverters in a NOR configuration is given in Fig. 8.16. As shown, M_1, M_2, and M_3 make up one NOR circuit, and M_4, M_5, and M_6 the other.

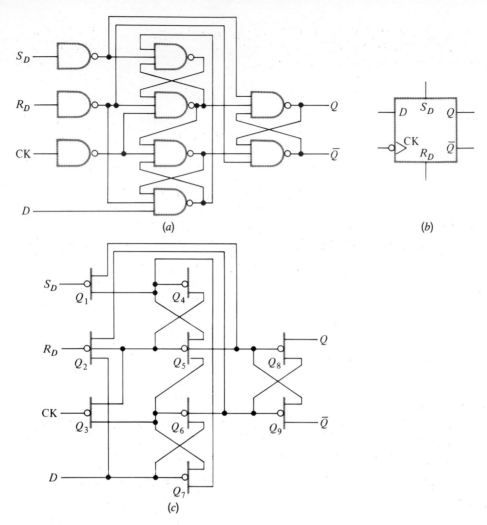

FIGURE 8.15
I^2L D flip-flop (negative edge-triggered). (a) Logic diagram. (b) Logic symbol. (c) Circuit diagram.

A more complicated NMOS circuit implemented with E-D inverters is shown in Fig. 8.17. This is a JK master-slave flip-flop, and from the logic symbol included in the figure we note that the Q and $\bar{Q}$ outputs change following the falling (or negative) transition of the clock pulse. From the circuit schematic note that both CK and $\overline{CK}$ are generated from buffer inverters on the chip.

The cross-coupled circuits forming the master and slave latches are readily identified. In the master latch, enhancement-mode inverters are stacked to form the AND of the J with the J feedback line, and similarly for the K and the K feedback line. The output from the master latch is labeled as Q_M and $\bar{Q}_M$. When $\overline{CK}$ goes high (following CK going low), the output from the master latch is

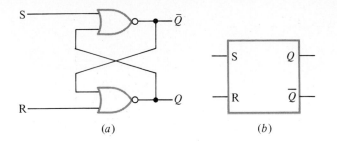

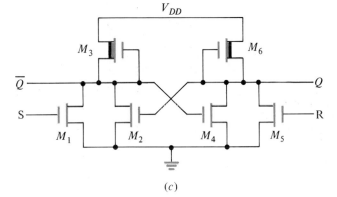

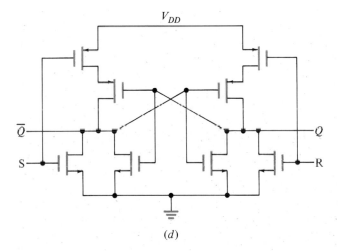

FIGURE 8.16
SR latch. (*a*) Logic diagram. (*b*) Logic symbol. (*c*) NMOS circuit. (*d*) CMOS circuit.

transferred to the slave latch via a stacked AND circuit, and the output of the flip-flop changes accordingly. The outputs Q and $\overline{Q}$ are derived from buffer inverters so that all other devices in the circuit can be minimum size. The output transistors are larger so that they can source and sink the necessary output currents.

Also included in the design are active high inputs for the direct set and reset,

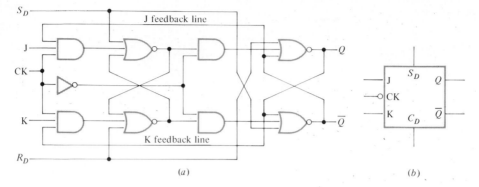

FIGURE 8.17
NMOS JK master-slave flip-flop. (*a*) Logic diagram. (*b*) Logic symbol.

which feed directly into the master and slave latches through buffer inverters. If more J and K inputs were required, they would simply be included by stacking transistors as series gates with the J and K input transistors.

The two NMOS circuits that have just been described are both examples of static logic in that there is no lower limit to the clock frequency. It has already been noted in Sec. 3.8 that NMOS is extensively used in dynamic logic — particularly for LSI, where it is desirable to design a given logic function with the minimum number of devices.

An example of dynamic logic used in a sequential operation is illustrated in Fig. 8.18. The circuit operation is similar to a simple D flip-flop. With CK at a high level, the transmission gate M_5 is on, and the state of the D input is transferred to the gate of M_1. Following the buffer inverters (M_1, M_2 and M_3, M_4), the Q output will have the same state as the D input. Note this is a transparent D flip-flop, in that the Q output follows the D input while CK is at a high level. When CK goes low, M_5 is turned off. The charge on the gate of M_1 will now leak away unless restored by periodic clocking of M_5. The clock can be stopped, but only with CK at a high level. The simplicity of this circuit should be compared with the complexity of the general-purpose JK flip-flop of Fig. 8.17.

Exercise 8.4. Assume that the simple NMOS D flip-flop of Fig. 8.18 is based on two NMOS inverters with specifications as given in Example 3.3, operating with $V_{DD} = 5$ V. Find the lowest value of voltage at the gate of M_1 for which the Q output remains at V_{DD}.

8.9   CMOS CIRCUITS

A simple SR latch formed with CMOS NOR gates is included in Fig. 8.16. A CMOS D-type positive edge-triggered flip-flop is shown in Fig. 8.19. The circuit diagram is simply derived from the logic diagram.

The cross-coupling in the input and output latches is completed through the

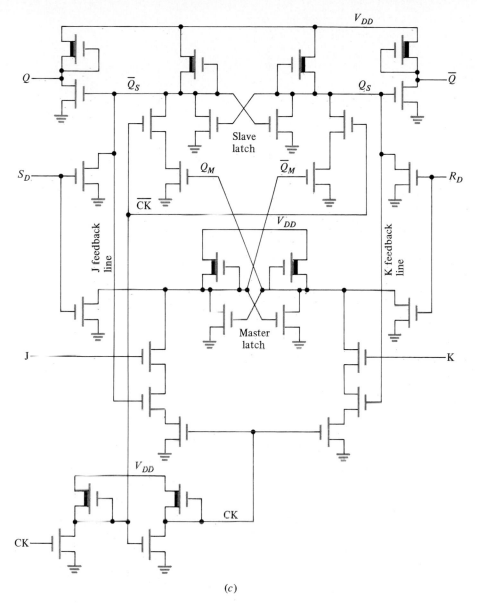

FIGURE 8.17 (*cont.*)
NMOS JK master-slave flip-flop. (*c*) Circuit schematic.

transmission gates TG_2 and TG_4, respectively. Transmission gates TG_1 and TG_3 are used to enter data into the input and output latches, respectively. As a result, when CK goes low TG_1 is on and data from the D input are entered into the input latch, but the input to the output latch is disabled since TG_3 is off. When CK goes high (and $\overline{CK}$ goes low) TG_1 turns off and TG_3 turns on. Consequently,

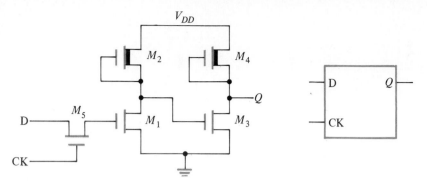

FIGURE 8.18
Simple NMOS D flip-flop used in LSI.

the D input is disconnected from the input latch, but the state of the D input one set-up time before the clock transition is transferred to the output latch and the output buffer inverters. These inverters allow for minimum-size devices to be used in the latch circuits. Buffer inverters are also used for CK and $\overline{CK}$, since it is required to quickly turn on and turn off the transmission gates. The

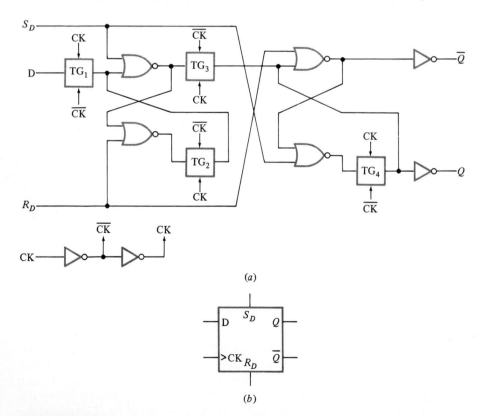

FIGURE 8.19
CMOS D flip-flop (positive edge-triggered). (*a*) Logic diagram. (*b*) Logic symbol.

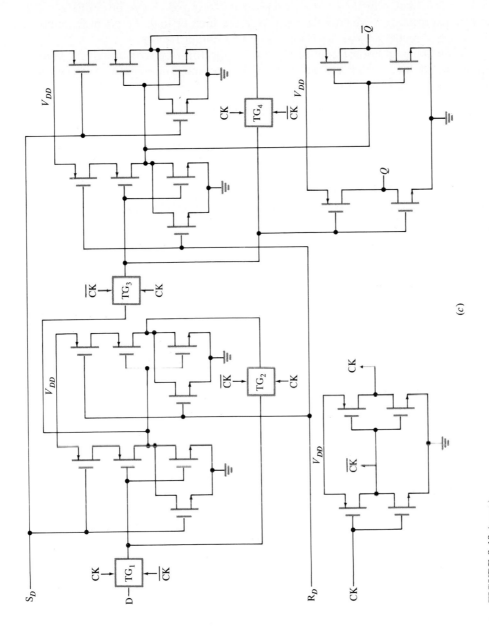

(c)

FIGURE 8.19 (*cont.*)

CMOS D flip-flop (positive edge-triggered). (c) Circuit diagram.

315

transmission gates provide a low-resistance path for rapid charge or discharge of the gate capacitances, whether the input is going high or low. Active high inputs are used to directly set and reset both input and output latches.

With some additional control logic a D-type flip-flop may readily be converted to a JK type. With the aid of a truth table and Karnaugh map it can be shown that for a JK flip-flop,

$$\overline{D} = \overline{J}\,\overline{Q} + KQ$$
$$= (J + Q) + KQ$$

hence

$$D = \overline{(J + Q) + KQ}$$

This combination of logic gates implemented with CMOS is illustrated in Fig. 8.20c.

Exercise 8.5. Show that a D-type flip-flop may be converted to a JK type with control logic consisting of two 2-input NAND gates and one 2-input OR gate.

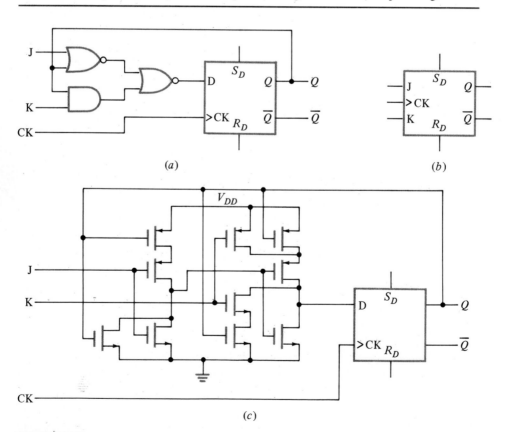

(a)

(b)

(c)

FIGURE 8.20
Conversion of a D flip-flop to a JK. (a) Logic diagram. (b) Logic symbol. (c) CMOS circuit diagram.

TABLE 8.1
Performance characteristics of IC flip-flops at 25°C

		Maximum frequency, typical/minimum	Power consumption,* typical/maximum
CMOS	$V_{DD} = 5$ V	7/3.5 MHz	0.1/5 μW
	$V_{DD} = 10$ V	16/8 MHz	0.2/20 μW
TTL	74LS	45/30 MHz	10/20 mW
	74S	110/75 MHz	75/125 mW
ECL	10K	200/140 MHz	230/350 mW
	100K	550/400 MHz	550/775 mW

*Power consumption is found from the average of I_{CCH} and I_{CCL}, with the clock input grounded.

8.10 SUMMARY OF FLIP-FLOP CIRCUITS

Flip-flop circuits designed with TTL, ECL, and CMOS are readily available as SSI circuits, usually as a dual circuit—that is, two D flip flops or two JK flip flops fabricated in one package, for example. Circuits of these same three technologies are also available as MSI circuits. Usually four flip-flops are contained in one package. Examples are four JK flip-flops interconnected as a 4-bit counter or four D flip-flops connected as a 4-bit shift register. For any particular technology in both SSI and MSI circuits, the input and output capabilities of the function block are compatible with the like capabilities of an individual logic gate. Indeed, as we have seen, the function block is essentially a connection of individual gates.

Continuing to create more complex ICs containing larger logic functions by connecting more and more gates is not advisable. The die size soon becomes too large for economic yields. However, LSI circuits are available in TTL, ECL, and CMOS; the first two are especially used in high-speed memory circuits, while CMOS is used where the lower power drain is of particular importance. Because of their higher packing density, NMOS and I²L are seldom, if ever, found in anything else but LSI circuits.

To conclude this section on flip-flops we include, for comparison purposes, Table 8.1 which shows some typical performance characteristics, namely, operating frequency and power consumption, for flip-flops of each of the SSI and MSI technologies discussed.

8.11 SCHMITT TRIGGER

Regenerative circuits are not limited to latches and flip-flops. Another regenerative circuit that is particularly useful in digital system is the *Schmitt trigger*.*

---

*Named after the originator of this type of circuit: O.H. Schmitt, "A Thermionic Trigger," *Journal of Scientific Instruments*, vol. 15, January 1938, pp. 24–26.

This circuit responds to a slowly changing input waveform with fast transition times at the output. Another important feature of the circuit is that the voltage transfer characteristic has different input thresholds for positive- and negative-going voltage signals. A simplified form of an emitter-coupled Schmitt trigger circuit and its voltage transfer characteristic are shown in Fig. 8.21.

8.11.1 Emitter-Coupled Schmitt Trigger

The operation of the circuit in Fig. 8.21a is as follows. Assume V_{in} is at a low level, say, near 0 V, on the voltage transfer characteristic of Fig. 8.21b. Then Q_1 is off and Q_2 is on and saturated, with $V_{out} = V_{OL}$. This establishes a certain voltage at V_E (the common-emitter voltage of Q_1 and Q_2). Next increase V_{in} to point b, where Q_1 is at the edge of conduction but Q_2 is still in saturation. Now a small increase in V_{in} causes Q_1 to conduct, and the voltage at its collector, V_{C1}, falls due to the collector current in Q_1. But $V_{C1} = V_{B2}$. Hence as Q_1 turns on, Q_2 turns off. The transition is rapid because of the regenerative action of the voltage at the base of Q_2 falling as I_{C1} increases and the voltage at the emitter of Q_2 increasing as V_{in} increases. There is therefore a steep change in the VTC from point b to point c at $V_{in} = V_{T+}$ and $V_{out} = V_{OH}$. Any further increase in V_{in} causes Q_1 to go into saturation, but Q_2 is off and hence V_{out} remains at V_{OH}.

Now suppose the level at V_{in} is steadily decreased from a high level near V_{CC}, which will cause V_E to fall, but Q_2 remains cut off until point d on the VTC is reached, where Q_1 is in the active region. Now decreasing V_{in} causes I_{C1} to decrease and V_{CE1} to increase. But $V_{CE1} = V_{BE2}$. Hence as Q_1 turns off, Q_2 turns on. Again, because of regeneration the transition is rapid. The voltage at

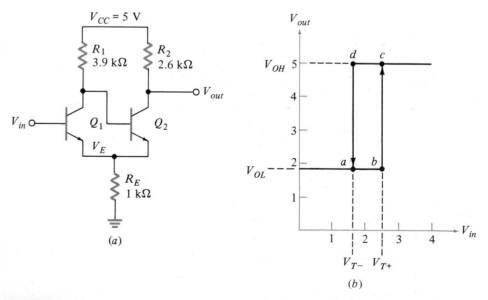

FIGURE 8.21
(a) Basic emitter-coupled Schmitt circuit. (b) Voltage transfer characteristic.

the base of Q_2 is increasing as I_{C1} decreases and the voltage at V_E is decreasing as V_{in} decreases. This causes the steep change in the VTC from point d to point a, where $V_{in} = V_{T-}$ and $V_{out} = V_{OL}$. The difference between the threshold levels, $V_{T+} - V_{T-}$, is known as the *hysteresis voltage*.

To determine the threshold voltages for the circuit of Fig. 8.21a, assume at the edge of conduction $V_{BE(EOC)} = 0.7$ V, at the edge of saturation $V_{BE(EOS)} = 0.8$ V, and in saturation $V_{CE(sat)} = 0.1$ V.

To calculate V_{T+}, assume V_{in} is at a low level, Q_1 is off, and Q_2 is on and saturated. Then

$$I_{E2} = I_{B2} + I_{C2}$$

that is,

$$\frac{V_E}{1\ k\Omega} = \frac{5 - (V_E + 0.8)}{3.9\ k\Omega} + \frac{5 - (V_E + 0.1)}{2.6\ k\Omega}$$

Solving,

$$V_E = 1.8\ \text{V} \quad \text{and} \quad V_{out} = V_{OL} = V_E + 0.1 - 1.9\ \text{V}$$

Now Q_1 turns on when $V_{BE1} = 0.7$ V. Therefore, $V_{T1} = V_E + 0.7 = 2.5$ V. That is, with $V_{in} < 2.5$ V, Q_1 is off and Q_2 is on as assumed.

To calculate V_{T-}, assume V_{in} is at a high level, Q_2 is off, and Q_1 is on and saturated. Then

$$V_{out} = V_{OH} = V_{CC} - 5.0\,\text{V}$$

and, neglecting base current,

$$I_{C1} = \frac{5 - 0.1}{(3.9 + 1)\ k\Omega} = 1\ \text{mA}$$

hence,

$$V_E = (1\ \text{mA})(1\ k\Omega) = 1.0\ \text{V}$$

Transistor Q_2 is off, since with Q_1 saturated $V_{BE2} = V_{CE1(sat)} = 0.1$ V. Now Q_2 turns on when $V_{CE1} = V_{BE2} = 0.7$ V. Then

$$I_{C1} = \frac{5 - 0.7}{(3.9 + 1)\ k\Omega} = 0.88\ \text{mA}$$

that is,

$$V_E = (0.88\ \text{mA})(1\ k\Omega) = 0.88\ \text{V}$$

Therefore,

$$V_{T-} = V_E + 0.7 = 1.6\ \text{V}$$

hence for this circuit the hysteresis voltage of 0.9 V.

The hysteresis feature of the Schmitt circuit is used to advantage in line receiver applications. Because of the fast transition times of high-speed digital systems and the parasitic series inductance and parallel capacitance of a bus line, the voltage pulse received at the end of a long line might be as in Fig. 8.22a. The output of a Schmitt buffer circuit with hysteresis would simply be a single step waveform, as in Fig. 8.22b. The output of a simple buffer with one threshold at V_{T+} would consist of three pulses, as shown dashed in Fig. 8.22b. Simply setting the threshold to V_{T-} may cause the circuit to trigger on noise, hence the advantage of the hysteresis in the Schmitt circuit.

Because of the fast transition times at the output, the Schmitt trigger circuit is also very useful in converting a sine-wave input to a pulse-train output in time-marker applications.

Preceded by an input diode AND circuit and followed by a totem-pole output circuit, the circuit in Fig. 8.21a becomes a standard Series 74 TTL part. Then, because of the voltage drop across the input diode, the input threshold levels V_{T+} and V_{T-} are 1.8 and 0.9 V, respectively. Similar circuits, but with Schottky-clamped transistors, are included in the 74S and 74LS Series of digital gates.

8.11.2   CMOS Schmitt Trigger

The Schmitt trigger circuit can also be implemented with CMOS transistors, where the different threshold voltage for the n-channel and p-channel transistors is used to advantage. The circuit of a CMOS Schmitt trigger is shown in Fig. 8.23. The Schmitt trigger proper consists of three p-channel devices, M_1 to M_3, and three n-channel transistors, M_4 to M_6. Devices M_3 and M_6 have minimum-sized geometries. The inverter M_7 and M_8 is a buffer for the output driver stage of M_{11}

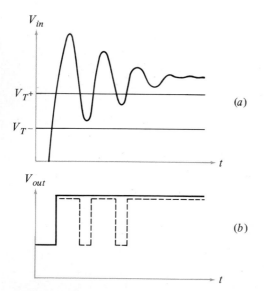

(a)

(b)

FIGURE 8.22
(a) Input voltage to Schmitt circuit. (b) Output voltage. Shown dashed is the output voltage due to one threshold at V_{T+}.

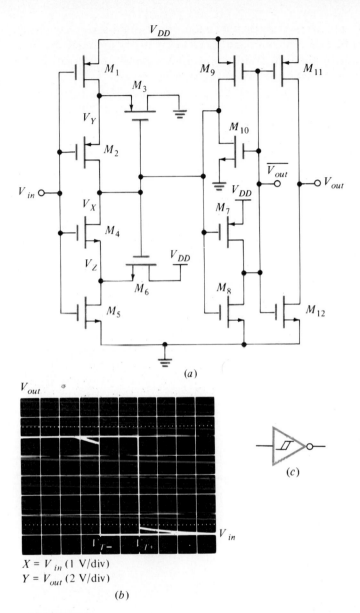

(a)

$X = V_{in}$ (1 V/div)
$Y = V_{out}$ (2 V/div)

(b)

(c)

FIGURE 8.23
(a) CMOS Schmitt trigger circuit. (b) Voltage transfer characteristic. (c) Logic symbol for an inverter with Schmitt trigger input.

and M_{12}. The transistors M_9 and M_{10} provide feedback to effect a rapid change of voltage at V_X.

To describe the circuit, assume $V_{DD} = 10$ V and the threshold voltages are $V_{TN} = +4$ V and $V_{TP} = -4$ V. Now with $V_{in} = 0$ V, the two stacked p-

channel transistors (M_1 and M_2) will be on but conducting negligible drain current, since M_4 and M_5 are off. Hence the voltage $V_Y \approx V_X \approx 10$ V; and after the two inverter stages, $V_{out} = V_{OH} = 10$ V. With $V_X = 10$ V, transistor M_6 will be at the threshold of conduction, and $V_Z = V_X - V_{TN} \approx 6$ V.

When V_{in} rises to V_{TN}, namely, 4 V, M_5 turns on, but M_4 is off because $V_Z = 6$ V. But now M_5 and M_6 form an inverting NMOS amplifier with a voltage gain of about -2. Thus, as V_{in} rises, V_Z is falling. With $V_{in} = 6$ V, $V_Z = 2$ V and M_4 turns on. With both M_4 and M_5 conducting, V_X rapidly goes to 0 V, turning off M_6. Following the two inverter stages, $V_{out} = V_{OL} = 0$ V. For the voltage transfer characteristic, $V_{T+} = 6$ V. Also, with $V_X = 0$ V, transistor M_3 turns on, which aids in turning off M_2 as V_Y goes from 10 V to $V_X - V_{TP} = 4$ V.

As V_{in} decreases from 10 to 0 V the operation is essentially similar. But now M_1 turns on when $V_{in} = 6$ V, and M_1 and M_3 form an inverting PMOS amplifier with a voltage gain of about -2. With $V_{in} = 4$ V, $V_Y = 8$ V and M_2 turns on. With M_1 and M_2 on, V_X rapidly goes to 10 V, turning off M_3 and turning on M_6. At the output, $V_{out} = 10$ V at $V_{in} = V_{T-} = 4$ V, and the hysteresis voltage is 2 V.

The voltage transfer characteristic with two threshold levels is particular to the Schmitt trigger circuit, so that including a stylized form of the VTC in a logic symbol, as in Fig. 8.23c, is the standard way of indicating a Schmitt trigger circuit at the input.

Exercise 8.6. For the basic emitter-coupled Schmitt trigger circuit shown in Fig. 8.21a, find values for R_1 and R_2 in terms of R_E so that $V_{T+} = 2.2$ V and $V_{T-} = 1.8$ V. Assume $V_{CC} = 5$ V, $V_{BE(EOC)} = 0.7$ V, $V_{BE(EOS)} = 0.8$ V, $V_{CE(SAT)} = 0.1$ V, and β_F is infinite.

8.12 MULTIVIBRATOR CIRCUITS

Another series of regenerative circuits that are especially useful in timing applications are the *multivibrator circuits*. These may be one of the three following forms:

1. The bistable circuit
2. The monostable circuit
3. The astable circuit

We are already familiar with the bistable circuit. Latches and flip-flops are examples of *bistable multivibrator* circuits, which have two stable states. The circuit will remain in the one stable state until it is triggered into the other stable state, where it will again remain stable until triggered back to its original stable state.

The *monostable multivibrator* has only one stable state. It will remain in this stable state until it is triggered into the other, a *quasi-stable state*. That is, it will eventually return to its original stable state in a time determined by the circuit parameters. This circuit, also termed a *one-shot*, is very useful in generating pulses of a known duration in time.

The *astable multivibrator* has no stable states. The output oscillates between two quasi-stable states. The time of each quasi-stable state is determined by circuit parameters. Thus this circuit can be used as a clock pulse generator.

8.12.1 CMOS Monostable Multivibrator

CMOS logic gates are very useful in implementing simple multivibrator circuits. This is because the CMOS gate exhibits output voltage levels that are essentially at V_{DD} or V_{SS} (0 V), and the input current at the gate of the device is essentially zero. It should be noted that if the protective diodes at the input of the gate become forward-biased, this will cause some modification of the basic operation. A simple monostable multivibrator circuit using two CMOS NOR gates is shown in Fig. 8.24a. The voltage waveforms at particular points in the circuit are given

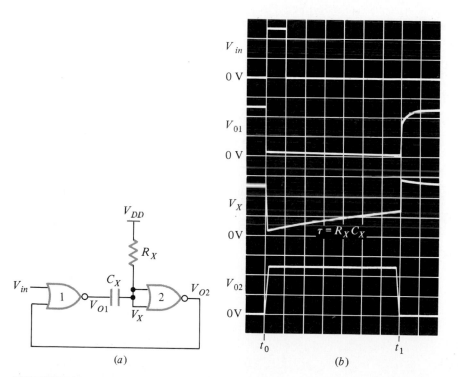

FIGURE 8.24
(a) Monostable multivibrator circuit with CMOS NOR gates. (b) Voltage waveforms with $V_{DD} = 5$ V, $R_X = 10$ kΩ, and $C_X = 0.001$ μF. Vertical scale = 2 V/division; horizontal scale = 1 μs/division.

in Fig. 8.24b. The switching threshold voltage V_{Th} is the value of V_X at which regenerative switching back to the original state occurs. Generally, little error is introduced in assuming $V_{Th} = V_T$, the threshold voltage for the device. In fact, V_X must cross the unity-gain point V_{IL} of gate 2 for this regenerative switching to occur.

In the stable state, both inputs of gate 1 are at a low level, because $V_{in} = 0$ V and V_{O2} is low since V_X at the input of gate 2 is connected to V_{DD} through the timing resistor R_X. Hence V_{O1} is at a high level.

A trigger pulse, as shown at time t_0, causes V_{O1} to go to 0 V. Now provided the output current I_{OL1} can provide the necessary voltage drop across the resistor R_X, V_X will go to 0 V. Then the output V_{O2} will go to V_{DD}, and the circuit is in the quasi-stable state. Once in the quasi-stable state the trigger pulse can be terminated. That is, it is only necessary that the width of the trigger pulse be a little wider than the propagation delay time through the two NOR gates. The output of gate 1 is coupled to the input of gate 2 by the timing capacitor C_X. Hence, following the initial transition, the capacitor changes from 0 V toward V_{DD}, with a time constant $\tau = R_X C_X$. But when V_X reaches the switching threshold voltage V_{Th} at time t_1, gate 2 turns on and V_{O2} goes to 0 V, and the quasi-stable state is terminated. Consequently V_{O1} returns to its quiescent state at V_{DD}. Therefore the output waveform at V_{O2} is a pulse whose time duration is determined primarily by R_X and C_X and the regenerative switching threshold voltage V_{Th}.

Following the initial transition, the voltage V_X is of the form

$$V_X(t) = A + Be^{-t/\tau} \tag{8.12.1a}$$

where

$$\tau = R_X C_X$$

but

$$V_X(0) = 0 = A + B \quad \text{and} \quad V_X(\infty) = V_{DD} = A$$

Therefore,

$$B = -V_{DD}$$

Hence,

$$V_X(t) = V_{DD}(1 - e^{-t/\tau}) \tag{8.12.1b}$$

but

$$V_X(t_1) = V_{Th}$$

Solving,

$$t_1 = \tau \ln \frac{V_{DD}}{V_{DD} - V_{Th}} \tag{8.12.2}$$

Hence with

$$V_{Th} = V_T = \frac{V_{DD}}{2} \qquad t_1 = 0.69 R_X C_X \qquad (8.12.3)$$

The threshold voltages in a CMOS gate are relatively unchanged by temperature; hence in this circuit the output pulse width stability can thus be reasonably good. However because of production tolerances there will be a spread in V_T from one production run to another. Consequently, for any given value of R_X and C_X there can be a substantial change in the output pulse width from unit to unit. Of course, the input trigger pulse width must be less than the output pulse width. (Also note, the monostable circuit can just as readily be implemented with two NAND gates. Then the trigger pulse would be a negative-going pulse, from V_{DD} to 0 V.)

Exercise 8.7. In the monostable circuit of Fig. 8.24, $V_{DD} = 10$ V, $V_{Th} = 5$ V, $R_X = 10$ kΩ, and $C_X = 0.001$ μF.
(a) Calculate the output pulse width.
(b) If the variation of V_{Th} is $+20\%$, what is the variation in the output pulse width?

8.12.2 CMOS Astable Multivibrator

A basic CMOS astable multivibrator circuit implemented with two CMOS NOR gates is shown in Fig. 8.25a with some pertinent voltage waveforms for the circuit in Fig. 8.25b. In the analysis we neglect the propagation delay time through the gates and assume that the output voltage of the gates changes levels instantaneously when the input voltage crosses the threshold level V_T.

Because of the inverting action of gate 2, V_{O2} is the complement of V_{O1}. So that at time t_0 with $V_X = V_T$ gate 1 turns on, gate 2 turns off, and V_{O2} goes from 0 V to V_{DD}. This step change of voltage ($= V_{DD}$) is coupled through C_X so that V_X changes from V_T to $V_T + V_{DD}$. Note that the output V_{O1} is at 0 V and V_{O2} is at V_{DD}, hence C_X discharges through R_X toward 0 V. But when V_X crosses V_T at time t_1, gate 1 turns off and gate 2 turns on. That is, V_{O1} goes to V_{DD}, and V_{O2} goes to 0 V. Consequently, V_X goes from V_T to $V_T - V_{DD}$. The timing capacitor now charges through R_X toward V_{O1}, which is at V_{DD}; when V_X crosses V_T at time t_2, the whole cycle repeats itself.

In this circuit there are no stable states, but the output voltages switch from one level to the other at a period determined by the passive components R_X and C_X and the threshold voltage V_T. With $V_T = V_{DD}/2$, the output waveform is symmetrical with $t_1 - t_0 = t_2 - t_1$, and the frequency of oscillation is given as

$$f = \frac{1}{2t_1} = \frac{1}{2.2\tau} \qquad \text{where } \tau = R_X C_X \qquad (8.12.4)$$

Also in this analysis we have assumed no protective diodes at the input of the logic gates. Including these diodes the input voltage is effectively clamped on the positive swing to one diode voltage drop above V_{DD} and on the negative swing to one diode voltage drop below V_{SS} ($=0$ V).

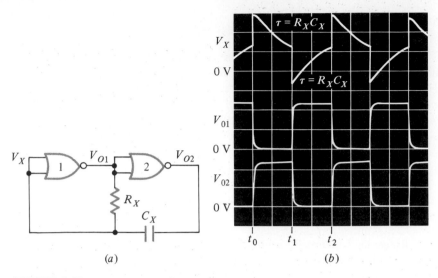

(a)   (b)

FIGURE 8.25
(a) Astable multivibrator circuit with CMOS NOR gates. (b) Voltage waveforms with $V_{DD} = 5$ V, $R_X = 10$ kΩ, and $C_X = 0.001$ μF. Vertical scale = 2 V/division; horizontal scale = 5 μs/division.

Similar to the monostable multivibrator, in this circuit the frequency of oscillation is relatively stable with temperature but does change from unit to unit as the threshold voltage changes. However, the circuit is a simple and effective way of generating a clock frequency of up to 1 MHz.

Exercise 8.8. For the CMOS astable multivibrator circuit shown in Fig. 8.25, prove that the frequency of oscillation is given as in Eq. (8.12.4).

8.12.3 TTL Monostable Multivibrator

Simple monostable multivibrator circuits may be constructed with TTL NOR or NAND gates, but because of their input characteristics, namely, I_{IL} and I_{IH}, the timing stability is inferior to that of the CMOS circuits. However, a very useful and popular digital IC is the 9600 Series of monostable multivibrators. The basic circuit of the 9600 is described here to illustrate the problems and solutions to the design of a general-purpose IC monostable multivibrator. The illustrated circuit is compatible to the familiar standard 74 Series, but S and LS versions of the circuit are also available.

As shown in Fig. 8.26, the input circuit (Q_1 to Q_3) is TTL-compatible. Transistors Q_4 and Q_5 form a bistable, Q_6 and Q_7 a monostable, and Q_8 and Q_9 an emitter-coupled Schmitt trigger circuit. Not shown are the buffer circuits that connect from the output of the Schmitt trigger to conventional TTL totem-pole output circuits so that both Q and $\overline{Q}$ outputs are available at the output of the IC.

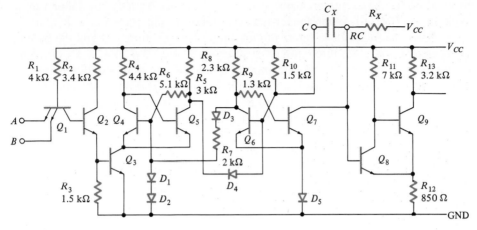

FIGURE 8.26
Basic circuit of the 9600 monostable multivibrator.

In the quiescent state, Q_3 is off so that there is no current in either Q_4 or Q_5. The monostable is in its stable state with Q_6 on and Q_7 off. This is because the resistance from V_{CC} to the base of Q_6 is only 1.5 kΩ, but to the base of Q_7 it is 3.6 kΩ. With Q_7 off, for the Schmitt circuit, Q_8 is on and Q_9 is off. The result is that the Q output is low and the $\overline{Q}$ output is high.

The circuit is triggered when Q_3 turns on, because the logic function at the input of the IC is true. With Q_3 conducting, Q_5 preferentially turns on and the bistable is set. Now Q_5 takes the base current away from Q_6 through the diode D_4, so Q_6 turns off. With Q_6 off, current through the diode D_3 turns on Q_4 and resets the bistable. Hence the output of the bistable at the collector of Q_5, which is the trigger pulse to the monostable via diode D_4, is a very narrow negative-going pulse. The width of this trigger pulse is due only to the time taken for the bistable to set and then reset. The bistable effectively differentiates the leading edge of the pulse applied to the base of Q_3.

With the input trigger, Q_6 turns off and Q_7 turns on, and the monostable is in its quasi-stable state. With Q_7 on, the base of Q_8 is pulled down, turning Q_8 off and Q_9 on. The emitter coupling of Q_8 and Q_9 provide a regenerative action—the base of Q_9 is rising as the emitter of Q_9 is falling. The change of state at the collector Q_9 is connected via buffers to the output circuits of the IC, so that the voltage levels at Q and $\overline{Q}$ also change state.

When Q_7 turns on, the negative transition at the collector is coupled to the base of Q_6 through C_X. But then C_X charges toward V_{CC} through R_{10} and the conducting Q_7. The monostable returns to its stable state in a time determined by the time constant $R_{10}C_X$, when the voltage at the base of Q_6 has risen sufficiently to turn Q_6 on, and then goes Q_7 off. Now the principal timing of the IC occurs as C_X charges toward V_{CC} in the opposite direction, namely, through R_X and the forward-biased base-emitter junction of Q_6. Since $R_X \gg R_{10}$, the latter timing cycle ($\tau_2 = R_X C_X$) is much longer than the initial timing cycle ($\tau_1 = R_{10} C_X$).

When the monostable returns to its stable state, Q_7 turns off and the base of Q_8 rises as C_X charges through R_X. At some threshold level (V_{T+}), Q_8 turns on, regeneratively turning Q_9 off (the base of Q_9 is falling as its emitter is rising). The result is a negative waveform at the collector of Q_9 with short fall and rise times and a duration due primarily to R_X, C_X, and the threshold level V_{T+}.

The voltage waveforms at particular points in the circuit are presented in Fig. 8.27. Note that if the circuit is retriggered after the initial timing cycle but before the primary timing cycle is completed, a complete new timing cycle is initiated. This is the retriggerable feature of the 9600, which is due to the trigger pulse at the collector of Q_5 that starts each new timing cycle.

The logic symbol for the 9600 is included in Fig. 8.27. Note that the circuit may be triggered with either active high inputs or active low inputs. The output pulse width is nominally

$$t = 0.32 R_X C_X \left(1 + \frac{0.7\,\text{k}\Omega}{R_X} \right) \qquad \text{ns} \qquad (8.12.5)$$

where R_X is in kΩ, C_X is in pF, and $0°C < T_A < 75°C$. Also, it is required that $5\,\text{k}\Omega < R_X < 50\,\text{k}\Omega$ and $C_X > 1000$ pF.

For smaller values of C_X, a graphical relation included in the data sheet must be used to determine the pulse width. But the minimum output pulse width is typically 75 ns.

Referenced to $T_A = 25°C$, the output pulse width of the 9600 varies from $+1.5$ to -1.5% as the ambient temperature is varied from 0 to 70°C. The pulse

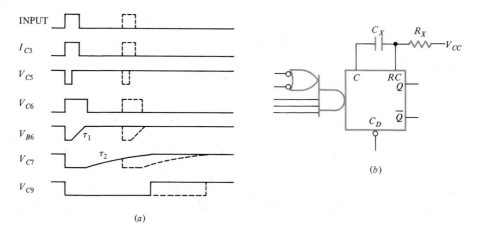

FIGURE 8.27
(a) Waveform for the 9600. (b) Logic symbol for the 9600.

width shows a variation of -1.5 to $+1.5\%$ as the V_{CC} is varied about 5 V, from 4.75 to 5.25 V. The power dissipation of the 9600 is typically 125 mW.

Exercise 8.9. Sketch a logic block diagram of the 9600 circuit given in Fig. 8.26.

8.13 IC TIMER

Another timing circuit that is widely used in digital systems, though genetically it is of the linear IC family, is the 555 timer. The circuit is useful where pulse widths of $\geq 1.0\ \mu s$ are required. Also only two or three external components are needed to operate the circuit as either a monostable or an astable multivibrator. While the 555, a single unit, is described here, the circuit is so popular that dual and quad units are also commercially available. Also while the 555, the original IC timer, is designed with bipolar junction transistors, there are now available CMOS versions of the circuit.

8.13.1 Timer Operated as a Monostable Multivibrator

A basic block diagram of the 555 connected as a monostable multivibrator is shown in Fig. 8.28, as well as the important voltage waveforms. The junction of the two external timing components R_X and C_X is connected to the *threshold input*, which is connected to the *discharge output*. The monostable is triggered by a negative-going voltage (V_{Trig}) at the *trigger input*.

The basic timer circuit consists of two *voltage comparators*, an SR latch, a discharge transistor, and a totem-pole output circuit that is capable of sourcing and sinking up to 200 mA. The supply voltage V_{CC} may vary from 4.5 to 16 V.

Voltage comparators are high-gain (typically ≥ 1000) voltage amplifiers with output circuits that may be directly connected to digital ICs, usually TTL. That is, they have linear input circuits but digital output circuits, and only a small voltage difference is required at the input terminals to switch the output level from V_{OH} to V_{OL}, or vice versa.

In this circuit the reference voltage for the comparators is obtained from a string of three equal-valued resistors. Recall that in the IC process the absolute value of a diffused resistor is not well-controlled, but the ratio of such resistors can be controlled to better than 2%. Hence the value of the resistor string in the 555 timer is not well-determined, but the reference voltages V_1 and V_2 will be $V_{CC}/3$ and $2V_{CC}/3$, irrespective of the value of V_{CC}.

In the quiescent state the latch is reset with the Q and output at the low level, equal to V_{OL}. Since $\overline{Q}$ is at the high level, the discharge transistor is on and saturated. The voltage across the capacitor C_X is therefore clamped at approximately 0 V. Assuming $V_{CC} = 5$ V, the voltage at the threshold input is less than the reference voltage $V_2 = 3.3$ V and the output of comparator 2 is

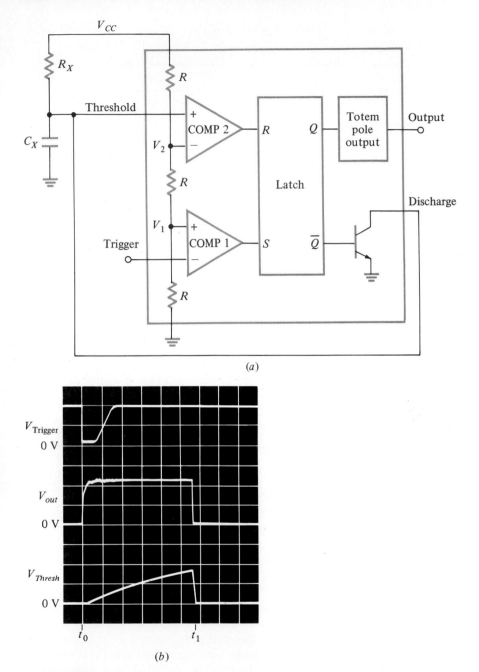

(a)

(b)

FIGURE 8.28
Basic block diagram of the 555 timer. (a) Connected as a monostable multivibrator. (b) Waveform of monostable multivibrator with $V_{CC} = 5$ V, $R_X = 10$ kΩ, and $C_X = 0.001$ μF. Vertical scale = 2 V/division; horizontal scale = 2 μs/division.

at the low level. With the voltage at the trigger input greater than the reference voltage $V_1 = 1.7$ V, the output of comparator 1 is also at the low level.

The circuit is triggered to the quasi-stable state when $V_{Trig} < V_1$; then the output of comparator 1 goes to the high level, which sets the latch, causing the Q and output to go to the high level and the discharge transistor to turn off. The timing capacitor C_X is now allowed to charge toward V_{CC} with a time constant $\tau = R_X C_X$. When the capacitor voltage crosses the threshold level $(= V_2)$, the output of comparator 2 goes to the high level, which resets the latch. Consequently, the Q and the output switch back to the low level, and with $\overline{Q}$ at the high level, the discharge transistor turns on, rapidly discharging C_X and thereby ending the quasi-stable state.

The output pulse width is dependent upon the voltage across the timing capacitor, and

$$V_{CAP}(t) = V_{CC}(1 - e^{-t/\tau}) \tag{8.13.1}$$

where $\tau = R_X C_X$. Solving Eq. (8.13.1) for t,

$$t = \tau \ln \frac{V_{CC}}{V_{CC} - V_{CAP}} \tag{8.13.2}$$

The latch is reset at time t_1 when

$$V_{CAP}(t_1) = V_2 = \frac{2V_{CC}}{3}$$

Therefore,

$$t_1 = \tau \ln \frac{V_{CC}}{V_{CC} - 2V_{CC}/3} \tag{8.13.3}$$

or

$$PW = \tau \ln 3 = 1.1 R_X C_X \tag{8.13.4}$$

Note that the output pulse width is independent of the value of V_{CC} and dependent only on the value of the external timing components R_X and C_X.

8.13.2 Timer Operated as an Astable Multivibrator

The connection of the 555 timer as an astable multivibrator is shown in Fig. 8.29. Note that the trigger input is now connected to the threshold input and the discharge output to the junction of the two external timing resistors R_A and R_B.

Assume at time t_0 that the latch has just been set, causing V_{out} to switch to V_{OH} and the discharge transistor to turn off. The timing capacitor C_X charges toward V_{CC} with a time constant $\tau_1 = (R_A + R_B)C_X$. At time t_1, $V_{CAP} = V_2$, causing the latch to reset. Hence V_{out} switches to V_{OL} and the discharge transistor turns on. The timing capacitor now discharges toward ground with a time constant

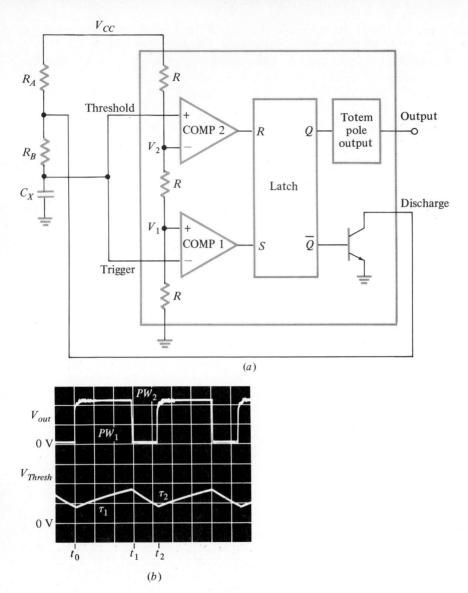

(a)

(b)

FIGURE 8.29
Basic block diagram of the 555 timer. (a) Connected as an astable multivibrator. (b) Waveforms of astable multivibrator with $V_{CC} = 5$ V, $R_A = R_B = 10$ kΩ, and $C_X = 0.001$ μF. Vertical scale = 2 V/division, horizontal scale = 5 μs/division.

$\tau_2 = R_B C_X$. But at time t_2, $V_{CAP} = V_1$, and then the output of comparator 1 causes the latch to set and the whole timing cycle repeats itself.

During the charge cycle,

$$V_{CAP}(t) = V_{CC}\left(1 - \tfrac{2}{3}e^{-t/\tau_1}\right) \qquad (8.13.5)$$

where $\tau_1 = (R_A + R_B)C_X$. The latch is reset at time t_1 when

$$V_{CAP}(t_1) = V_2 = \frac{2V_{CC}}{3}$$

Therefore,

$$t_1 - t_0 = \tau_1 \ln\left(\frac{2}{3} \frac{V_{CC}}{V_{CC} - 2V_{CC}/3}\right) \tag{8.13.6}$$

or

$$PW_1 = \tau_1 \ln 2 = 0.69(R_A + R_B)C_X \tag{8.13.7}$$

During the discharge cycle,

$$V_{CAP}(t) = \tfrac{2}{3}V_{CC}e^{-t/\tau_2} \tag{8.13.8}$$

where $\tau_2 = R_B C_X$. The latch is set at time t_2 when

$$V_{CAP}(t_2) = V_1 = \frac{V_{CC}}{3}$$

Therefore,

$$t_2 - t_1 = \tau_2 \ln\left(\frac{2}{3} \frac{V_{CC}}{V_{CC}/3}\right) \tag{8.13.9}$$

or

$$PW_2 = \tau_2 \ln 2 = 0.69R_B C_X \tag{8.13.10}$$

Notice during both timing cycles that the output pulse width is dependent only on the value of the external timing components. Also since $R_A + R_B > R_B$, $PW_1 > PW_2$.

The frequency of oscillation for the astable is the reciprocal of the time period. That is,

$$f_0 = \frac{1}{t_p} = \frac{1}{PW_1 + PW_2} \tag{8.13.11}$$

Therefore, $t_p = 0.69(R_A + 2R_B)C_X$.

The duty cycle of the output waveform is defined as

$$D = \frac{PW_2}{t_p} = \frac{0.69R_B C_X}{0.69(R_A + 2R_B)C_X} = \frac{R_B}{R_A + 2R_B} \tag{8.13.12}$$

The timing characteristics of the 555, which are typical of most IC timers, are given in Table 8.2.

For the 555 with $V_{CC} = 5$ V, the supply current is typically 3 mA; but with the CMOS versions at $V_{DD} = 5$ V, the supply current is typically only 80 μA.

8.14 SUMMARY

This chapter has presented an analysis and description of the design of regenerative logic circuits implemented with:

TABLE 8.2
Characteristics of type 555
integrated circuit timer

Timing accuracy	±1%
Temperature stability	50 ppm/°C
Supply stability	0.05%/V
Output rise/fall times	100 ns

- Bipolar junction transistors (BJT)
- MOS field-effect transistors (NMOS)
- Complementary field-effect transistors (CMOS)

Examples are given of:

- SR latches
- JK flip-flops
- D flip-flops

The difference between level-triggered and edge-triggered flip-flops is described. Also described in this chapter are examples of:

- Schmitt trigger circuits
- Monostable multivibrator circuits
- Astable multivibrtor circuits

Finally, the very popular IC timer is described.

REFERENCES

1. V. H. Grinich and H. G. Jackson, *Introduction to Integrated Circuits*, McGraw-Hill, New York, 1975.
2. H. Taub and D. Schilling, *Digital Integrated Electronics*, McGraw-Hill, New York, 1977.
3. *The TTL Data Book*, Texas Instrument Inc., Dallas, Tex., 1976.
4. *MECL Integrated Circuits*, Motorola Inc., Phoenix, Ariz., 1978.
5. *COS/MOS Integrated Circuits*, RCA Corp., Somerville, N.J., 1977.

DEMONSTRATIONS

D8.1 CMOS Monostable Multivibrator

The objective of this experiment is to demonstrate the timing characteristics of a simple monostable multivibrator made from two CMOS gates.

1. (a) Connect the circuit shown in Fig. 8.24, using the 4001B—a quad 2-input NOR gate. Let $R_X = 10$ kΩ and $C_X = 0.001$ μF.
   (b) With $V_{DD} = 5$ V, trigger the circuit with a pulse 1 μs wide and compare the voltage waveform at the nodes, as illustrated in Fig. 8.24.
   (c) Record the output pulse width V_{O2} as V_{DD} is varied from 5 to 15 V in 1-V steps.
2. Based on the given values of R_X and C_X and the data from Demonstration D2.1, use Eq. (8.12.2) to calculate the output pulse width for $V_{DD} = 5$, 10, and 15 V, and compare with the observed results of part 1(c).

D8.2   CMOS Astable Multivibrator

In this experiment two CMOS gates are connected to yield an astable multivibrator and the timing characteristics are demonstrated.

1. (a) Connect the circuit shown in Fig. 8.25 using the 4001B. Let $R_X = 10$ kΩ and $C_X = 0.001$ μF.
   (b) Compare the voltage waveforms with those illustrated in Fig. 8.25, and comment on any differences.
   (c) Record the output pulse period as V_{DD} is varied from 5 to 15 V in 1-V steps.
2. Repeat part 2 of D8.1 using Eq. (8.12.4).

D8.3   555 Timer: Monostable Multivibrator

The objective of this experiment is to demonstrate the superior timing characteristics of a 555 timer connected as a monostable multivibrator.

1. (a) With $R_X = 10$ kΩ and $C_X = 0.001$ μF, connect the 555 as in the circuit of Fig. 8.28.
   (b) With $V_{CC} = 5$ V, trigger the circuit with a pulse 1 μs wide, and compare the voltage waveforms at the nodes as illustrated in Fig. 8.28.
   (c) Record the output pulse width as V_{CC} is varied from 5 to 15 V in 1-V steps. What are the reasons for the smaller variations in the pulse width as compared to the CMOS monostable of D8.1?
2. Use Eq. (8.13.4) to calculate the output pulse width, and compare with the observed results of part 1(c).

D8.4   555 Timer: Astable Multivibrator

In this experiment the 555 is connected as an astable multivibrator and the timing characteristics are demonstrated.

1. (a) With $R_A = R_B = 10$ kΩ and $C_X = 0.001$ μF, connect the 555 as in the circuit of Fig. 8.29.
   (b) Compare the voltage waveforms with those illustrated in Fig. 8.29, and comment on any differences.
   (c) Record the output pulse period as V_{CC} is varied from 5 to 15 V in 1-V steps.
2. Use Eq. (8.13.11) to calculate the pulse period, and compare with the observed results in part 1(c).

PROBLEMS

P8.1. The voltage waveforms shown in Fig. P8.1 are applied to the JK master-slave flip-flop illustrated in Fig. 8.5. With the flip-flop initially reset, show the resulting waveform at the Q output of the master and slave latches.

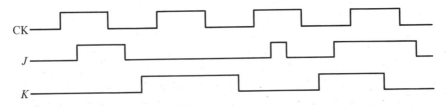

FIGURE P8.1

P8.2. Repeat P8.1 for the JK edge-triggered flip-flop shown in Fig. 8.6. Assume the flip-flop is initially set.

P8.3. For the TTL D flip-flop of Fig. 8.9a, number the NAND gates 1 through 6.
  (a) With CK = D = low and S_D = R_D = high, determine the output state of each gate (H or L). Assume the flip-flop is initially set.
  (b) Repeat part (a) after CK = high.

P8.4. For the ECL D flip-flop of Fig. 8.12a, number the gates 1 through 8.
  (a) With CK = S_D = R_D = low and D = high, determine the output state of each gate (H or L). Assume the flip-flop is initially reset.
  (b) Repeat part (a) after CK = high.

P8.5. For the RTL flip-flop shown in Fig. P8.5, assume $V_{BE(sat)}$ = 0.8 V and $V_{CE(sat)}$ = 0.1 V. Determine values for R_2 and R_3 such that (i) for the off transistor, V_{BE} = −0.5 V, and (ii) for the on transistor, $I_{C(sat)}/I_{B(sat)}$ = 10.

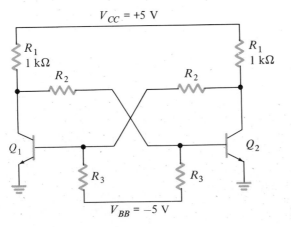

FIGURE P8.5

P8.6. Shown in Fig. P8.6 is a simple ECL latch. The output emitter-followers have been omitted for clarity. Neglect base currents and determine values for R_1, R_2, and R_3 such that (i) $V_{BE(on)} = 0.8$ V, $V_{BE(off)} = 0.2$ V, and (ii) $V_{\bar{Q}} = -0.1$ V, $V_Q = -0.9$ V. Assume $V_S = V_R = -5.2$ V.

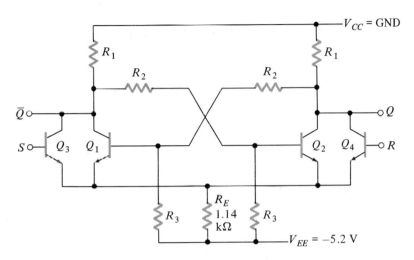

FIGURE P8.6

P8.7. Design a BJT Schmitt trigger circuit similar to Fig. 8.21, with $V_{T+} = 3$ V and $V_{T-} = 2$ V. $V_{CC} = 10$ V and $I_{C(max)} = 5$ mA. The output voltage swing is 5 V, and transistor Q_2 must not saturate. Assume $V_{BE(EOC)} = 0.7$ V, $V_{BE(EOS)} = 0.8$ V, $V_{CE(sat)} = 0.1$ V, and β_F is infinite. (You will need to add a resistor from the base of Q_2 to ground to meet these specifications.)

P8.8. In Fig. 8.21 a resistor R_{E1} in series with the emitter of Q_1 will cause the hysteresis voltage to go to zero. Similarly, a resistor R_{F2} in series with the emitter of Q_2 will result in $V_{T+} - V_{T-} = 0$. Use the same transistor data as in P8.7.

(a) Find the value of R_{E1} to eliminate the hysteresis. What is V_{T+} and V_{T-}?

(b) Similarly, solve for the value of R_{E2}, V_{T+}, and V_{T-}.

P8.9. Refer to the CMOS Schmitt trigger circuit shown in Fig. 8.23. Let $V_{DD} = 5$ V. CMOS data:

$$V_{TP} = -2 \text{ V}, \ k_p' = 10 \ \mu\text{A/V}^2, \ (W/L)_p = 20, \ \text{but}(W/L)_3 = (W/L)_9 = 2$$

$$V_{TN} = +2 \text{ V}, \ k_n' = 20 \ \mu\text{A/V}^2, \ (W/L)_n = 10, \ \text{but}(W/L)_6 = (W/L)_{10} = 1$$

(a) Sketch the VTC including estimates for V_{T+} and V_{T-}.

(b) Verify with SPICE.

P8.10. Shown in Fig. P8.10 is an NMOS Schmitt trigger circuit. Assume $V_{TD} = -2$ V, $V_{TE} = +1$ V, $k' = 20 \ \mu\text{A/V}^2$, and $\gamma = \lambda = 0$.

(a) Sketch the VTC including values for V_{T+} and V_{T-}.

(b) Verify with SPICE.

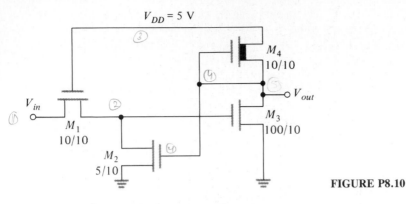

FIGURE P8.10

P8.11. Shown in Fig. P8.11 is an alternative circuit for a CMOS monostable multivibrator.
  (a) Sketch the voltage waveforms at V_1, V_2, and V_X, including a suitable trigger signal.
  (b) Develop a simple equation for the output pulse width at V_1.

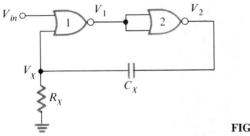

FIGURE P8.11

P8.12. A simple NMOS monostable multivibrator is shown in Fig. P8.12. Assume the NMOS data as in P8.10.
  (a) Sketch the voltage waveforms at V_{DS2}, V_{GS4}, and V_{DS4}, including a suitable trigger signal.
  (b) Calculate the value of C_X for an output pulse width of 10 μs.

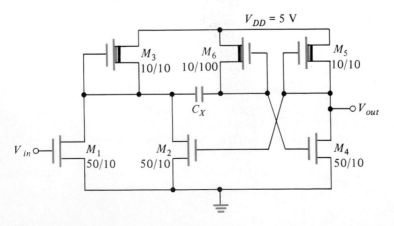

FIGURE P8.12

P8.13. Calculate the frequency of oscillation for a CMOS astable multivibrator circuit similar to Fig. 8.25a but with input protective diodes included. Let $V_{DD} = 10$ V, $V_{Th} = 5$ V, and $V_{D(on)} = 0.7$ V. For the timing components $R_X = 10$ kΩ and $C_X = 0.001$ μF. Assume the gates switch instantaneously.

P8.14. Repeat P8.13 but with $V_{DD} = 5$ V, $V_{TN} = +1$ V, and $V_{TP} = -2$ V.

P8.15. Refer to the monostable multivibrator shown in Fig. 8.26. Transistor data:

$$V_{D(on)} = 0.7 \text{ V}, \ V_{BE(on)} = 0.7 \text{ V}, \ \beta_F = 50, \ V_{BE(sat)} = 0.7 \text{ V}, \ V_{CE(sat)} = 0.2 \text{ V}$$

(a) Determine the threshold voltages V_{T+} and V_{T-} for the Schmitt circuit. ($V_{CC} = 5$ V and $R_X = 10$ kΩ. In the stable state Q_7 is off.)

(b) Show that the pulse width is given as $PW \approx 0.34 R_X C_X$.

P8.16. The complementary outputs of an ECL OR/NOR gate may be used to advantage in constructing a simple monostable multivibrator, as illustrated in Fig. P8.16.

(a) Sketch the voltage waveforms at V_1, V_2, and V_X, including a suitable trigger signal.

(b) Show that the output pulse width is given as $PW \approx 0.69 R_X C_X$.

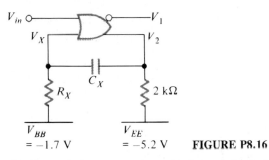

FIGURE P8.16

P8.17. A BJT monostable multivibrator circuit is shown in Fig. P8.17. Assume each transistor instantly saturates when turned on.

(a) Sketch and label the voltage waveforms at each base and collector, including a suitable trigger signal.

(b) Show that the output pulse width is given as $PW \approx 0.69 R_3 C_1$.

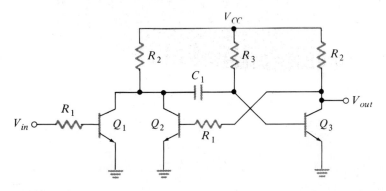

FIGURE P8.17

P8.18. A monostable multivibrator similar to that shown in Fig. P8.17 is to be designed using RTL NOR gates. Calculate the values of R_1, R_2, R_3, and C_1 if the output pulse width is to be 10 μs. The power dissipation in the quiescent state is not to exceed 20 mW. Use the transistor data from P8.7, but $\beta_{sat} = 20$. Fan-out = 5.

P8.19. An emitter-coupled monostable multivibrator circuit is shown in Fig. P8.19. Use the transistor data from P8.7. Also, $V_{SBD(on)} = 0.4$ V.

    (a) Compute the quiescent operating point (I_C and V_{CE}) for each transistor and hence determine the minimum trigger amplitude for V_{in}.

    (b) With a 1-V narrow pulse at V_{in}, what is the output pulse width and amplitude?

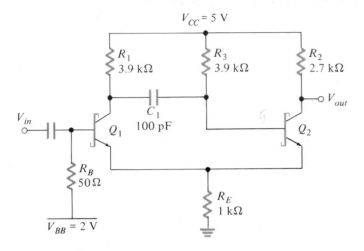

FIGURE P8.19

P8.20. A complementary transistor monostable multivibrator circuit is shown in Fig. P8.20. Assume the same absolute data values for the *pnp* transistor and for the *npn* transistor as in P8.15.

    (a) Describe the operation of this circuit and determine a suitable trigger signal at V_{in}.

    (b) Sketch the voltage waveform at the collector of Q_1 and Q_2 including values for all voltages and times.

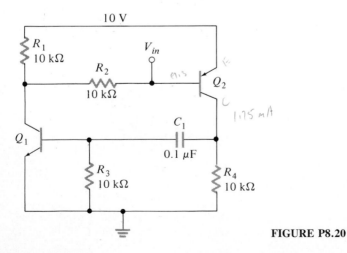

FIGURE P8.20

P8.21. The BJT astable multivibrator circuit shown in Fig. P8.21 is used to generate positive clock pulses at a repetition rate of 100 kilopulses per second. The clock pulse itself is 1 μs wide. Calculate all resistor and capacitor values and sketch the clock pulse waveform, indicating all voltages and times. Assume the transistors instantly saturate when $V_{BE} = 0.7$ V, and are nonconducting when $V_{BE} < 0.7$ V. $\beta_{sat} = 20$, $I_{C(sat)} = 10$ mA, and $V_{CE(sat)} = 0.1$ V.

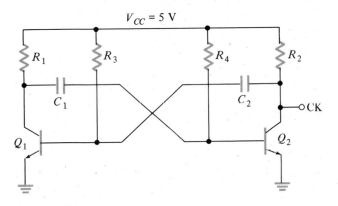

FIGURE P8.21

P8.22. Shown in Fig. P8.22 is an emitter-coupled astable multivibrator circuit. Assume $V_{BE(on)} = 0.8$ V and β_F is infinite.

(a) Sketch the voltage waveforms at V_{E1}, V_{E2}, V_1, and V_2.

(b) Calculate the frequency of oscillation.

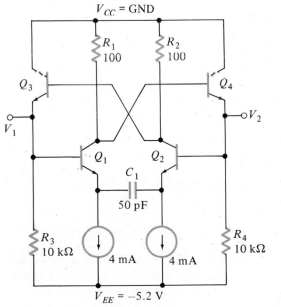

FIGURE P8.22

CHAPTER
9

SEMICONDUCTOR MEMORIES

9.0   INTRODUCTION AND DEFINITIONS

Modern digital systems require the capability of storing and retrieving large amounts of information at electronic speeds. *Memories* are devices or systems that store digital information in large quantity. This chapter addresses the analysis and design of large-scale integrated circuit memories, commonly known as *semiconductor memories*. Magnetic core and magnetic bubble memories are not covered, nor are moving-surface memories such as discs and tapes.

Electronic memory capacity in digital systems ranges from fewer than 100 bits for a simple four-function pocket calculator, to 10^5 to 10^7 bits for a personal computer, and up to 10^{10} bits for the largest commercial computers. Circuit designers usually speak of memory capacities in terms of bits, since a separate flip-flop or other circuit is used to store each bit. On the other hand, system designers usually state memory capacities in terms of *bytes* (8 or 9 bits); each byte represents a single alphanumeric character. Very large scientific computing systems often have memory capacity stated in terms of *words* (32 to 80 bits). Each byte or word is stored in a particular location that is identified by a unique numeric *address*. A key characteristic of memory systems is that only a single byte or word, at one single address, is stored or retrieved during each cycle of memory operation.

Except for the smallest systems, memory storage capacity is usually counted in units of *kilobytes*, or *K bytes*. Because memory addressing is based on binary codes, capacities that are integral powers of 2 are most common. Thus the convention that, for example, 1K byte = 1,024 bytes and 64K bytes = 65,536 bytes.

The memory capacity of computers in bits is usually 100 to 1,000 times the number of logic gates employed in the central processing unit. Clearly the memory cost per bit must be kept very low. Design methods used to achieve high density (and therefore low cost), low power consumption, and fast memory operation are described in the following sections.

Figure 9.1a shows a data storage register for storing 1 byte or word of several bits. Such a register can be designed using an SR or D flip-flop for each bit.

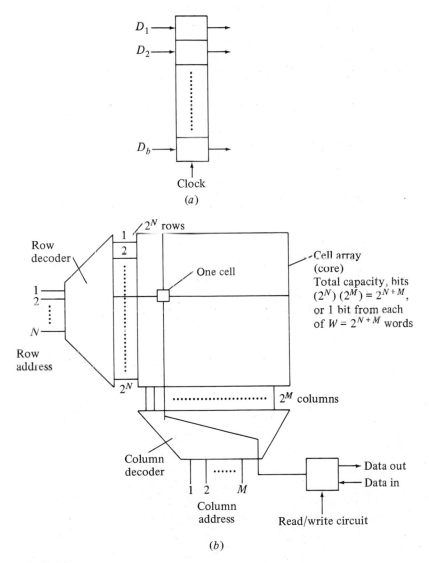

(a)

(b)

FIGURE 9.1

(a) Single data register; b bits. (b) Random access memory organization.

Storage capacity can be increased by using many such registers. However, it is not economical to build large memories this way. Each flip-flop or bit storage cell (henceforth, *cell*) of a register requires a number of transistors and has two or more connections for data access. Additional logic gates would be required to select one register to store or retrieve (*write* or *read*) data. A total of 10 to 30 transistors per bit are required for register arrays. The chip area and cost of a large memory (greater than 64 bits) designed in this way would be grossly excessive.

Memory cell circuits are simplified compared to register circuits by sacrificing most of the desired properties of digital circuits listed in Sec. 1.1. While the logic function is preserved, other properties including quantization of amplitudes, regeneration of logic levels, input-output isolation, and fan-out may be sacrificed for individual storage cell circuits. Thus the complexity of a single cell is reduced to the range from one to six transistors. At the level of a memory chip, the desired properties are recovered through use of properly designed *peripheral circuits*. These circuits are designed so that they may be shared among many memory cells.

The preferred organization for most large memories is shown in simplest form in Fig. 9.1*b*. This is termed *random-access memory* (*RAM*) organization. The name derives from the fact that memory locations (*addresses*) can be accessed in random order, at a fixed rate independent of physical location, for reading or writing. The storage array, or *core*, of a RAM is made up of simple cell circuits arranged to share connections in horizontal rows and vertical columns. The horizontal lines, which are driven only from outside the storage array, are sometimes called *word lines*, while the vertical lines, along which data flow into and out of cells, in this parlance are called *bit lines*. A cell is made available (*accessed*) for reading or writing by selecting one row and one column. (Some memory components simultaneously select four or eight columns in one row so that a group of four or eight cells are simultaneously accessed.) The row and column (or group of columns) to be selected are determined by decoding binary-encoded address information. For example, a *decoder* for row selection as shown in Fig. 9.1*b* has 2^N output lines, a different one of which is selected for each different N-bit input code.

Read-write (R/W) circuits determine whether data are being retrieved or stored, and perform any necessary amplification, buffering, and translation of voltage levels. Specific examples are presented in the following sections.

Read-write random-access memories may store information in flip-flops, or simply as charges on capacitors. Approximately equal delays (usually in the range 10 to 500 ns) are encountered in reading or writing data. Because read-write memories store data in active circuits, they are *volatile*; that is, stored information is lost if the power supply is interrupted. The natural abbreviation for read-write memory would be RWM. However, pronunciation of this acronym is difficult. Instead, the term RAM is commonly used to refer to read-write random-access memories. If we were consistent, both read-only (see below) and read-write memories would be called RAMs.

Read-only memories (ROMs) store information according to the presence or absence of diodes or transistors joining rows to columns. ROMs employ the organization shown in Fig. 9.1*b* and have read speeds comparable to those for read-write memories. All ROMs are nonvolatile, but they vary in the method used to enter (write) stored data. The simplest form of ROM is written when it is manufactured by formation of physical patterns on the chip; subsequent changes of stored data are impossible. These are termed *mask-programmed*, or simply *masked*, ROMs. *Programmable* read-only memories (PROMs) have a data path present between every row and column when manufactured, corresponding to a stored 1 in every data position. Storage cells are selectively switched to the zero state (the PROM is programmed) once after manufacture by applying appropriate electrical pulses to selectively open (blow out) row-column data paths. Once programmed, or *blown*, a 0 cannot be changed back to a 1.

Erasable programmable read-only memories (EPROMs) also have all bits initially in one binary state. They are programmed electrically (similar to the PROM), but all bits may be erased (returned to the initial state) by exposure to ultraviolet (UV) light. The packages for these components have transparent windows over the chip to permit the UV irradiation. Finally, *electrically erasable programmable read-only memories* ("E squared," or E^2 PROM) may be written and erased by electrical means. These are the most advanced and most expensive form of PROM. Unlike EPROMs, which must be totally erased and rewritten to change even a single bit, E^2PROMs may be selectively erased. Writing and erasing operations for all PROMs require times ranging from 10 μs to many milliseconds. However, all PROMs retain stored data when power is turned off; thus they are termed *nonvolatile*.

A nonvolatile memory that could read or write any location in a few microseconds or less would be termed a *nonvolatile RAM*. Nonvolatile semiconductor RAMs based upon MNOS (metal-nitride-oxide-semiconductor) technology are available but do not offer the bit density or short access times of more conventional RAMs. They are used in such applications as television tuners with preset channel selections.

Memories based on magnetic materials, e.g., ferrite cores, can be designed to retain stored information when power is off. Despite the disadvantage of volatility, semiconductor memories are preferred over magnetic core memories for most applications because of their advantages in cost, operating speed, and physical size.

The principal time-dependent parameters of random-access memories are illustrated in Fig. 9.2. *Read access time* is the delay from presentation of an address until data stored at that address are available at the output. Maximum read access time is an important memory parameter. *Cycle time* is the reciprocal of the time rate at which address information is changed while reading or writing at random locations. A minimum value of cycle time, below which errors may occur, is specified. Minimum cycle times for reading and writing are not necessarily the same, but for simplicity of design, most systems employ a single minimum cycle time for both reading and writing. For semiconductor read-write memories, the read access time is typically 50 to 90% of read cycle time.

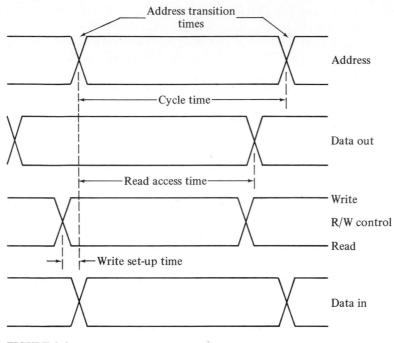

FIGURE 9.2
Definition of memory transient parameters.

9.1 READ-ONLY MEMORIES

Read-only memories are used to store constants, control information, and program instructions (*firmware*) in digital systems. They may be thought of as components that provide a fixed, specified binary output for every binary input. The cartridges for most of the popular home video games each contain a 16K or 32K MOS ROM chip. In this section we will study the internal circuits in ROMs based on MOS and bipolar integrated circuit technologies.

9.1.1 MOS ROM Cell Arrays

A bit is stored in a ROM by the presence or absence of a data path from a row to a column. The absence of a path is achieved simply by having no circuit element joining row and column. Paths must be established using nonlinear elements (transistors or diodes) to ensure that data flow only in the desired routes from input to output. Figure 9.3 shows the two basic forms of MOS ROM cell arrays. In each array, bits are stored according to the presence or absence of a transistor switch at each row-column intersection.

The NMOS array of Fig. 9.3a implements the NOR function in the sense that a column goes low when any row, joined to the column with a transistor, is

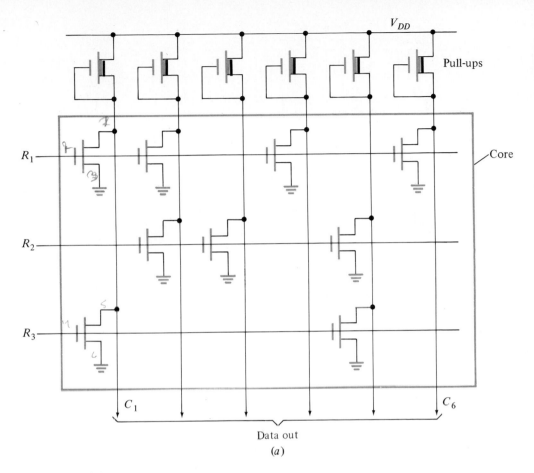

FIGURE 9.3
(a) NOR array.

raised to a high level. In normal operation, all but one row conductor is held low. When a selected row line is raised to V_{DD}, all transistors with gates connected to that line turn on. The columns to which they are connected are pulled low. The remaining columns are held high by the pull-up or load devices shown at the top in this example. Using positive logic, a stored 1 is defined as the absence of a transistor. Usually the array is formed with transistors at every row-column intersection. The desired bit pattern is placed in the array by omitting the drain or source connection, or the gate electrode, at locations where a 1 is desired.

The array shown in Fig. 9.3b is usually called a NAND ROM because the column output goes low only when all series bit locations provide a conducting path toward ground. In this case, all but one of the row conductors are normally held at V_{DD}. When the selected row line is pulled low, all transistors with gates

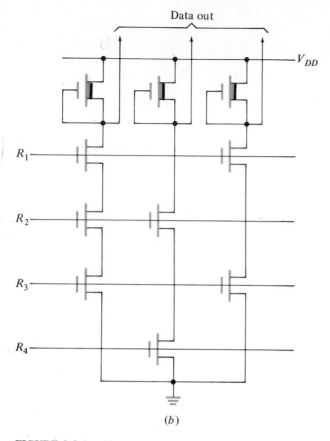

FIGURE 9.3 (cont.)
(b) NAND array.

connected to that line turn off. Above these transistors, the columns to which they are connected are pulled high by the pull-up devices shown. The data output is taken at the top. In positive logic, a stored 1 is defined as the presence of a transistor, while a stored 0 is achieved with a direct connection in place of the transistor. The array is formed with transistors at every intersection, but the source-drain paths of transistors at the desired 0 locations are shorted out by an implant or diffusion.

The performance characteristics of ROMs depend on the details of the MOS fabrication process. In any case, the NOR array usually has a faster access time and has the advantage that the stored bit pattern can be determined by the mask defining contacts to the transistors or the metal interconnection layer. Therefore these ROMs may be kept in inventory with most of the fabrication processing completed, then customized quickly to a particular bit pattern by preparing and using a mask that makes contact only to transistors at 0-bit locations.

The NAND-based ROMS have longer access time and must be customized (usually with a masked implant to form electrical connections between source and drain where stored 0s are desired) near the beginning of the manufacturing process. Their advantage is that bit density per unit area is considerably higher than that for a NOR-based ROM using the same process and design rules.

The access time of a ROM is limited by the resistance and capacitance of row and column lines and the currents available to drive these lines. For large ROMs with a low cost per bit, emphasis is on high circuit layout density. The result is cell and decoder devices with small W and L, resulting in relatively small driving currents. Decoder drive current is limited by power consumption considerations as well as by device area considerations.

In silicon-gate MOS technology, rows are usually formed in polycrystalline silicon and must be driven through a voltage excursion of several volts. The sheet resistance of polysilicon ranges from about 20 to 50 Ω per square. Thus a row line 6 μm wide and 3,000 μm long (that is, 500 squares) could have as much as 25,000 Ω of series resistance. Supposing that such a line forms a 6- by 6-μm gate for each of 128 transistors, the total row line capacitance will be about 2 pF. The distributed RC time constant can become a serious limitation on access time under conditions like these.

Closed-form analysis of the transient response of distributed RC circuits is complex. Furthermore, some circuit simulators do not provide for simulation of distributed circuits. A useful rule of thumb for the case of a uniformly distributed RC line with a step-function input and open-circuit conditions at the far end of the line can be stated as follows. The propagation delay time to the 50 output voltage transition is $0.38RC$, where R and C are the total resistance and total capacitance for the distributed circuit. This result should be compared to the corresponding delay time to the 50% point for a classical RC charging problem, which is $0.69RC$. Note that if the length of a uniformly distributed RC line is doubled, the delay is multiplied by 4.

Current begins to flow in the columns as soon as the row voltage rises above V_T of the cell transistors. Even when the row line reaches its maximum value (usually 5 V), the column current from a small array transistor is usually in the range 10 to 100 μA, while the column capacitance is usually 1 to 5 pF. The resulting time rate of change of column voltage can be rather small. These data correctly suggest that careful design is necessary to achieve short access time with low power consumption.

Example 9.1. A simplified layout for a NOR ROM cell in silicon-gate NMOS is shown in Fig. 9.4. Assume that 1s are programmed in this ROM by using a channel implant that raises threshold voltage to a value $V_{TI} > 5$ V.

(a) Use the given process and device data to calculate transistor parameters and row and column C and R per memory cell.

(b) Make hand calculations to compare the access time of storage arrays for 32K-bit ROMs organized as 128 rows by 256 columns, and as 256 rows by 128 columns.

The following data are available:

Name	Symbol in text	Value	SPICE name
Threshold voltages	V_{TE}	+1.0 V	VTO
	V_{TD}	–3.0 V	VTO
	V_{T1}	+6.0 V	VTO
Transconductance	k	20 μA/V^2	KP
Body factor	γ	0.37	GAMMA
Body doping	N_A	5×10^{14} cm^{-3}	NSUB
Source, drain doping	N_D	1×10^{20} cm^{-3}	
Source, drain sheet resistance		20 Ω per square	
Polysilicon sheet resistance		40 Ω per square	
Gate oxide thickness	t_{ox}	0.1 μm	TOX
Junction depth	X_j	1.0 μm	XJ
Lateral diffusion		1.0 μm	LD
Metal sheet resistance		0.2 Ω per square	

Solution

(a) The drawn value of transistor W/L is $\frac{7}{7}$ μm. Due to lateral diffusion, the final value of L is 5 μm. Therefore, device transconductance is given by

$$k = k'\frac{W}{L} = 20 \times \frac{7}{5} = 28 \ \mu\text{A/V}^2$$

Capacitances per unit area are calculated as they were for Example 3.4.

Symbol	Value	Where from?	SPICE name
ϕ_0	$=0.86$ V	Eq. (3.7)	PB
C_{j0}	$= 70 \ \mu$F/m^2	Eq. (3.6)	CJ
C_{jsw}	$= 220$ pF/m	$X_j\sqrt{10}\ C_{j0}$	CJSW
C_{ox}	$= 345 \ \mu$F/m^2	Eq. (2.8)	
$C_{ox}LD$	$= 345$ pF/m		CGSO = CGDO

Areas, perimeters, and capacitances are calculated from Fig. 9.4.

Row (gate) capacitance and resistance:

$$\text{Area of thin oxide} = 7 \times 7 = 49 \ \text{pm}^2/\text{bit}$$

$$\text{Capacitance} = 49 \times 345 = 16.9 \ \text{fF/bit}$$

$$\text{Poly resistance} = \frac{14}{7} \times 40 = 80 \ \Omega/\text{bit}$$

Column (drain) capacitance and resistance:

$$\text{Area} = \frac{20 \times 7}{2} = 70 \ \text{pm}^2/\text{bit}$$

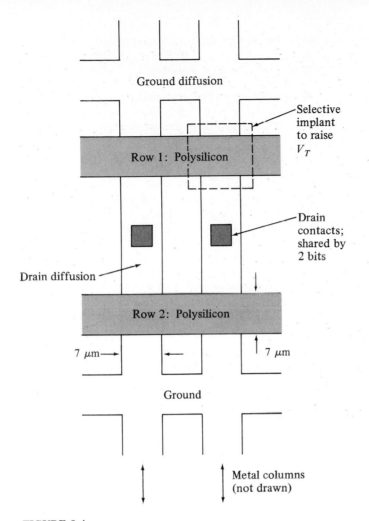

Ground diffusion

Selective
implant
to raise
V_T

Row 1: Polysilicon

Drain
contacts;
shared by
2 bits

Drain diffusion

Row 2: Polysilicon

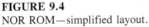 7 μm→   ←   7 μm

Ground

Metal columns
(not drawn)

FIGURE 9.4
NOR ROM—simplified layout.

$$\text{Oxide sidewall perimeter} = \frac{2 \times 20}{2} = 20 \ \mu\text{m/bit}$$

$$\text{Gate overlap area} = 1 \times 7 = 7 \ \mu\text{m}^2/\text{bit}$$

$$\text{Junction capacitance at zero bias} = 70 \times 70 + 20 \times 220$$

$$= 4.9 + 4.4 = 9.3 \ \text{fF/bit}$$

$$\text{Overlap capacitance} = 7 \times 345 = 2.4 \ \text{fF/bit}$$

$$\text{Metal resistance} = \frac{24}{7} \times 0.2 = 0.7 \ \Omega/\text{bit (negligible)}$$

Ground (source):

Always grounded; capacitance has no influence on performance.

$$n^+ \text{resistance} = \frac{14}{7} \times 20 = 40 \ \Omega/\text{bit}$$

In the above calculations, the capacitances between elements separated by thick dielectrics (e.g., polysilicon to field, metal to polysilicon) are omitted because these capacitances are 5 to 10 times smaller per unit area than those considered. The gate overlap capacitance is added to column capacitance because each of the 127 or 255 unselected cells contribute such a capacitance to the total column capacitance.

(b) Several assumptions must be made to calculate access time. Provided the same (reasonable) assumptions are made in both cases, the comparison will be qualitatively correct. The hand analysis is not expected to be accurate in an absolute sense. SPICE should be used for accurate evaluation of absolute access time.

- Assume 5-V operation; $V_{OH} - V_{OL} \approx 5$ V.
- Note that (as expected for NMOS) longest delay occurs for t_{PLH} on column; assume row line is driven with an ideal step from 5 to 0 V.
- Calculate row delay t_R to furthest cell as $0.38RC$.
- Calculate column delay t_C while the load transistor charges the column capacitance; assume a depletion load transistor is used, with $K_R = 4$. A conservative approximation is to use the zero-bias value of junction capacitances.
- Access time $t_a \approx t_R + t_C$

First complete the design of the load transistor. For $K_R = 4$, it should have final $W/L = \frac{7}{20}$, which will provide a saturated drain current

$$I_{DL} = \frac{7}{20} \times \frac{20}{2}[0 - (-3)]^2 = 31.5 \ \mu A$$

The column decoder must be designed so that only a small number of load devices (say, one or four or eight) conduct simultaneously. If 128 or 256 loads could conduct at once, the voltage drops in the diffused ground line parallel to a selected row would be unacceptably large.

For the array of 128 rows and 256 columns, the row delay time is

$$t_R = 0.38(256 \times 80 \ \Omega)(256 \times 16.9 \ \text{fF}) = 34 \ \text{ns}$$

The column delay time is calculated in the manner shown in Sec. 3.4. The column capacitance is $(128)(9.3 + 2.4 \ \text{fF}) = 1.5$ pF. The average current for the transition from 0 to 2.5 V is found to be 28 μA.

$$t_C = \frac{C \Delta V}{I_{av}} = \frac{1.5 \ \text{pF} \times 2.5 \ \text{V}}{28 \ \mu A} = 134 \ \text{ns}$$

The access time for the 128 by 256 array is then

$$t_a = t_R + t_C = 168 \ \text{ns}$$

The corresponding calculations for the 256 row and 128 column array give

$$t_R = 8.5 \text{ ns} \qquad t_C = 280 \text{ ns} \qquad t_a = 289 \text{ ns}$$

The conclusion to be drawn in this case is that performance of the 128 by 256 array is considerably better. A complete comparison would include a study of the comparative characteristics of the row and column decoders for the two arrays, analysis for the worst-case variations in device parameters, and consideration of the buffer amplifier between the output of the column decoder and chip output. Possible buffer amplifiers are described in Sec. 10.6. Note that if ΔV on the columns could be reduced, say from 2.5 to 0.5 V, the access time for both arrays would be greatly reduced.

9.1.2   MOS EPROM and E^2 PROM Cells

The most widely used form of erasable programmable ROM is based on the special MOS device structure shown in Fig. 9.5. This storage structure is used in the NOR cell array of Fig. 9.3a. Two layers of polysilicon form a double gate, as seen in Fig. 9.5a. Gate 1 is a "floating gate" that has no electrical contact. Gate 2 is used for cell selection, taking the role of the single gate of the MOS ROM cell in the NOR array. A circuit symbol for the structure with critical capacitances drawn in is shown in Fig. 9.5b.

Operation of this EPROM cell relies on being able to store charge on the floating gate. Initially, we assume no charge on the floating gate so that with

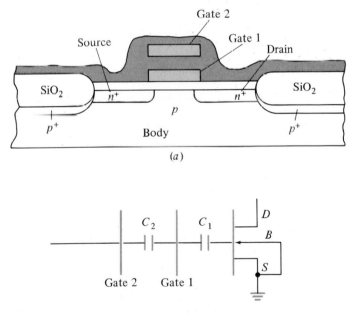

(a)

(b)

FIGURE 9.5
MOS erasable programmable read-only memory.

gate 2, drain, and source all grounded the potential of gate 1 is 0 V. As the voltage on gate 2 is increased, gate 1 voltage rises also, but at a lower rate as determined by the capacitive divider $C_2 - C_1$. The effect of this is to raise the threshold voltage of this transistor as seen from gate 2. However, when gate 2 voltage is raised sufficiently (above about twice the normal V_T), a channel forms. Under these conditions the device provides a positive logic stored 0 when used in a NOR array.

To write a 0 in this cell, both gate 2 and drain are raised to about 25 V while source and substrate remain grounded. A relatively large drain current flows due to normal device conduction characteristics. In addition, the high field in the drain-substrate depletion region results in avalanche breakdown of the drain-substrate junction, with a considerable additional flow of current. The high field in the drain depletion region accelerates electrons to high velocity such that a small fraction traverse the thin oxide and become trapped on gate 1.* When gate 2 and drain potentials are reduced to zero, the negative charge on gate 1 forces its potential to about -5 V. If gate 2 voltage for reading is limited to $+5$ V, a channel never forms. Thus a 1 is stored in the cell.

Gate 1 is completely surrounded by silicon dioxide (SiO_2), an excellent insulator, so charge can be stored for many years. Data may be erased, however, by exposing the cells to strong ultraviolet (UV) light. The UV radiation renders the SiO_2 slightly conducting by direct generation of hole-electron pairs in this material. These EPROMs must be assembled in packages with transparent covers so that they may be exposed to UV radiation.

Electrically erasable PROMS employ a somewhat different structure and a different mechanism for writing and erasing. A portion of the dielectric separating gate 1 from body and drain is reduced in thickness to about 100 Å. When about 10 V is applied across this thin dielectric, electrons flow to or from gate 1 by a conduction mechanism known as *Fowler-Nordheim tunneling.*† This mechanism is reversible, so erasing is achieved simply by reversing the voltage applied for writing. For reading, operation is similar to the EPROM. The threshold voltage as seen at gate 2 is raised if electrons have been placed on gate 1.

9.1.3 MOS Decoders

The row and column decoders identified in Fig. 9.1*b* are essential elements in all random-access memories. Access time and power consumption of memories may be largely determined by decoder design. Similar designs are used in read-only and read-write applications.

---

*D. Frohman-Bentchkowsky, "FAMOS—A New Semiconductor Charge Storage Device," *Solid-State Electronics,* vol. 17, 1974, pp. 517–529.

†E. H. Snow, "Fowler-Nordheim Tunneling in SiO_2 Films," *Solid-State Communications,* vol. 5, 1967, pp. 813–815.

Figure 9.6 shows NMOS decoders based on NOR and NAND gates. Input to the decoder is a binary address of N bits. The input signals are restored to uniform levels, inverted (to provide each bit and its complement), and provided with adequate drive capability using *address buffer* circuits. Two inverters per address bit provide the simplest implementation of these functions.

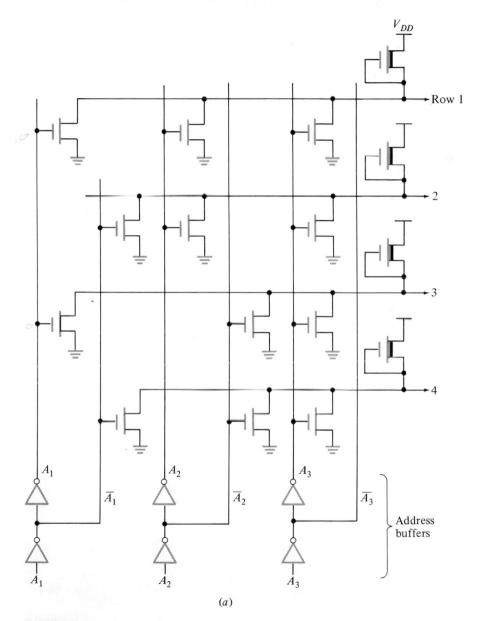

(a)

FIGURE 9.6
(a) NOR decoder in array form.

The decoders select one of 2^N output lines for each address input using 2^N logic gates, each with N inputs. In the NOR decoder of Fig. 9.6a, one of 2^N output lines goes high when selected, so this decoder is suitable for direct connection to the row lines in the MOS NOR ROM of Fig. 9.3a. In the NAND decoder of Fig. 9.6b, one output line goes low when selected, so this decoder is suitable for direct connection to rows in the MOS NAND ROM of Fig. 9.3b. Of course, a single inverter may be used between each decoder output and the ROM array row lines if it is desired to use a NOR decoder with a NAND array, or vice versa.

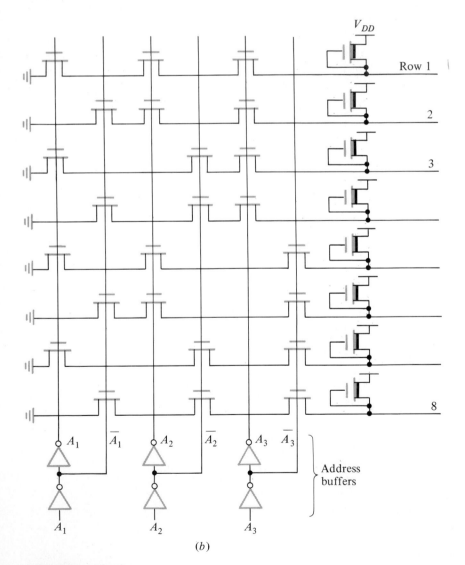

(b)

FIGURE 9.6 (*cont.*)
(b) NAND decoder.

The geometrical arrangement of the logic gates in the address decoder may be designed to be the same as in the ROM array. Then the decoder and ROM array are two similar structures, differing only in the data pattern. Figure 9.7 shows the decoder of Fig. 9.6a and the ROM array of Fig. 9.3a redrawn together to make this conclusion obvious. The programmable logic array described in Sec. 10.4 is derived from this type of ROM.

A different decoder design is often used on the columns of ROMs and RAMs (henceforth we revert to popular usage of RAM to denote random-access read-write memory). In the ROMs of Fig. 9.3, one column, or perhaps a group of four or eight columns, selected from a larger number of columns, must be given an output path. One possibility would be to use a NOR decoder driving transmission gate, as shown, for example, in Fig. 9.8a. However, the tree decoder shown in Fig. 9.8b provides the same function at the expense of a smaller number of transistors. The selected column is given a direct conducting path to the output through transistor channels. Improved load-driving capability is obtained by using one or two inverters to drive the data output node, as shown in Fig. 9.8.

9.1.4   MOS Sense Amplifiers

The maximum voltage at the labeled *data* outputs from the column decoders shown in Fig. 9.8 approaches one threshold below the supply in the high state. As was seen in Example 9.1, memory cell transistors with minimum length and width are normally chosen ($W/L \approx 1$) to minimize chip area. Using the parameters of Example 9.1, the low state output voltage will be less than V_T. In this case, a simple inverter (or more likely, a cascade of two or three inverters) may be used to buffer the output data and deliver conventional logic levels of $V_{OH} = 5$ V and $V_{OL} = 0.3$ V.

This simple approach results in long memory access times, as seen in Example 9.1. Access time can be shortened greatly by reducing the voltage change on the highly capacitive column line. Figure 9.9 shows NMOS and CMOS amplifiers (these are commonly called *sense amplifiers*) that when properly designed restrict the voltage swing at their inputs to 0.5 V or less. Commonly, ROMs use eight identical sense amplifiers to provide simultaneous output of 8 data bits. When using sense amplifiers, the pull-up transistors (shown at the top of Fig. 9.3) are removed. Column current is provided by M_4 within each sense amplifier. For simplicity, with each amplifier in Fig. 9.9 we show only a single memory cell transistor M_1 and a single transistor M_2 representing the combined effect of all series-connected transistors in the column decoder. These sense amplifiers are purely static circuits that require no clock signals or precharge time for correct operation.

Both circuits are designed to minimize the voltage swing on highly capacitive nodes 1 and 2 while producing a larger voltage swing at node 3. The maximum value of V_2 is one NMOS threshold voltage below the fixed reference voltage V_4, corresponding to zero current in the selected memory cell, i.e., positive logic 1. Under these conditions, V_3 rises to the full supply voltage V_{DD}. When a cell

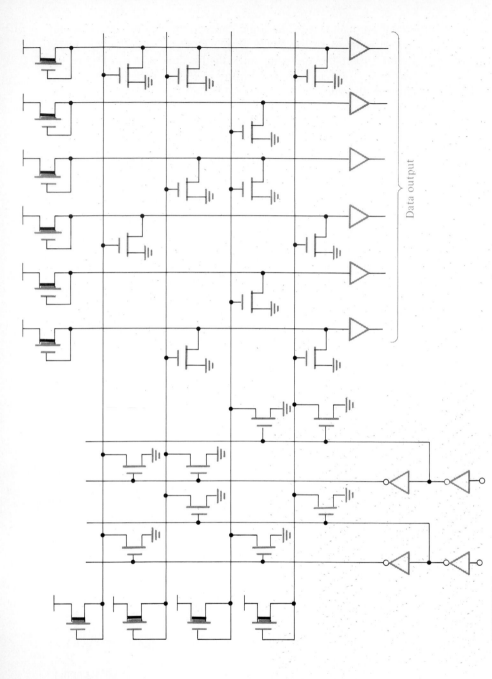

FIGURE 9.7
Decoder plus ROM as two arrays.

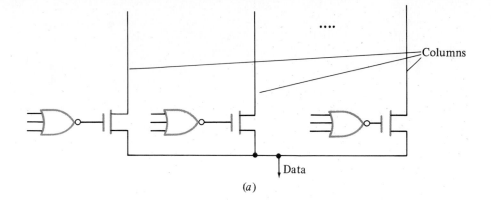

(a)

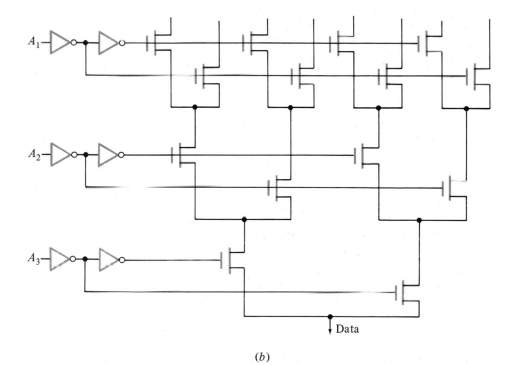

(b)

FIGURE 9.8
(a) NOR plus pass gate for column decode. (b) Tree decoder.

programmed to logic 0 is selected, V_1 and V_2 should fall only slightly, while V_3 falls from V_{DD} until it approaches the value of V_2. This result is achieved by designing such that M_2 and M_3 have much larger values of W/L or, in other words, be *much stronger* devices than M_1. On the other hand, M_4 must be designed as a weaker device with less conduction capability than M_1. To obtain the desired large voltage swing at node 3, the maximum value of V_2 should be less than

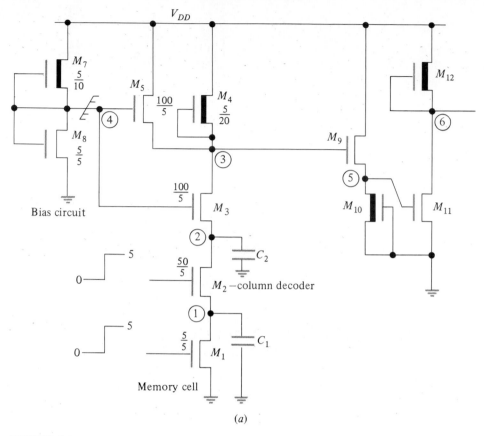

(a)

FIGURE 9.9
(a) NMOS sense amplifier.

half of V_{DD}. However, V_2 should not approach zero because this would reduce the current available from minimum-size memory device M_1. Note that M_1 when conducting will be operating in the linear region, since it will have $V_{GS} = V_{DD}$ and V_{DS} less than half of V_{DD}. With $V_{DD} = 5$ V, a reasonable compromise is achieved with $V_4 \approx 3$ V. The bias circuit, composed of M_7 and M_8, produces reference voltage V_4. It may be shared by all eight sense amplifiers.

In the NMOS sense amplifier, M_5 provides additional current that prevents V_2 from falling more than a few tenths of a volt below the initial level achieved with zero current. This speeds up the recovery after reading a logic 0. A level shifter, composed of M_9 and M_{10}, produces an output that swings between about 0.5 and 3.4 V. It is needed because the lowest level for V_3 is only about 1.8 V, not low enough to turn off M_{11}. The output inverter M_{11} and M_{12} must be designed with a larger-than-normal value of K_R in order to achieve an acceptable value of V_{OL} with an input of about 3.4 V. In the CMOS sense amplifier, M_5 and M_6 cause node 6 to swing nearly the full V_{DD} voltage without additional level shifting.

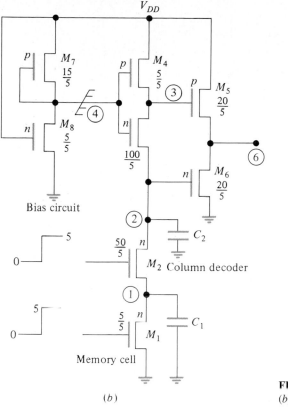

V_{DD}

Bias circuit

Column decoder

Memory cell

(b)

FIGURE 9.9 (*cont.*)
(*b*) CMOS sense amplifier.

The device sizes shown in Fig. 9.9 are representative. In normal operation, V_1 will fall by a fraction of a volt and there will be small voltage drops across M_2 and M_3. If the sizes of M_2 and M_3 are increased to reduce these nonideal voltage drops, total capacitances on nodes 2 and 3 will increase, slowing memory operation. Many other sense amplifier designs, dynamic as well as static, are possible.

9.1.5 Bipolar ROM and PROM Cell Arrays

The shortest memory access and cycle times are achieved with bipolar transistor technology. The cells used in bipolar ROMs and PROMs are shown in Fig. 9.10. The Schottky diode and emitter-follower ROM arrays shown in Fig. 9.10*a* and *b* are programmed by selectively omitting a contact at the contact mask step of the fabrication process. The emitter-follower PROM array of Fig. 9.10*c* incorporates a small fusible link in series with each emitter. These fusible links are formed using nichrome, polysilicon, or another conductor (aluminum doesn't work) and require extra steps in fabrication for deposition and patterning of this material. The memory is programmed by selectively blowing fuse links in the desired data pattern. These PROMs are designed for operation with a 5-V supply while reading. The memory circuits are designed so that higher voltages (10 to 15 V)

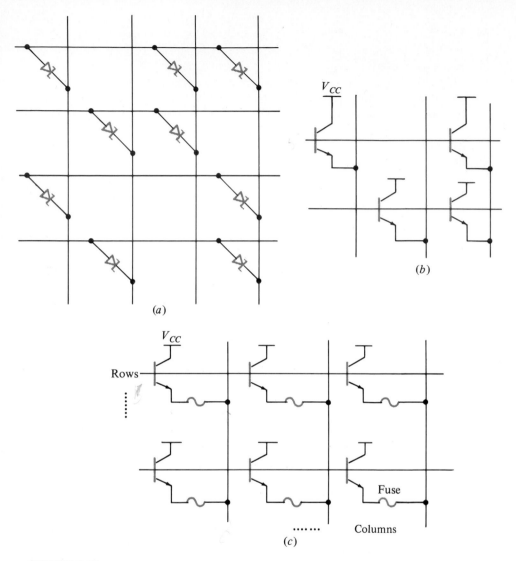

FIGURE 9.10
Bipolar ROM and PROM arrays. (*a*) Schottky ROM. (*b*) Emitter-follower ROM. (*c*) Emitter-follower PROM.

are necessary to force the 10 to 30 mA necessary to blow fuses. Thus there is no danger of blowing fuses during reading.

9.1.6   Bipolar Decoders and Read Circuits

Decoders for bipolar transistor ROM and RAM are often based on standard TTL or ECL gate circuits. Power consumption and complexity may often be reduced if the designer takes advantage of the particular features of a specific application. For instance, row decoders for the ROM and PROM arrays of Fig. 9.10 do not require strong pull-down capability. A simple AND gate shown in Fig. 9.11*a* is perfectly satisfactory as a row decoder in this application.

Column selection for these ROMs requires routing of a selected column current to the input of a read amplifier buffer. Bipolar transistors operated as transmission gates are suitable for the current routing operation provided the base drive current is controlled. An example is shown in Fig. 9.11*b*. Resistor R_4 is chosen to set the base current of column switch transistor Q_4 at a magnitude much smaller than the current available from a column in the array. When reading a 0, no current flows down the column, so the current flowing after column selection is only the base current of Q_4. When reading a 1, the corresponding current after

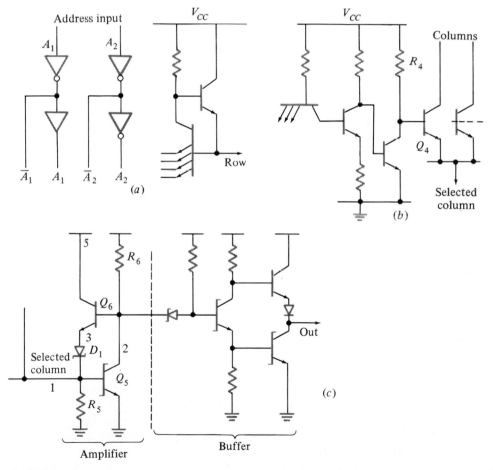

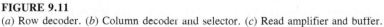

FIGURE 9.11
(*a*) Row decoder. (*b*) Column decoder and selector. (*c*) Read amplifier and buffer.

column selection is the sum of base current and column current, which should be several times larger than the zero current.

A read amplifier and buffer as shown in Fig. 9.11c can be used to sense the data and restore it to TTL logic levels. Resistor R_5 in this circuit must be chosen in relation to R_4 of the previous circuit so that the zero current through R_5 does not exceed about half of $V_{BE(on)}$ of transistor Q_5.

9.2   STATIC READ-WRITE MEMORIES

Memories and other sequential circuits are said to be *static* if no periodic clock signals are required to retain stored data indefinitely. Read-only memory cell arrays are inherently static. Read-write memory cell arrays based on flip-flop circuits are static. These cells and their associated peripheral circuits are described in this section. The simpler read-write memory cells that store information only as charge on a capacitor require periodic clock signals for *refreshing* of stored data. These *dynamic* memories are described in Sec. 9.3.

9.2.1   Static MOS Memory Cells

Popular static memory cells based on NMOS and CMOS technology are illustrated schematically in Fig. 9.12. Because their modes of operation are very similar, they will be described together. Both cells employ a pair of cross-coupled inverters, M_1, M_5 and M_2, M_6, as the storage flip-flop. Major design efforts are directed at minimizing the chip area and power consumption of these cells so that 16,384 or more may be fabricated on a single chip of reasonable dimensions.

The steady-state power consumption of the CMOS circuit is very small because it is determined by junction leakage currents only. In the NMOS circuit, one inverter is always on, drawing current from V_{DD}. This cell standby current can be reduced only by designing M_5 and M_6 with W/L much smaller than unity, at the expense of increased cell area. If the standby current of an NMOS cell must be reduced to the range of 1 μA or less, this may be achieved in a small area by replacing load devices M_5 and M_6 with resistors formed in undoped polysilicon. Sheet resistance of these resistors is 10 MΩ per square or higher. Thus power and area may be saved at the expense of extra processing complexity to form the undoped polysilicon resistors.

These cells both employ a pair of access devices, M_3, M_4, to provide a switchable path for data into and out of the cell. Row select line R is held low except when cells connected to it are to be accessed for reading or writing. Two column lines, C and $\overline{C}$, provide the data path. While this is not used as a true differential signal path, normal operation is symmetrical in a more limited sense, as explained below.

In principal, it should be possible to achieve all memory functions using only one column line and one access device. Attempts have been made to do this, but due to normal variations in device parameters and operating conditions, it is difficult to obtain reliable operation at full speed using a single access line

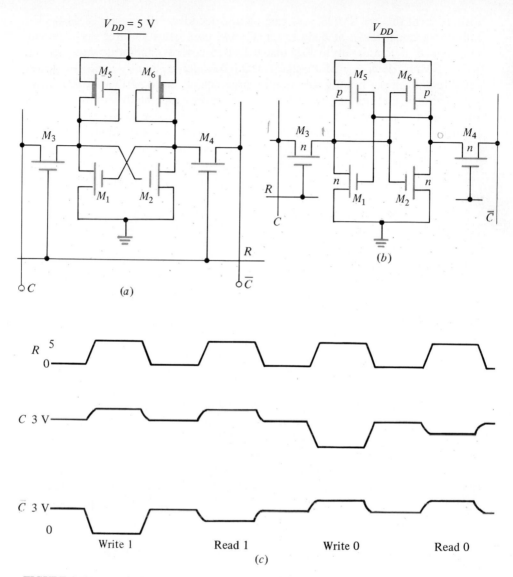

FIGURE 9.12
Static MOS memory cells.

to flip-flop cells. Therefore, the symmetrical data paths C and $\overline{C}$ as shown in the figure are almost always used.

Row selection in NMOS memory may be accomplished using the NOR decoder of Figs. 9.3a or 9.6a or a CMOS NOR decoder. Column selection in NMOS may be achieved using NORs plus pass gates or a full tree decoder, in the forms shown in Fig. 9.8. Twice as many pass devices are required as are shown in Fig. 9.8, because C and $\overline{C}$ require separate transistors. Note that the highest positive level that an NMOS pass gate or tree decoder will convey is $V_{DD} - V_T$, with due account taken of body effect. CMOS pass gates and tree decoders can be designed to conduct up to V_{DD} (using PMOS devices) and/or down to ground (using NMOS devices). Multistage CMOS decoder designs, which use fewer transistors when performing selection in large arrays, are described in Sec. 10.6.

The cell circuits of Fig. 9.12a and b are best operated with voltage and timing conditions approximately as shown in Fig. 9.12c. Assume that a stored 1 is defined as the state in which the left side of the flip-flop is high; that is, M_1 is off. In Fig. 9.12c we introduce a graphical convention that signals originating from external row and column circuits are drawn with sharp trapezoidal shapes, while the typically weaker signals coming from the memory cell are drawn with gradual rise and fall times.

Operation of the static NMOS and CMOS memory cells proceeds as follows. The row selection line, held low in the standby state, goes to V_{DD}, turning on access transistors M_3 and M_4. Writing is accomplished by forcing either C or $\overline{C}$ low, while the other remains at about 3 V. To write a 1, $\overline{C}$ is forced low. The cell must be designed such that the conductance of M_4 is several times larger than that of M_6 so that the drain of M_2 and gate of M_1 may be brought below V_T. M_1 turns off and its drain voltage rises due to the currents from M_5 and M_3. Within about 10 ns, M_2 turns on and the row line may return to its low standby level, leaving the cell in the desired state.

To read a 1, C and $\overline{C}$ are initially biased at about 3 V. When the cell is selected, current flows through M_4 and M_2 to ground and through M_5 and M_3 to C. The gate voltage of M_2 does not fall below 3 V, so it remains on. However, to avoid altering the state of the cell when reading, the conductance of M_2 must be about 3 times that of M_4 so that the drain voltage of M_2 does not rise above V_T. The operations of writing and reading a 0 are complementary to those just described.

Several factors contribute to a limit on the maximum speed of operation. Delays in address buffers and decoders naturally increase as the number of inputs and outputs increase. Row lines are typically formed in polysilicon and may have substantial delays due to distributed RC parameters. Column lines are usually formed in metal, so resistance is not significant, but the combined capacitance of the line and many paralleled access transistors connected to them results in a large equivalent lumped capacitance on each of these lines.

A typical layout for an NMOS silicon-gate static cell is shown in Fig. 9.13. Every effort is made to minimize the area of memory cells in the interest of economy. The load transistors have a small value of W/L to minimize power consumption. The cross-coupling connections are made with *buried contacts,* which are direct contacts between polysilicon and drain diffusions, to reduce cell area. The cell transistors drive only a small current onto the highly capacitive column during a read operation. To reduce read access time, the memory is designed so that only a small voltage change on the column line is needed during a read operation. Two or more amplifying stages are used to give a valid logic output when the voltage difference between C and $\overline{C}$ is about 0.5 V. Thus the column delay is only the time needed to achieve this small voltage change.

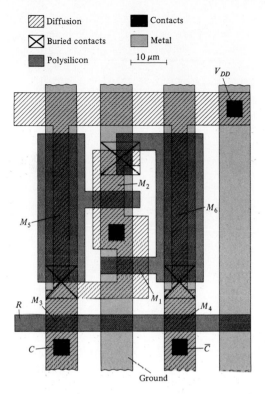

Diffusion   Contacts

Buried contacts   Metal

Polysilicon   10 μm

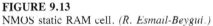

FIGURE 9.13
NMOS static RAM cell. *(R. Esmail-Beygui.)*

Figure 9.14 shows a simplified version of the read amplifier and read and write buffers for an NMOS static memory. The circuits of Fig. 9.14a are provided for each column. The columns are biased at 3 to 3.5 V by transistors M_7, M_8. A number of these circuits are connected to a shared read bus RB, $\overline{RB}$ for data output, and shared write bus WB, $\overline{WB}$ for data input. A NOR gate is used for column decoding, with transistor switch M_{11} used to provide a current return path only for the selected column. Thus only one of these column circuits is operating on each memory cycle.

During a read cycle, WB and $\overline{WB}$ are both held low due to the presence of a high level on $\overline{W}$. Thus M_{13}, M_{14} remain off. When reading a stored 1, $\overline{C}$ is pulled low. M_{10} tends to turn off, so the main current path is from $\overline{RB}$ through M_9 and M_{12} to ground. This current can be several times larger than the cell readout current because M_9 through M_{12} can be larger than cell transistors.

Since the read bus is shared by many columns, it is highly capacitive. In the interests of fast operation, the voltage swing should be kept small here too. The read buffer of Fig. 9.14b uses a common gate input stage M_{15}, M_{16} to provide high gain with a small voltage difference between RB and $\overline{RB}$. Capacitance is relatively low, for the first time along the readout signal path, at the drains of M_{15}, M_{16} so high gain can be achieved in this stage with a minimum speed penalty. Transistors M_{19} through M_{22} provide level shifting and differential to single-ended conversion. M_{23} through M_{26} provide thresholding, additional gain,

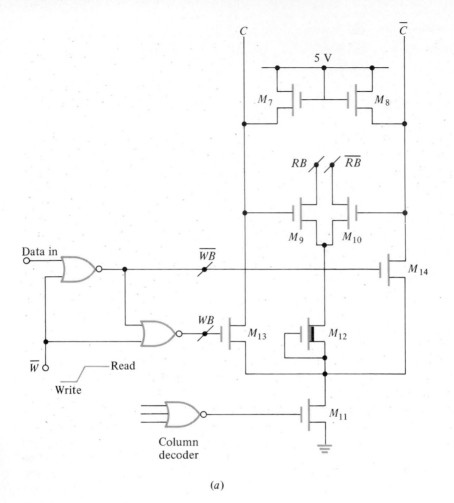

FIGURE 9.14
(a) Read amplifier and write buffer.

and off-chip drive capability. Memory chips are commonly designed with one, four, eight, or nine parallel data input-output paths. There must be an independent read buffer, read bus, and write bus for each parallel data path.

For a write cycle, $\overline{W}$ goes low, allowing C or $\overline{C}$ to be pulled low according to the data input. Additional gating needed to provide a chip enable function is not shown in Fig. 9.14.

9.2.2 Bipolar Cell Arrays

The first semiconductor memories employed bipolar technology. In recent years, NMOS and CMOS memories have achieved lower cost and lower power per bit. Bipolar memory is now used primarily where highest speed operation is required.

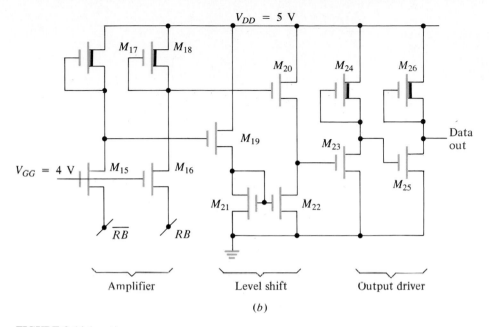

FIGURE 9.14 (*cont.*)
(*b*) Read buffer.

Memory cells based upon I^2L received much attention in the 1970s but eventually proved to have poorer performance without commensurate advantages in density compared to alternative cells. Two widely used bipolar transistor cells, both based on normal (not inverted) transistor action, are shown in Figs. 9.15 and 9.16. These may be termed the *emitter-coupled cell* and *diode-coupled cell*, based upon the means of coupling to the columns in an array. Numerous variations on these two basic cells have been used. The choice between these two cells is primarily determined by the availability of Schottky diodes. If they are available, the diode-coupled cell is usually preferred. One important bipolar LSI process eliminates lightly doped *n*-type silicon from the transistor structure, making it impossible to form Schottky diodes. In this case, the emitter-coupled cell is used. Both row and column lines must have low series resistance if high-speed operation is to be achieved. Therefore two layers of metallization are a virtual necessity.

Both cells are operated at a low voltage, about 1 V, in the unselected or standby condition in order to minimize power consumption. Row and column selection voltages are chosen so that current available from the cell during a read cycle is larger than the standby current. The emitter-coupled cell requires two row lines per cell to achieve this result.

Operation of the emitter-coupled cell proceeds as follows, assuming a logic 1 is stored with Q_1 on. Row selection requires that both R and R^* go to the positive levels shown in Fig. 9.15. To write a 1, column line C must be held low. Regardless of the previous state of the cell, this will forward bias the emitter of Q_1 connected to C. Collector-emitter voltage of Q_1 falls quickly, removing

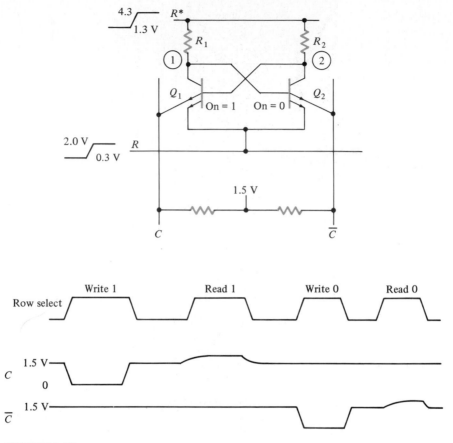

FIGURE 9.15
Emitter-coupled memory cell.

the base drive from Q_2. When the row voltages return to standby levels, Q_1 remains on with its base current coming from R_2. Cell current flows through Q_1 and returns to ground through line R. Emitters connected to C and $\overline{C}$ are reverse-biased in the standby condition.

To read a stored 1, the cell is selected in the same way as for writing. Emitters connected to R become reverse-biased, and the current flowing in Q_1 transfers to the emitter connected to C. The resulting rise in the voltage on C indicates the presence of a stored 1. The operations of writing and reading a 0 are complementary to those just described.

Operating speed of this cell is limited by the rate at which a load resistor can charge node 1 or 2 during writing and the time required to charge highly capacitive line C or $\overline{C}$ during reading.

Operation of the diode-coupled cell of Fig. 9.16 proceeds as follows, assuming a stored 1 corresponds to the state with Q_2 on. Row selection requires that the row voltage be pulled low. To write a 1, the voltage on line C is raised, forward-

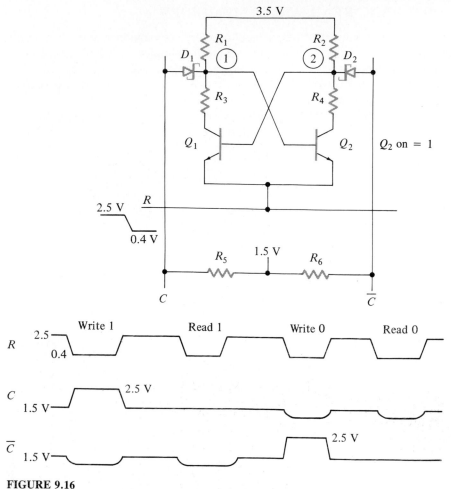

FIGURE 9.16
Diode-coupled memory cell.

biasing diode D_1. Sufficient current must be forced so that voltage across R_3 and Q_1 is adequate to turn on Q_2. The current gain of Q_1 is likely to be high and it will remain saturated, so most of the voltage drop appears across R_3. When Q_2 turns on, its collector voltage drops rapidly, turning off Q_1. The currents in R_1 and R_2 must always be much smaller than the current used for writing so that the drops these currents cause across R_3 and R_4 are much less than $V_{BE(on)}$. In the standby condition, D_1 and D_2 are reverse-biased.

To read a stored 1, the row is pulled low and current flows from $\overline{C}$ through D_2, R_4, and Q_2 to R. The resulting drop in the voltage on $\overline{C}$ indicates the presence of a stored 1. The most important limitations on the operating speed of this cell are the time required to change the row voltage for selection and the time to discharge the column line during reading.

9.3 DYNAMIC READ-WRITE MEMORIES

The importance of reducing cost per bit of memory, as mentioned at the beginning of this chapter, has led to a continuing search for simpler, smaller-area memory cells that could be more densely packed on a chip. The static memory cells described in Sec. 9.2 all require four to six transistors per cell and four or five lines connecting to each cell.

In the late 1960s it was realized that memory cells with reduced complexity, area, and power consumption could be designed if dynamic MOS circuit concepts, similar to those introduced in Sec. 3.8, were used. Static memory cells store data as a stable state of a flip-flop, and data are retained as long as dc power is supplied. In contrast, dynamic cells store binary data as charge on a capacitance. Normal leakage currents can remove stored charge in a few milliseconds, so dynamic memories require periodic restoration, or *refreshing*, of stored charge, typically every 2 or 4 ms. For memories of 64K bytes and larger, the cost of a complete dynamic memory system including provision for refresh cycles is lower than the cost of a system based on static memory components.

9.3.1 Three-Transistor Dynamic Cell

The first widely used dynamic memory cell is shown schematically in Fig. 9.17. Note that this three-transistor (3-T) cell circuit can be derived by eliminating M_1, M_5, and M_6 from the static NMOS cell shown in Fig. 9.12. This cell, unlike the static cell, does not require internal device conductance ratios for proper operation. Transistors M_2, M_3, and M_4 can all be small devices to minimize cell area. Parasitic node capacitance C_1 is drawn in explicitly because it is essential to normal operation. Charge stored on C_1 represents stored binary data. Selection lines for reading and writing must be separated because the stored charge on C_1 would be lost if M_3 turned on during reading. Although separate column lines are shown here for data in (D_{in}) and data out (D_{out}), these two may be combined at the expense of some extra complexity in the read-write circuits.

The cell operates in two-phase cycles. The first half of each read or write cycle is devoted to a precharge phase during which columns D_{in} and D_{out} are charged to a valid high logic level via M_{Y1} and M_{Y2}. A 1 is assumed to correspond to a high level stored on C_1 and is written by simply turning on M_3 after D_{in} is high. The D_{in} line is highly capacitive because it joins many cells. The read-write circuits do not need to hold D_{in} high because sharing the charge on D_{in} with C_1 does not significantly reduce the precharged high level. A 0 is written by turning on M_3 after the precharge phase is over, then simultaneously discharging D_{in} and C_1 via a grounded-source pull-down device (not shown) in the read-write circuit.

Reading is accomplished by turning on M_4 after the precharge is over. If a 1 is stored, D_{out} will be discharged through M_2 and M_4. If a 0 is stored, there will be no conducting path through M_2, so the precharged high level on D_{out} will not change. The cell may be read repeatedly without disturbing the charge stored on C_1. The drain junction leakage of M_3 depletes the stored charge over the span

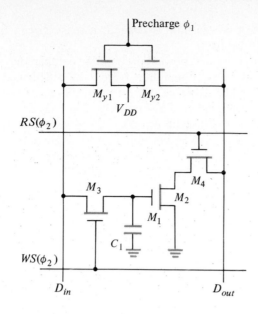

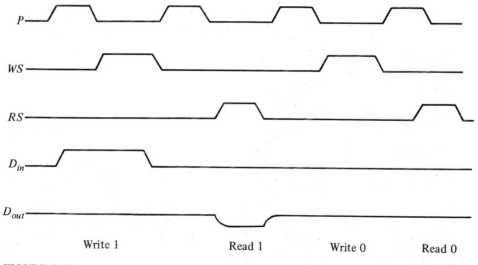

FIGURE 9.17
Three-transistor dynamic cell.

of many milliseconds. Refreshing is performed by reading stored data before they all leak away, inverting the result, and writing back into the same location. This is done simultaneously for all the bits in a row once every 2 or 4 ms.

The level on the D_{out} line in principle can be detected with a simple inverter, but considerable delay would be encountered in achieving the needed 2- or 3-V swing on the D_{out} line. If a short access time is desired, a sense circuit of the type shown in Sec. 9.1.4 would be used.

Use of the dynamic precharge reduces power consumption compared to static NMOS operation. For instance, if the D_{out} high level were to be established only through a static pull-up device, higher average current drain would be required for D_{out} to recover a high level after reading a 1. Similarly, fast changes of state on D_{in} would require excessive power if static drivers were used. Thus, the use of dynamic techniques sharply reduces power consumption for a given operating speed. Some additional cost of using dynamic techniques arises from the need to provide several clocking and timing signals with closely controlled timing specifications. For large memories, however, memory cost per bit is lowest for dynamic designs.

For this 3-T cell, output data are inverted compared to input data. However, memory component data input and output will have the same logic polarity if one extra inversion is included in either the read or write data path.

9.3.2   One-Transistor Dynamic Cell

Most modern dynamic RAMs of capacity 16K bits and more all use the one-transistor (1-T) cell shown schematically in Fig. 9.18a. There are many variations in the detailed realization of this cell, depending on the number of polysilicon layers, method of capacitor formation, conductors used for row and column, etc.*
For simplicity in this introductory evaluation we overlook the many differences and focus on the schematic representation that is common to all.

Selection for reading or writing is accomplished by turning on M_1 with the single row line. Data are stored as a high or low level on C_1. In the interests of minimum cell area, C_1 is very small, on the order of 30 to 100 fF (10^{-15} F). The selection signal on the row line is the same for reading or writing, as shown in Fig. 9.18b. Data are written into the cell by forcing a high or low level on the column when the cell is selected.

When reading, the charge stored on C_1 is shared with the 10 to 20 times larger capacitance C_2 of the column line. Considering that after 2 ms the difference in stored voltage between a 1 and a 0 may be as little as 2 V, the data output signal may be as small as 100 mV. Stored data must be regenerated every time they are read, in addition to refreshing them every 2 ms even if they are not read. Read amplifier design for reliable detection of the small column signal is one of the most difficult aspects of 1-T dynamic RAM design.

The simplified schematic for a read-refresh circuit for a 1-T RAM is shown in Fig. 9.19a. The regenerative switching of a dynamic flip-flop detects the small data signal and restores the high or low signal level. The storage array is split

*V. L. Rideout, "One-Device Cells for Dynamic Random-Access Memories: A Tutorial," *IEEE Transactions on Electron Devices,* vol. ED-26, June 1979, pp. 839–852.

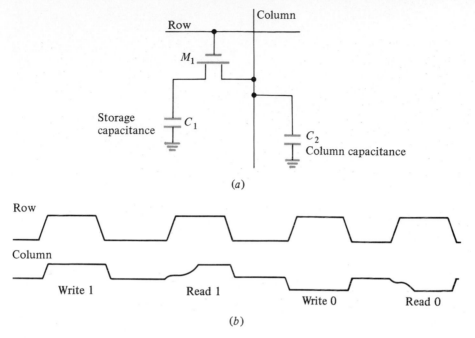

(a)

(b)

FIGURE 9.18
One-transistor storage cell. (a) Circuit diagram. (b) Timing diagram.

in half so that equal capacitances are connected to each side of the flip-flop. As shown in Fig. 9.19a, each half column in the array has a single additional *dummy cell* that will be used as described below.

A possible sequence of signals for reading from a 1-T array proceeds as follows. A precharge clock phase ϕ_p sets the voltages on all column lines near the supply level V_{DD} and sets the voltage in all dummy cells to zero. One row (either in the left or right array of storage cells in Fig. 9.19a) is then selected with a signal ϕ_R. The dummy cell on the opposite side of the sense amplifier is selected simultaneously with a signal ϕ_D. The column voltage on the side connected to the selected dummy cell drops slightly as the column charge is shared with the dummy cell capacitance. The column voltage on the side connected to the selected storage cell drops twice as much (if a 0 was stored) or does not change (if a 1 was stored). The resulting small voltage difference between the two sides of the array determines the final state of the flip-flop when latching signal ϕ_s is applied. (Note that if no dummy cell were used here, there would be no voltage difference between the two sides for a stored 1. Circuit noise or unbalance then would determine the latched result.) The data are taken out through a column decoder circuit to a final amplifier and output buffer. The timing diagram in Fig. 9.19b shows the sequence of signals that appear when a stored 0 level is read from the left half of the array.

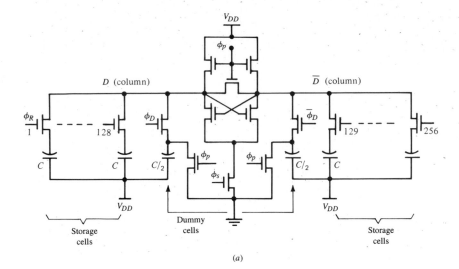

(a)

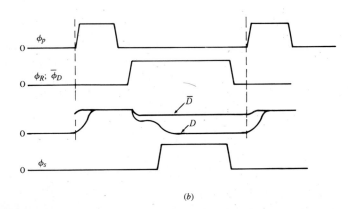

(b)

FIGURE 9.19
(a) Sense amplifier for 64K-bit dynamic RAM. (b) Possible sense amplifier timing.

9.3.3 External Characteristics of Dynamic RAMs

The internal circuit design and internal timing diagram for the widely used 64K and 256K dynamic RAMs are much more complicated than shown above, involving, for instance, a total of more than 20 internally generated clock phases for read, write, and refresh functions. These matters are beyond the scope of this book. To conclude our study of dynamic RAMs, we will now step back to an external or user's point of view and consider the terminal characteristics of these widely used components.

The block diagram for the Texas Instruments TMS 4164, a 65,536-bit dynamic RAM, is shown in Fig. 9.20a. Although internal process and circuit details vary, this component has similar external characteristics to 64K RAMs from other vendors. Features common to all include the following. A single 5-V power supply is specified. All inputs and outputs are TTL-compatible. The chip is encapsulated in a 16-pin dual in-line package. The one-transistor cell described above is used, and the 65,536 cell storage array is organized in two halves, each containing 32,768 cells. The 256K-bit chips are very similar. In order to limit the row and column delays, the Texas Instruments TMS4256 is organized as four blocks of 64K bits, each block having its own sense amplifiers. The row and column address buffers and the output buffers are shared by the four blocks. One additional address pin is multiplexed to bring in the needed additional bit of row address and column address.

A read cycle for the TI TMS 4164 proceeds as follows, with specific reference to the internal design of this particular chip. First, row selection is achieved using eight row address bits that are latched into eight row address buffers. These buffers provide true and inverted outputs for each input bit. Seven address bits are routed to either the upper or lower set of row decoders, as determined by the state of the eighth row address bit. The upper or lower group of row decoders performs a 1 out of 128 selection, driving a single row line to a high level.

The high or low level stored in each of 256 cells along the selected row line is transferred to its column line, with an inevitable capacitive attenuation of voltage level. One regenerative flip-flop *sense amplifier* (similar to that shown in Fig. 9.19a) for each of the 256 columns is used to capture the relatively small change in column voltage and regeneratively restore it to a full 1 or 0 voltage level. The correct levels are restored to all selected cells without any involvement by column address selection circuits.

Address information for column selection comes on to the chip via the same eight pins used for row addresses. This is accomplished by time-sharing these pins. After the row addresses are latched they are disconnected from the address pins. Address changes at the pins do not affect row selection for the remainder of the cycle. Therefore the pins may be used for eight additional bits for column selection. (The purpose of this *address multiplexing* is to reduce the package pin count, size, and cost.) The column address is latched into eight column address buffers. The control of all these events is described below. Six column address bits are used to select one of 64 groups of four sense amplifiers for connection

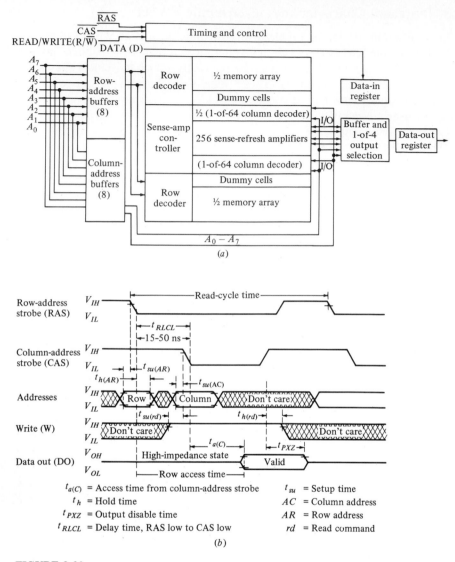

FIGURE 9.20
(a) Block diagram. (b) Timing diagram.

to four data buffers and a 4 to 1 multiplexor (using the last two address bits). A single selected bit is passed to the data output buffer and to an off-chip data bus.

A write cycle proceeds with many similarities, except that a single data input bit is used to set the state of one sense amplifier flip-flop, overriding the effect of the data previously present in the selected cell. An externally applied write enable signal W specifies the write operation. Regardless of any other activity, every storage cell must be *refreshed* at least once each 4 ms. This can be achieved by ensuring that every row in each half of the array is accessed at least this often. During each row access, the sense amplifiers perform their function of regenerating stored signal levels. If memory cycle time is 200 ns, selection of $(2)(256) = 512$ columns requires $(0.2 \ \mu s)(512) = 102.4 \ \mu s$ every 4 ms, or 2.56% of total elapsed memory operating time. The digital timing and control functions needed to provide the refresh functions can be provided by additional hardware on the memory board or by additional software executed by the central processing unit. Most systems (but not all!) can be designed to allow the necessary refresh cycles without serious effects.

9.3.4   Timing Requirements for Dynamic RAMs

The static memories described in Sec. 9.2 operate correctly at any cycle time from a specified minimum value up to infinity, i.e., indefinitely long selection of any address, without loss of stored data. As already pointed out, dynamic memories require periodic refresh operations while cycling through all row addresses to avoid loss of stored data. Other functions in dynamic RAMs that cannot be performed with static signals (i.e., signals that may remain unchanged for arbitrarily long periods) are the time-multiplexing of row and column addresses through one set of pins, the precharging of column lines, and the actuation *(strobing)* of sense amplifiers just after stored data are transferred to column lines.

Modern dynamic RAMs derive the signals needed to perform these dynamic functions from two externally generated timing signals, known by the names *row address strobe (RAS)* and *column address strobe (CAS)*. In fact, the complements of these signals, known as $\overline{RAS}$ (RAS bar) and $\overline{CAS}$ (CAS bar), are actually applied to the component pins. The time relationship of $\overline{RAS}$, $\overline{CAS}$, and $\overline{W}$ to address and data signals in 64K and 256K RAMs is illustrated in the simplified read cycle timing diagram shown in Fig. 9.20b. Complete specification of the read cycle timing relationships involves approximately 20 timing parameters, all of which must be held within specified limits to assure proper operation. Write cycle timing (not shown) is similarly complex. Manufacturer's data and application notes are the best source of detailed information for any particular component.

Through extensive use of dynamic circuit design in decoders and sense amplifiers as well as in the storage cells, the power consumption of 64K NMOS DRAMs such as the TMS 4164 is typically only 125 mW in full-speed operation and 20 mW with just the necessary refresh cycles taking place. (Because of the block organization of 256K and larger DRAM chips, with only a fraction of the array accessed on any single memory cycle, total chip power consumption is only slightly larger than for 64K chips.) In all DRAMS, large peak currents flow when $\overline{RAS}$ and $\overline{CAS}$ change states, because many capacitive nodes are charged at these times. For the TMS 4164, the peak current supply current can reach 60 mA, compared to 25 mA average current at full speed. These "current spikes" can generate excessive noise in power and ground lines. Care must be taken to include proper bypass capacitors and in the design of printed wiring boards to prevent such noise from affecting system operation.

At the level of 1M- and 4M-bit DRAMs, several manufacturers have changed from NMOS to CMOS technology. This change does not bring any major reduction in power consumption because NMOS DRAMs have always been designed to eliminate static power consumption. (That is, virtually all power consumption is dynamic, determined by CV^2f.) The advantages of using CMOS for DRAM include wider operating margins for V_{DD}, simpler circuit design for zero static power, and greater tolerance of process variations. The main speed and power consumption characteristics are similar for NMOS and CMOS DRAMs; indeed, they are designed to be interchangeable in most digital systems.

9.4 SERIAL MEMORIES

The block diagram for a serial or shift register memory is shown in Fig. 9.21a. Data entered into the memory at the left shift one location to the right for each cycle of the clock. A recirculate-write control line determines whether stored data are recirculated or new data are entered. Obviously data output are in the same time sequence as data input.

9.4.1 MOS Transistor Shift Register

Shift registers were a popular early form of LSI memory because they are easy to design and test and because the internal circuits are tolerant of considerable variations in device parameters. The schematic for a two-phase ratioed dynamic shift register cell is shown in Fig. 9.21b. This will be recognized as a cascade of two inverters based on the two-phase ratioed logic circuit shown in Fig. 3.18a. During ϕ_1, the data present at the input are inverted and transferred to the following inverter. During ϕ_2, these data are inverted again and transferred to the output. Logic levels are regenerated in each inverter, while fan-in and fan-out are limited to 1, so device parameters can vary considerably without degrading the performance.

For large storage capacity, shift registers are quite limited in comparison to RAMs. First, the inherently serial access to stored data is inferior to random access for most applications. Secondly, six transistors per bit are an economic penalty compared to a total of between two and three transistors per bit of 1-T RAM with all row and column circuits. Third, every cell of the SR is clocked with every cycle. This imposes heavy capacitive loads on clock drivers and results in a large component of dynamic power consumption, CV^2f. For these reasons, shift registers as shown in Fig. 9.21b are rarely used today for large-capacity memories.

9.4.2 Bucket Brigades and Charge-Coupled Devices

A considerably less complex serial memory is achieved by eliminating the logic inversions and level regenerations found in every stage of the 6-T shift register

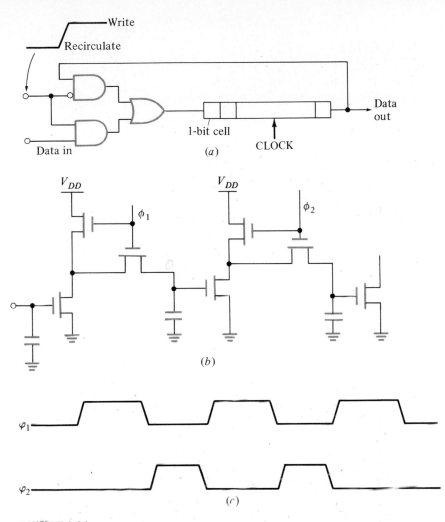

FIGURE 9.21
(*a*) Block diagram of a shift register. (*b*) One stage of a two-phase dynamic shift register. (*c*) Clock for two-phase shift register.

above. Bucket brigades (BB) and charge-coupled devices (CCD) provide inherently serial memories that convey packets of charge—mobile electrons or holes—along a defined path under the control of a two-, three-, or four-phase clock.

Charge-coupled devices cannot be adequately represented with conventional schematic circuit diagrams. Instead, the cross-sectional view shown in Fig. 9.22*a* is useful as an aid to understanding. The clocking electrodes are termed *transfer gates*.

When D_{in} is low (logic 0), electrons are drawn in under the first ϕ_1 electrode from the n^+ input electrode when ϕ_1 goes high. This process is similar to

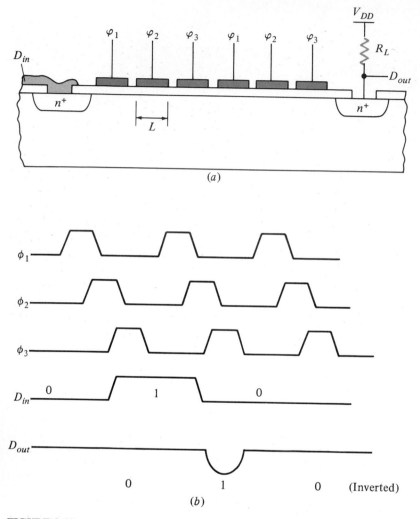

FIGURE 9.22
Charge-coupled device (CCD). (*a*) 2-bit CCD serial memory. (*b*) Timing diagram for above CCD.

formation of a conducting channel in the MOS transistor, as described in Sec. 2.3. When ϕ_2 goes high and ϕ_1 goes low, these electrons move to the right as a packet. The third clock phase is necessary to ensure the desired flow to the right. Clock phases should overlap slightly in time to ensure that almost all of the mobile charge moves as desired.

When the charge packet reaches the output electrode, it is collected by the n^+ output electrode. A brief pulse of current flows from V_{DD}, giving a momentary dip in the output voltage. This signifies a binary 0 as was originally input. For a binary 1 at the input, the n^+ input electrode is held high and no electrons flow to the right. Therefore the output voltage never dips; this is interpreted as a 1 at the output. The timing diagram of Fig. 9.22b shows operation for an input sequence of 0, 1, 0.

A number of practical problems intrude. Less than 100% of the charge packet is transferred from stage to stage, so the binary signal must be regenerated every 128 to 1024 stages. The clock electrodes must be placed very close together (within about 1 μm), so that the depletion regions created by two adjacent electrodes overlap when both are at the high clock voltage level. Even if etching 1-μm gaps is feasible, leakage or breakdown between electrodes is likely when they are at different voltages. Therefore, the simple structure diagrammed in Fig. 9.22a is unsatisfactory, and more complex structures with two or three overlapping levels of clock electrodes, separated by solid insulators, are necessary.

Charge-coupled and bucket-brigade devices as serial digital memories do overcome some of the limitations of 6-T shift registers. Practical designs with two or three overlapping clocking electrodes can provide large-capacity memories in somewhat less area than required for 1-T RAM. Due to the more limited market for serial memory as compared to RAM, CCDs and bucket brigades have not been widely used for digital memory.

9.5 SUMMARY

This chapter begins by defining a number of special terms used in the description of semiconductor memories. The student should be familiar with the following ideas:

- Bit, byte, word
- Memory address; row and column selection
- Read and write operations; timing diagrams
- Memory cells, decoders, read and write circuits
- Volatile versus nonvolatile information storage
- Read-write memories: random access, serial access
- Read-only memories: masked, PROM, EPROM, E^{2}PROM
- Features of static and dynamic memory circuit design

The analysis of static and dynamic RAM performance follows the techniques presented in the earlier chapters. Elements such as sense amplifiers require careful analog circuit design.

REFERENCES

1. D. A. Hodges (ed.), *Semiconductor Memories*, IEEE Press, New York, 1972.
2. G. Luecke, J. P. Mize, and W. N. Carr, *Semiconductor Memory Design and Application*, McGraw-Hill, New York, 1974.
3. M. I. Elmasry (ed.), *Digital MOS Integrated Circuits*, IEEE Press, New York, 1981.

4. *IEEE Journal of Solid-State Circuits,* special issues on semiconductor memory and logic, October of each year since 1970.
5. L. C. Sood et al., "A Fast 8K × 8 CMOS SRAM With Internal Power Down Design Techniques," *IEEE Journal of Solid-State Circuits,* vol. SC-20, no. 5, October 1985, pp. 941–949.
6. A. Mohsen et al., "The Design and Performance of CMOS 256K Bit DRAM Devices," *IEEE Journal of Solid-State Circuits,* vol. SC-19, no. 5, October 1984, pp. 610–618.
7. T. Furuyama et al., "An Experimental 4-Mbit CMOS DRAM," *IEEE Journal of Solid-State Circuits,* vol. SC-21, no. 5, October 1986, pp. 605–611.

PROBLEMS

Use the MOS device parameters from Example 9.1. Use the following bipolar transistor device data:

Transistors

$I_{ES} = 10^{-16}$ A	$I_{CS} = 4 \times 10^{-16}$ A
$\alpha_F = 0.98$	$\alpha_R = 0.25$
$C_{je0} = 1$ pF	$C_{je0} = 0.5$ pF
$\phi_e = 0.9$ V	$\phi_c = 0.8$ V
$m_e = 0.5$	$m_c = 0.33$
$\tau_F = 0.2$ ns	$\tau_R = 10$ ns
$C_{js0} = 3.0$ pF	$r_b = 50$ Ω
$\phi_s = 0.7$ V	$r_c = 20$ Ω
$m_s = 0.33$	$r_e = 1$ Ω

Schottky diodes

$I_S = 10^{-12}$ A	$\phi_0 = 0.7$ V
$C_{j0} = 0.2$ pF	$m = 0.5$

P9.1. Use SPICE to simulate the access time for the NOR ROM array described in Example 9.1, and compare the simulation results for row, column, and overall delays with the results of hand analysis.

P9.2. Draw the layout for a NAND ROM in silicon gate NMOS with a level of detail similar to that of Fig. 9.4. Calculate row and column capacitances per bit. This ROM is programmed with the depletion implant; do not neglect the capacitance from this channel implant to body. Also calculate the nominal current available when reading a stored 0 (low column voltage).

P9.3. Consider an EPROM cell as shown in Fig. 9.5, assuming that the transistor characteristics if gate 1 could be used as an input are those given above. (Of course, gate 1 is not directly accessible in the EPROM device.) The oxide under gate 1 is 0.1 μm thick and that under gate 2 is 0.12 μm thick. Drawn channel dimensions are $W = L = 10$ μm.

   (*a*) Calculate the saturated drain current for this device with source grounded and 5 V applied to drain and gate 2, assuming that the potential on gate 1 is zero with 0 V on all other electrodes.

   (*b*) Suppose that during writing a current of 10^{-11} A flows from gate 1 to drain, induced by the high field in the drain depletion region. How long will it take for the potential on gate 1 to change by 5 V due to this current?

P9.4. Calculate the threshold voltage seen from gate 2 for the EPROM cell of the previous example, assuming the potential on gate 1 is -5 V with 0 V applied to all other electrodes.

P9.5. Calculate the number of transistors required for N-bit decoders in the two forms shown in Fig. 9.8. Compare results for $N = 4$, 8, and 12 bits.

P9.6. Find the transfer characteristic (output voltage as a function of input *current*) for the bipolar transistor read amplifier of Fig. 9.11c, comprising Q_5, Q_6, D_1, R_5, and R_6. Use the transistor data given for the problems in Chap. 5.

P9.7. Measure the device W/L ratios for the NMOS static RAM cell with (true scale) drawn layout as shown in Fig. 9.13. Assume that source and drain each diffuse 0.5 μm into the channel so that the electrical channel length is 1 μm less than the channel length measured on Fig. 9.13.

(a) Use the device data on page 384 to calculate the nominal power consumption at standby, the maximum initial bias voltage on the column lines so that the internal cell voltage does not exceed V_T on the low side, the readout current when column lines are biased at this voltage, and the column voltage needed to change the state of a selected cell.

(b) Find the needed W/L ratio for the cross-coupled flip-flop transistors such that the column lines can be biased at 4 V without exceeding V_T at the internal low node. For this value of W/L (other devices unchanged), find the power consumption, readout current, and column voltage to change the state of a selected cell.

P9.8. For the cell of the previous problem, calculate the row and column capacitances and resistances per memory cell. Suppose that a read operation requires that column voltage change from 3.0 to 2.5 V. Calculate the row and column delay times for a 4,096-bit (64 $\times$ 64) array of these cells.

P9.9. Consider the diode-coupled bipolar memory cell of Fig. 9.16 with $R_1 = R_2 = 50$ kΩ, $R_5 = R_6 = 1$ kΩ. Schottky diodes D_1 and D_2 each have 100 Ω of series resistance. Find the maximum and minimum values of $R_3 = R_4$ such that reading and writing can be achieved with the voltages shown.

P9.10. Assume that a CCD transfer gate has length L in the direction of charge flow and width W perpendicular to L in the surface plane. The gate insulator is SiO_2, 0.1 μm thick. Assume that transfer gates are clocked between 0 and 10 V and that the threshold voltage (transfer gate voltage at which electrons begin to collect under the gate) is 2 V.

(a) What is the maximum charge representing a binary 0?

(b) With $V_{DD} = 5$ V and $R_L = 50$ kΩ, find the shaded area over the dip in D_{out} voltage (see Fig. 9.22b).

P9.11. For the NMOS sense amplifier shown in Fig. 9.9a, find the W/L for transistor M_7 so that 2.8 V appears at node 4. Use the device data from Example 9.1. Do not neglect body effect.

P9.12. Use SPICE to simulate the characteristics of the sense amplifier of Fig. 9.9a (M_1 through M_5).

(a) Apply a voltage sweep from 0 to 5 V at the gate of M_1. Find the voltage swing at nodes 1, 2, and 3.

(b) Assume the total capacitance C_1 on the column line (node 1) is 1 pF, C_2 on node 2 is 0.3 pF, and C_3 on node 3 is 0.2 pF. No other capacitances need be considered. Use SPICE to simulate the transient performance when a step (0 to 5 V) is applied to the gate of M_1. How long does it take for node 3 to drop from 5 to 2 V? Compare this with the corresponding time when nodes 2 and 3 are connected together so that the amplifying function of M_3 is eliminated.

P9.13. For the CMOS sense amplifier shown in Fig. 9.9*b*, find the bias voltage established at node 4 by the combination of M_7 and M_8. Use the NMOS data from Example 9.1. The PMOS device has $V_{TP} = -1.0$ V, $k_p' = 8$ μA/V^2, $\gamma = 0.6$.

P9.14. For the CMOS sense amplifier shown in Fig. 9.9*b*, assume that $V_4 = 3$ V. Find W/L of M_4 such that its saturated drain current is half the saturated drain current of M_1 when M_1 has $V_{GS} = 5$ V. Use SPICE to find the dc transfer characteristics from the gate of M_1 (sweep 0 to 5 V) to nodes 2, 3, and 6.

P9.15. For the CMOS sense amplifier shown in Fig. 9.9*b*, assume the total capacitance C_1 on the column line (node 1) is 1 pF, C_2 on node 2 is 0.3 pF, and C_3 on node 3 is 0.2 pF. No other capacitances need be considered. Use SPICE to simulate the transient performance when a step (0 to 5 V) is applied to the gate of M_1. How long does it take for node 3 to drop from 5 to 2 V? Compare this with the corresponding time when nodes 2 and 3 are connected together so that the amplifying function of M_3 is eliminated.

CIRCUIT
DESIGN
FOR LSI
AND VLSI

10.0 INTRODUCTION

The gates described in Chaps. 3 and 7 are in principle sufficient to design any
digital system. In fact, almost all small-scale and medium-scale integrated (SSI
and MSI) digital integrated circuits (1 to approximately 100 gates per chip)
are simply made up from combinations of these circuits with minor variations.
NAND and NOR gates with varying fan-in are common; more complex gate
functions such as AND-OR-INVERT, decoders, and multiplexors are often used;
and various flip-flops and registers are composed from gates as shown in
Chap. 8.

Transistor-transistor logic (TTL) and complementary MOS (CMOS) are the
most popular technologies at SSI and MSI complexity levels. Several hundred
different circuit combinations are available as individual components in each of
these two most popular families. Emitter-coupled logic (ECL) is used for highest
speed. N-channel MOS (NMOS) is essentially unused at SSI or MSI complexities,
because at low levels of integration NMOS is inferior in performance to TTL and
CMOS without offering any economic advantage.

Most SSI and MSI digital circuits perform standardized logic functions.
These circuits are manufactured in large quantities. They are not specialized to
particular applications but rather find diverse use in a variety of digital systems.
These components are designed to have large noise margins and the capability
of driving large fan-out and heavy loads in order to maximize their versatility. The

overall function for a chosen application is determined by the interconnections among the SSI and MSI components.

For most applications it is now desirable to increase the number of logic circuits on a chip above the 10 to 100 range typical of SSI and MSI circuits. Best results are achieved if the gate and flip-flop circuits described in Chaps. 3, 7, and 8 are modified for use on LSI and VLSI chips. For example, circuits that are interconnected only with other circuits on the same chip can be safely designed with smaller noise margins and reduced load-driving capability compared to SSI and MSI circuits. Circuit density can be increased, power consumption reduced, and performance and reliability of the complete system improved if intelligent decisions are made in design for LSI and VLSI.

Engineers frequently face the problem of selecting from among various approaches to LSI and VLSI design. In most cases two or more design approaches can yield the desired technical specifications. In such circumstances, choices must be based on economic and project development scheduling considerations. The following sections describe several alternative approaches to building digital systems employing LSI and VLSI components.

10.1 GATE ARRAYS

A conceptually simple form of LSI logic, known as the *gate array* or *uncommitted logic array (ULA)*, is one example of a larger category known as *semicustom* integrated circuits. Gate arrays typically incorporate between 100 and several thousand NOR or NAND gate circuits on a chip, arranged in rows and columns. Four to six masking levels needed to make transistors and other circuit elements are standardized, independent of the final application. Hence integrated circuit wafers can be processed in advance with these standard pattern levels.

One to five additional pattern levels are used to determine the overall function of a gate array chip. These levels, which are unique to the specific application, provide the desired interconnections. They can be designed and applied to a standard gate array wafer quickly and inexpensively compared to the time and cost of developing a totally unique LSI circuit for the same overall function.

Gate arrays find use in places where they can replace 5 to 50 separate digital SSI and MSI chips. Often substantial reductions in physical size, power consumption, and total cost of a complete system are achieved with gate arrays. Reliability and high-speed performance can be improved because the number of off-chip connections is sharply decreased. Also, it is more difficult for a competitor to copy a product using gate arrays than one built from standard SSI and MSI components.

10.1.1 CMOS Gate Arrays

Circuit design and logic design for gate arrays is quick and simple because only one or a small number of standard gate circuit designs are used. CMOS gate arrays typically use the NAND and NOR circuits shown in Sec. 3.7. Figure 10.1

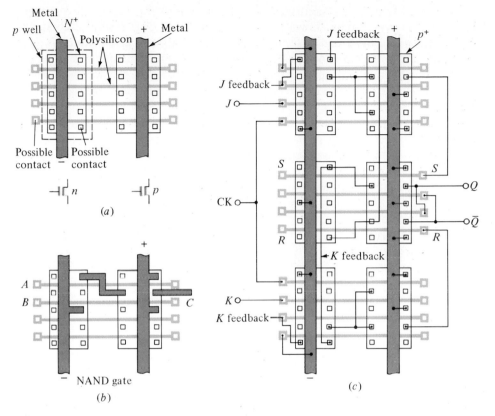

FIGURE 10.1
CMOS gate arrays. (*a*) Basic block design. (*b*) Physical array interconnection. (*c*) JK flip-flop.

shows four levels of detail for a silicon-gate CMOS gate array. The circuit block shown in Fig. 10.1*a* comprises four NMOS and four PMOS transistors. The solid black lines are polysilicon, and the diagonally striped lines carry power, ground, and logic interconnections. The dashed line denotes the outer boundary of the *p*-type well diffusion in which the NMOS transistors are formed. The clear areas inside the rectangles are heavily doped n^+ or p^+ regions that can serve as sources and drains. There are two possible contact points to each diffusion region, on the left and right of the vertical metal lines, and two possible contact points to each polysilicon line, at the left and right ends. Transistor channels are the areas under the polysilicon and inside the diffusion rectangles; source-drain diffusions do not penetrate the polysilicon gates.

The transistors in these circuit blocks may be connected in various ways to provide inverters or NAND or NOR gates with 2 to 4 inputs. Figure 10.1*b* shows the interconnections needed to provide 1 two-input NAND gate (see schematic of Fig. 3.14*b*). Note that another identical gate could be realized using the bottom half of this block.

Figure 10.1c shows sketched-in connections for three blocks to produce a JK flip-flop of the type shown schematically in Fig. 8.4. Note the complex route followed by the J and K feedback lines. The very top and bottom NMOS transistors are forced to a permanent off state by wiring their gates to ground. This is done so that diffused regions may be used to carry the feedback lines under the metal ground line. Although this circuit has been fully interconnected using only a single metal layer, more complex circuits often require customized patterns for two layers of metal and the connection paths between them in order to achieve efficient routing of interconnections.

Figure 10.2 is a photomicrograph of one corner of a complete silicon-gate CMOS gate array chip that employs two layers of metal for interconnections. Overall size of this complete chip is 4.55 by 5.9 mm; 960 copies of the block seen in Fig. 10.1a are arranged in 16 columns, 60 blocks per column. A total of 960 four-input gates or 1,920 two-input gates are available for interconnections. The spaces between columns of circuit blocks, called *wiring channels,* are used for interconnections. The squares along each edge are bonding pads to which bonding wires are attached when the chip is installed in its package. The dark area adjacent to each bonding pad is a special input-output (I/O) circuit block that may be connected as an input buffer, as a binary or tristate output buffer, or as a bidirectional input-output buffer.

10.1.2 Bipolar Transistor Gate Arrays

The gate arrays that offer the shortest logic propagation delays employ variations on the ECL circuits described in Sec. 7.4. Because these circuits are complex and dissipate 10 to 50 mW each, only a few hundred may be included on a single chip. Some very large high-speed computers are built with this form of gate arrays.

When 1,000 or more gates are to be incorporated on array chips, very simple, dense gate circuits are necessary. Standard TTL or ECL circuits are used at input and output nodes of LSI chips to provide standard logic levels and noise margins for external connections. Some of the alternative simple circuits for bipolar LSI are described in Secs. 7.5 and 7.6. Several more are shown below. Common features of all are the small number of devices per gate, the use of a supply voltage lower than 5 V, and a logic voltage swing between 0.2 V and 1 V.

Bipolar transistor resistor-transistor logic (RTL) circuits as shown in Fig. 10.3a are one option for gate arrays. The simplified form of Schottky TTL shown in Fig. 10.3b is also used. A single-ended form of current-switch or current-mode logic (CML) as shown in Fig. 10.3c has been used for operation of up to at least 20-MHz clock rates. (Operation of current-switch logic is explained in Sec. 7.4.) Integrated injection logic (see Sec. 7.5.1) is the highest density bipolar logic technology and is attractive for micropower operation. The other bipolar array gates provide better high-speed operation because they avoid the use of slow inverted transistors.

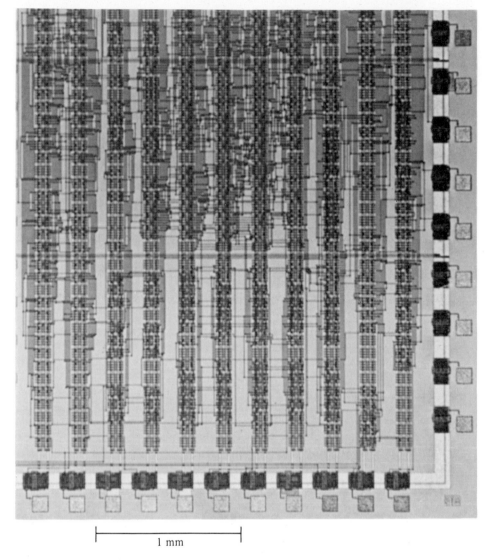

1 mm

FIGURE 10.2
Portion of a CMOS gate array chip. *(STC Microtechnology, Inc.)*

Integrated Schottky logic (ISL) as shown in Fig. 10.3*d*, and its close relative, Schottky transistor logic (STL) of Fig. 10.3*e*, have been used for gate arrays. These gate circuits are described in Sec. 7.5. As in the case of I²L, logic is performed by using multiple outputs. High-speed operation is possible because a normal downward *npn* transistor with a Schottky or *pnp* transistor clamp is employed. Either a *pnp* transistor or two different types of Schottky diode are necessary for high-speed operation; these requirements increase the complexity of fabrication processes.

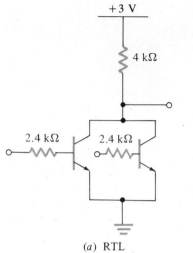

(a) RTL

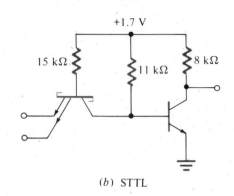

(b) STTL

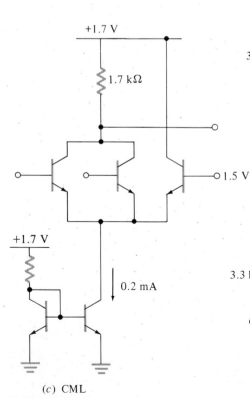

0.2 mA

(c) CML

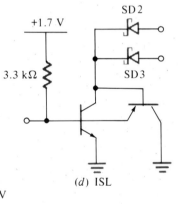

(d) ISL

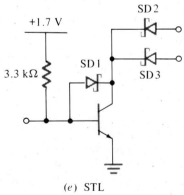

(e) STL

FIGURE 10.3
Bipolar circuits for LSI gate arrays.

10.1.3   Limitations of Gate Arrays

Drawbacks of gate arrays result from their simplicity and high degree of standardization. Complex digital functions, especially ROM and RAM, are not efficiently realizable using individually connected transistors or elementary gates as building blocks. The lack of memory functions may result in requirements for very large numbers of connections (100 to 200 or more) to array chips. The *standard cell* design approach to LSI, described below, addresses this problem by permitting use of more complex circuit blocks, including memory functions, on the same chip with elementary gates.

Good computer-based tools for designing interconnections are essential to successful application of logic arrays larger than a few hundred gates. *Logic simulators* are used to confirm that a specified network of gates and interconnections will perform the desired logic functions. *Placement and routing* programs are used to establish the relative locations of specific gates on a chip and to design the exact paths of all interconnections. Parasitic capacitance and resistance of interconnections must be calculated and included in a detailed *timing simulation* to ensure that the final circuit will operate at the desired speed. The final pattern of interconnections must be automatically translated into the format for the pattern generator that will generate the masks for interconnecting metallization.

Examination of Fig. 10.2 makes clear one of the limitations of gate arrays. Wiring channels toward the center of the chip may fill up before completing all necessary connections. Placement and routing programs are not sufficiently advanced to provide optimum utilization of channel space. In practice, it is difficult to make use of more than about 60 to 80% of the gates on a large gate array chip. If placement and routing cannot be completed, it may become necessary to use a larger (more costly) array chip, or to use two chips instead of one.

10.2   STANDARD CELLS

This technique is based on use of a library of standard logic cells, many of which are more complex than a basic gate. The standard cell design approach has found use in NMOS and CMOS technologies. In addition to basic NOR and NAND circuits, cells such as exclusive-OR, AND-OR-INVERT, D and JK flip-flop, binary full adder, ROM, and RAM are included in the cell library. Design of an LSI chip proceeds by selecting the most suitable cells for the overall functional requirements, then placing and interconnecting these cells using computer aids. There is an additional degree of freedom in interconnecting standard cells compared to gate arrays. The width of the interconnecting channel may be adjusted to accommodate the necessary number of lines.

Figure 10.4 shows a chip designed by the standard cell approach. Columns of cells, each of uniform width but varying height, are separated by wiring channels. The latter are devoted exclusively to interconnections, with metal running vertically and diffusion running horizontally. A densely packed section of memory is evident on one side of this chip.

FIGURE 10.4
CMOS standard cell chip that provides echo cancellation for a data communications modulator-demodulator (modem). *(Zymos, Inc.)*

The layout of devices and connections within each cell is the same every time that cell is used, but placement of cells is unique to each chip. Therefore circuits designed by the standard cell method are unique at every mask level, and fabrication can begin only after the specific design is completed. Compared to gate arrays, LSI chips developed by this approach have more digital functions per unit of chip area. However, fabrication is more costly and time-consuming because all masking layers are unique to a chip.

10.3 PROGRAMMABLE LOGIC ARRAY (PLA)

Arbitrary combinational logic functions may be realized in the so-called *sum-of-products* form described in Sec. A.2 (Appendix). Based on conventional gate circuits, a logic function sum of products may be implemented using any of the equivalent two-level logic configurations: AND-OR, NAND-NAND, NOR-OR, OR-NAND, AND-NOR, NOR-NOR, NAND-AND, or OR-AND. The *programmable logic array (PLA)* provides an alternative to gates for implementation of sums of products.

The advantage of PLA compared to gate logic is that PLA has a highly regular geometric structure similar to ROM. The logic function of a PLA is determined by the presence or absence of contacts or connections at fixed, predefined positions in a single conducting layer. Physical means of forming or omitting contacts are identical to those described for ROMs in Sec. 9.1. Both mask programming and electrical programming are feasible for PLAs. Electrical programming by means of fuse links is offered in commercially available standard chips called *field-programmable logic arrays (FPLA)* and *programmable array logic (PAL)*. These chips provide varying numbers of inputs and outputs and 20 to 60 product terms, depending upon the particular part chosen. Empirical observations of completed designs for control logic functions show that each PLA product term is the functional equivalent of 3 to 10 logic gates.

Schematic diagrams for PLAs in NMOS and bipolar technology are shown in Fig. 10.5*a* and *b*. The NMOS example implements the NOR-NOR form for sum of products, while the bipolar example implements the AND-OR. Regardless of the particular implementation, the outputs of the first array in the signal path are known as *product terms*, or *products*.

An N-input PLA reaches a limiting case when it has 2^N product terms. In this case, the PLA is equivalent to a ROM with N address bits, and the first array would be identified as the ROM address decoder. However, the PLA finds most effective use as a replacement for gate logic when the number of product terms is much smaller than 2^N. Such a requirement is often found in the control units of computers. For instance, a 32-bit VLSI computer instruction decoding unit uses a PLA of 26 inputs, 206 product terms, and 22 outputs for decoding of instruction operation codes.[*] A ROM with 26 address bits would have more than 67 million addresses, compared to only 206 product terms used in this example!

The advantages of PLA over gate logic for such applications arise from its regularity of design. With good computer-based design tools, the PLA can be designed and laid out more compactly and with less likelihood of errors than gate logic for the same function.

10.4 MICROPROCESSORS AND MICROCOMPUTERS

Microcomputers provide a different means of producing digital systems for a variety of applications using standardized LSI components. These components are tailored for a specific function by writing a program that is stored in memory. The programming of microcomputers is beyond the scope of this book. The internal design of microprocessor circuits follows the ideas introduced in the next section.

A major limitation of microcomputers is that they are relatively slow in performing many specialized digital functions compared to circuits designed specif-

---

[*]W. S. Richardson et al., "The 32b Computer Instruction Decoding Unit," *1981 ISSCC Digest of Technical Papers*, pp. 114–115.

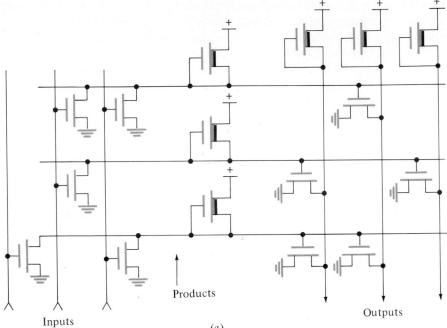

Products

Inputs

Outputs

(a)

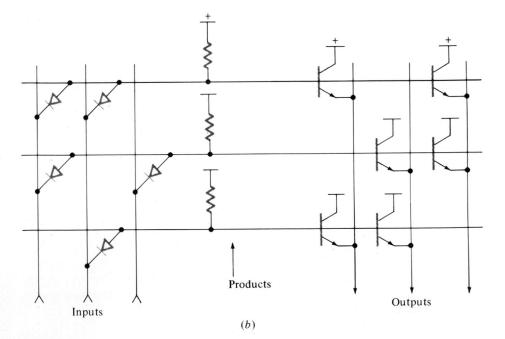

Products

Inputs

Outputs

(b)

FIGURE 10.5
Programmable logic array circuit design. (a) NOR-NOR PLA in NMOS. (b) AND-OR PLA in
bipolar.

ically for the application. The reason for this performance limitation is that computers execute even relatively simple logic or mathematical functions as a sequence of very elementary operations. In microcomputers, typical operations are load a byte of data into a register, add the contents of a register to a memory location, compare 2 bytes of data, etc. Each of these operations requires 0.2 to 2 μs in a microcomputer. On the other hand, a circuit uniquely designed to perform any one of these operations will produce a result in 10 to 100 ns, and additional circuits may be used to simultaneously perform other operations.

Microcomputers should be used for performing digital system functions wherever they can meet the system requirements. The fact that a very large fraction of a microcomputer's circuitry is not used in a particular application is seldom sufficient to justify design of any form of custom circuit. High-volume production and use of microcomputers has reduced their cost to as little as $1 to $2 apiece. Requirements that frequently cannot be met by microcomputers include those for high speed and those for specialized signal levels at inputs and outputs.

10.5   CIRCUIT DESIGN FOR VLSI

The design techniques described in Secs. 10.2 and 10.3 are attractive for quick low-cost design of digital circuits and subsystems of moderate complexity. However, very complex LSI circuits, such as memories, microprocessors, and specialized microprocessor peripheral circuits, that are manufactured by the millions, cannot be developed successfully using the above circuit design approaches. For complex components, circuit design and chip layout must be optimized to the particular application to achieve competitive circuit density, high-speed performance, and low power consumption.

Circuit design for the most complex components usually involves careful, selective compromises of some of the important properties of ideal logic elements that were described in Chap. 1. When this is skillfully done, the result is a severalfold increase in the circuit density, plus faster operation with less power consumption.

Layout design for these very complex components requires detailed understanding of the interactions between layout geometries and circuit performance. Carefully defined elements such as adders, decoders, multiplexors, and registers are repeated many times within a single chip. A measure of good design is the ratio on a chip of the total number of transistors to the number of individually designed transistors. This ratio, which was only slightly greater than unity for the first microprocessors, is now in the range 10 to 100. The remainder of this section focuses on the circuit design of the elements of LSI and VLSI.

10.5.1   Logic Circuits for VLSI

The number of devices (transistors or other circuit elements) that can be successfully manufactured on a single chip has grown steadily from about 10 in 1965 to about 1 million in 1987. As first mentioned in Sec. 3.0, considerations of heat removal place a limit on the amount of power that can be consumed by a chip. As the number of devices and circuits on a chip grows, power limitations create increasingly difficult constraints on circuit design.

There are advantages to reducing power consumption beyond the matter of heat removal. The cost of power supplies and regulators will be reduced. If power consumption of a chip can be kept below about 500 mW, low-cost plastic packages (which have only about half the heat dissipation capability of expensive ceramic packages) can be used. Chip reliability improves as operating temperature is reduced.

Popular 8-bit and 16-bit microprocessors use 5,000 to 20,000 static NMOS gates with depletion loads and typically dissipate about 1 W. More complex static NMOS designs will sacrifice high-speed performance if forced to stay in this range of power consumption. Ratioless dynamic NMOS is not a popular choice for logic operating from 5 V because of the design difficulties when operating with widely varying loads and parasitic capacitances. Dynamic NMOS is used for memories designed for 5-V operation, but only at the burden of generating complex, multiphase clocks on the chip. The regularity of memory array structures leads to uniform, predictable loads on address buffers, decoders, and read-write circuits. Only one row in the array is active on any cycle. These factors make it feasible to design 5-V dynamic NMOS memory circuits. The practicality of 5-V dynamic NMOS logic and microprocessors is questionable.

An obvious solution to the power problem is to use CMOS gate logic as described in Sec. 3.7. However, because standard CMOS gates require two transistors per input (one NMOS, one PMOS), isolation of NMOS and PMOS devices, and metal interconnections between drains of opposite conductivity, the gate count per unit chip area is much lower than for NMOS. This density penalty for CMOS gate logic is on the order of a factor of 2 if NMOS and CMOS are compared using the same design rules. CMOS domino logic and CMOS PLA-based designs can provide the low power characteristics of CMOS with the high density of NMOS. These alternatives are briefly described below.

REDUCING POWER CONSUMPTION OF NMOS LOGIC. An obvious way to reduce the power consumption of NMOS logic is to reduce the supply voltage V_{DD} below the 5-V level that is almost universal today. The most important reason for the continued popularity of 5-V operation is because it provides compatibility with standard TTL supply voltage and logic levels. In all but a few unique applications, such as calculators, wristwatches, and other battery-powered products, integrated circuits and input-output devices based on a variety of technologies are used. TTL voltage levels are by far the most common standard for interconnection of digital components, even if no TTL at all is used in a given system. Any departure from these levels would significantly increase system cost.

Only the input and output nodes of LSI circuits need be compatible with TTL levels. Internal circuits could be operated at a lower voltage and with smaller logic swings to save power. Two separate supply voltages would be used. While this has been done in some bipolar LSI circuits, it has never been common for NMOS.

Reducing the dimensions, or scaling, of MOS devices in order to increase density and performance, while simultaneously reducing power consumption, is described in Sec. 3.9. When MOS transistor gate length is reduced below about 1 μm and gate oxide thickness is reduced below about 20 nm, it is necessary to reduce the supply voltage to 2 or 3 V to avoid possible destructive electrical breakdown. A departure from TTL compatibility will have to be accepted when device scaling reaches this stage. These matters are not considered further in this text.

In Sec. 3.4 we saw that t_{PLH} is the largest propagation delay for a conventional static NMOS gate such as the one shown in Fig. 10.6a. For specified voltage swing and load capacitance, t_{PLH} is inversely proportional to the load current when the output is low. This load current in turn determines the power dissipation.

Power consumption of NMOS can be substantially reduced without increasing propagation delay by driving load devices with the logic complement of the gate input, as shown in Fig. 10.6b. This configuration, termed *push-pull* or sometimes "poor man's CMOS," requires one load device for each input. It is used in applications such as NMOS memory decoders, where logic complements are available and where minimizing the power-delay product is important.

When a logic input is high, V_{GS} for the corresponding load device is approximately zero, and biasing conditions are similar to those in the conventional NMOS gate. However, during t_{PLH}, the load devices have a much larger value of V_{GS} than in the NMOS conventional circuit because their gates are driven to V_{DD} rather than remaining tied to their sources. Consequently the peak load current during this transition is much higher than the steady-state load current with the output low. For reduced power consumption, W/L for the load devices must be smaller than for the conventional gate of Fig. 10.6a.

Another variation in NMOS gate design is represented in the multi-output gate shown in Fig. 10.6c. Translation of logic designs from multi-input to multi-output form is as described for I^2L in Sec. 7.5. The advantage of this circuit is that only the internal node x has a full voltage swing of almost V_{DD} in magnitude. The connections between gates, which are sometimes long and heavily loaded with capacitance (represented by C_T here), only have a voltage swing of $V_{TD} - V_{OL}$. For a given operating current or power, propagation delays are directly proportional to voltage swing, hence the advantage of this circuit.

INCREASING CMOS DENSITY. The use of CMOS in an unconventional circuit configuration that employs a clock provides an excellent combination of high density and low power. Figure 10.7 shows a single gate implementing the logic AND-OR function $(A \cdot B \cdot D) + (E \cdot F)$. A single-phase clock ϕ controls operation. When ϕ is low, M_1 is off, so no current can flow. PMOS device M_2 is on, operating in a common-source mode; it charges node 2 to V_{DD}. Parasitic capacitance C_1 holds this voltage until discharged. Logic inputs are changing while ϕ is low but then must remain steady while ϕ is high. Node 3, the output, is forced low while ϕ is low because the inverter $M_3 - M_4$ has a high input from node 2.

When ϕ goes high, M_1 turns on and M_2 turns off. Node 1 is pulled down to

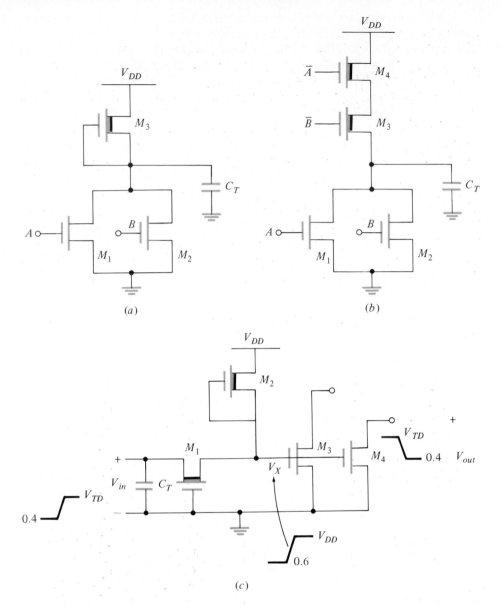

FIGURE 10.6
Standard and improved NMOS gates. (a) Standard NMOS NOR gate. (b) Push-pull NMOS NOR gate. (c) Multi-output NMOS gate.

ground level. If A, B, and D are all high, or E and F are both high, a conducting path is available from node 2 to node 1, and C_1 will be discharged. The output at node 3 will go high. On the other hand, if no conducting path is present between nodes 2 and 1, node 2 remains high and the output remains low. Thus the valid logic output is present at node 3 shortly after ϕ goes high.

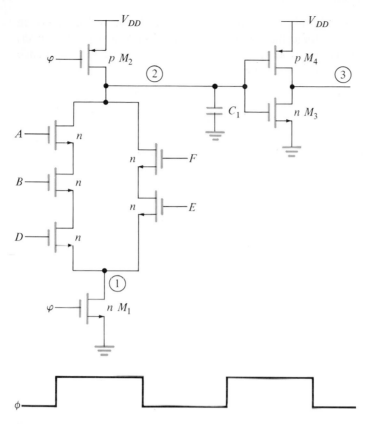

FIGURE 10.7
CMOS domino logic circuit.

Several aspects of this gate circuit deserve comment. Only one transist
is added for each additional logic input. All devices except M_3 and M_4 may be
minimum size, since they are required only to charge or discharge C_1. No steady-
state dc current flows for any state of inputs or clock. The inverter $M_3 - M_4$ serves
as an output buffer; it can be designed to drive the capacitance associated with a
large fan-out or a long bus with little loss of speed. Transistors M_1 through M_4
are "overhead" in the sense that they are needed whether the circuit has 1, 2,
or many inputs. Therefore overall device count will be minimized by designing
with complex logic functions. Two-input logic gates should be avoided.

Figure 10.8a is a simplified illustration of this technique as it might be
applied to perform a portion of the logic for an internal arithmetic unit in a
VLSI processor. The clock signals needed for operation are shown in Fig. 10.8b.
Register R_1, composed of static master-slave flip-flops similar to those shown
in Sec. 8.9, is assumed as the source of input data. The outputs of R_1 are
continuously valid except during the transition time that occurs on the rising edge
of clock ϕ_2. Output from the combinational logic is strobed into another static
master-slave register R_2 on the falling transition of ϕ_2, or, equivalently, on the

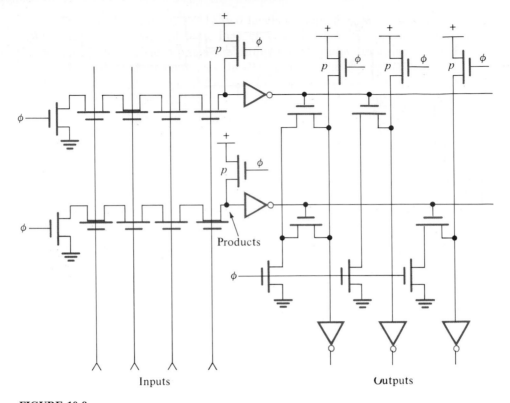

FIGURE 10.9
Domino PLA.

ROM or PLA is present only in a single masking layer, typically as the presence or absence of a contact or other feature in a designated fixed location.

An inherent drawback of array structures is that rows and columns contact many devices and hence are highly capacitive. Propagation delays can be reduced without excessive power consumption by driving highly capacitive lines with push-pull drivers. These have peak output currents considerably larger than their average dc supply current. Of course, a normal CMOS inverter is a nearly ideal push-pull driver. Figure 10.10a shows an NMOS realization of a push-pull driver circuit, often called a *superbuffer*. The similarity to the gate of Fig. 10.6b should be evident.

Further reductions in propagation delays through arrays can be achieved by detecting the desired logic output when there has been only a small voltage change, much less than the normal voltage swing of $V_{OH} - V_{OL}$. Normal NMOS or CMOS inverters operating from 5 V require a 2- to 2.5-V voltage change at the input, starting from a steady-state level at the input of V_{OL} or V_{OH}, to cause a change of state at the output. This problem has been described earlier in Sec. 9.1.4 in the context of ROM and EPROM design. Memory sense amplifiers for

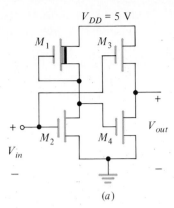

(a)

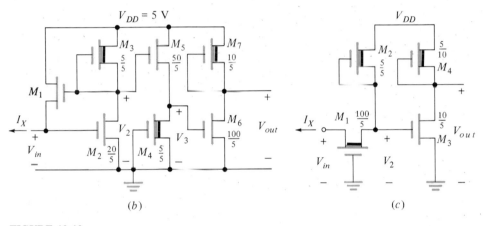

(b)   (c)

FIGURE 10.10
NMOS circuits for improved power delay product. (a) NMOS superbuffer. (b) Threshold detector
with output buffer. (c) Simple threshold detector.

NMOS and CMOS are shown there. Two additional circuits for faster detection
of valid output data are shown in Fig. 10.10b and c. If used outside the memory
context, these circuits may be called *threshold detectors* or *line receivers*. They
require a voltage change of only a few tenths of a volt at the input to quickly
give a much larger output voltage change.

Consider the use of the circuit of Fig. 10.10b in conjunction with either
of the MOS ROM arrays shown in Fig. 9.3. Transistor M_1 provides current to
the column lines, so the separate column pull-up devices shown in Fig. 9.3 are
unnecessary. When there is no conducting path from the column line to ground,
the steady-state input voltage is just higher than V_T for transistor M_2, and the
drain voltage of M_2 is slightly more than $2V_T$. Current is flowing in M_2 and M_3,
while M_1 is at the edge of conduction.

When current I_x flows to ground via the column in the array, the column
voltage drops. Only a small reduction in column voltage is needed to turn off M_2,

allowing its drain voltage to rise toward V_{DD}. A following level shifter $(M_4 - M_5)$ and inverter $(M_6 - M_7)$ are used to achieve an output logic low level well below V_T, restoring normal noise margins.

The simpler threshold detector of Fig. 10.10c is derived from the multi-output gate shown in Fig. 10.6c. It operates as follows. In the absence of input current I_x, the input voltage rises until depletion transistor M_1 is at the edge of conduction. This is achieved when its gate-source voltage $V_{GS} \simeq -3$ V, or when $V_{in} = +3$ V. The gate of M_3 rises to V_{DD}, giving a normal V_{OL} from inverter $M_3 - M_4$. When I_x begins to flow, V_2 drops much faster than V_{in} because M_1 is operating as a saturated common-gate amplifier with high voltage gain. The drawback to this circuit is that if V_{in} falls close to 0 V, the small device M_2 may recharge it only very slowly to the steady-state value, about $+3$ V. The circuit of Fig. 10.10b is better in this respect, since the gate voltage of M_1 of that circuit rises when V_{in} is low.

10.5.3 Decoders and Multiplexors

The array-structured circuits such as ROM and RAM that are an increasing part of VLSI design make extensive use of decoders and multiplexors. Decoders typically select and drive one line from a large group (e.g., row decoders), while multiplexors (e.g., column decoders) convey a signal from one of many lines to a single output node.

The complexity of single-stage NOR and NAND decoders such as those shown in Fig. 9.6 becomes excessive for large arrays. For example, NOR decoders to select 1 of 256 rows would require 256 eight-input NOR gates. Not only are the transistor count and power consumption very high, but the capacitance on the output node of an 8-input NOR would likely be a significant contributor to propagation delay. The same objections would apply to the use of NORs plus pass gates for column decoding, as shown in Fig. 9.8a, if applied to a 256-column array. On the other hand, the tree decoder of Fig. 9.8b would introduce a very long, slow series path for the desired signal, although transistor count and power consumption are not problems for this configuration.

Decoders and multiplexors for large arrays are best designed by using two or three stages of gating rather than the single stage considered in the above examples. Device count and power consumption can be reduced, and speed increased, by careful choices in circuit design. The last stage of decoding usually should be tailored to the specific case in hand.

An example of two-stage decoder design is illustrated in Fig. 10.11a. Only a 1-of-16 selection is shown for simplicity; the technique would be used in preference to a single-stage decoder only for much larger decoders. The address $A_1 - A_4$ is decoded in two halves, each of which comprises a decoding from 2 bits to a 1-of-4 selection. A total of eight 2-input NOR gates are adequate for this purpose. For convenience, we denote the outputs from the first stage as x_i and y_j, where in this example each index ranges from 1 to 4. By forming the logic ANDs of all possible combinations $x_i y_j$, we create the desired 1-of-16 selection. A total of sixteen 2-input AND gates are required for this stage. The

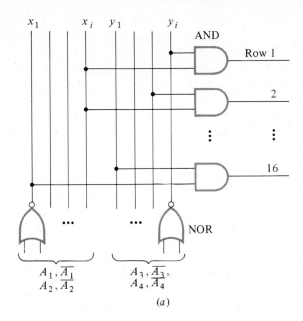

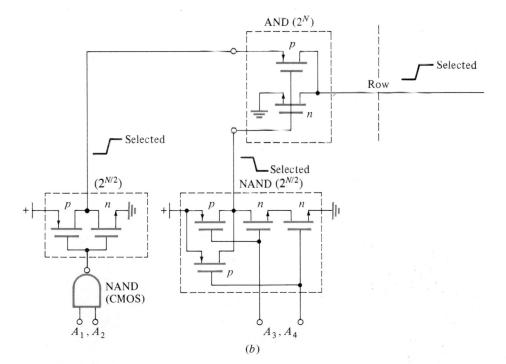

FIGURE 10.11

(a) Logic design for a two-stage row decoder. (b) Circuit design for a two-stage row decoder for $N = 4$-bit selection.

circuit implementation of this logic design requires some ingenuity to achieve a good compromise among complexity, speed, and power consumption. One possible design, in CMOS, is shown in Fig. 10.11b. NAND gates are preferred over NOR gates in CMOS because of the better conduction characteristics of NMOS devices. The AND function can be implemented without dc power consumption and with only two transistors, as shown, if one input is allowed to be inverting. The added inverter provides an economical way to drive row charging current, in addition to giving the needed inversion.

The number of instances of each circuit is indicated on Fig. 10.11b in terms of N, the number of bits being decoded. For complex decoders, the total number of transistors required for selection is sharply reduced compared to a single-stage decoder.

The design of column decoders or multiplexors can employ a combination of NOR and multibranch tree circuits to achieve a good compromise among speed and complexity.

10.5.4   Output Buffers

Internal gates in modern VLSI circuits employing 1.5- to 2-μm minimum feature sizes have input capacitances of 50 fF or less. CMOS and NMOS gates achieve propagation delays of 0.3 to 1 ns when driving moderate nearby fan-outs on the same chip. While device dimensions and capacitances on VLSI chips have decreased rapidly in recent years, unfortunately the dimensions and capacitances of chip packages and printed wiring connections have remained about the same. Commonly VLSI components are required to drive external loads of 50 pF or more at their outputs. Since for MOS circuits the propagation delay is directly proportional to capacitance load, using a normal gate or inverter to drive an off-chip node would result in a propagation delay of hundreds of nanoseconds. Consequently, special output buffer circuits employing a cascade of several progressively larger stages are designed to reduce this delay.

A simplified representation of an output buffer, shown in Fig. 10.12, can be used to calculate the optimum number of stages for minimum overall delay.[*]

---

[*]A. M. Mohsen and C. A. Mead, "Delay Time Optimization for Driving and Sensing of Signals on High-Capacitance Paths of VLSI Systems," *IEEE Journal of Solid-State Circuits,* vol. SC-14, 1979, pp. 462–470.

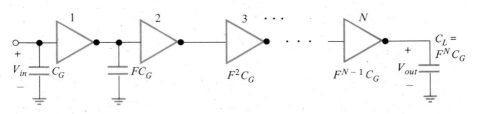

FIGURE 10.12
N-stage output buffer. Capacitors shown represent the input capacitances looking into each stage.

Small-size gates and inverters used for internal logic are assumed to have an input capacitance C_G. The capacitance load at the output is C_L. Denote the ratio C_L/C_G as Y. Assume the number of stages is N and the *fan-out factor* (capacitance seen by each stage relative to that seen by the preceding stage) is F. Either N or F is the independent design parameter. Define the propagation delay of one stage driving another identical stage ($F = 1$) as t_P. Then the overall delay for the buffer is

$$t_B = NF t_P \qquad (10.1)$$

The overall capacitance ratio is related to N and F by

$$Y = F^N \qquad \text{or} \qquad \ln Y = N \ln F \qquad (10.2)$$

By solving this for N and substituting the result in (10.1), we obtain

$$t_B = \ln Y \frac{F}{\ln F} t_P \qquad (10.3)$$

By differentiating this equation with respect to F and equating to zero, we can find that the overall delay t_B is minimized for $F = 2.72 = e$, the base of natural logarithms. At the optimum, the total delay in the output buffer is $2.72 \ln Y t_P$. For an example in which $Y = 1,000$, representative of real situations today, the minimum delay is $18.8 t_P$, while the delay would be $1,000 t_P$ if no buffer were used. The optimum is quite broad so that the delay is increased only 14% above the minimum by using $F = 5$. For $Y = 1,000$, 12 or 13 stages are required to approximate $F - 2.72$, while four stages suffice if we choose $F = 5$.

A five-stage CMOS output buffer with device sizes for an assumed 2-μm technology with $C_{ox} = 1 \text{ fF}/\mu m^2$ is shown in Fig. 10.13. For the moment, ignore the inductances in the power and ground leads. The input capacitance is

$$C_G = \text{gate area} \times C_{ox} = (5 + 12.5)(2)(1 \text{ fF}) = 35 \text{ fF} \qquad (10.4)$$

Then for a load capacitance of 50 pF, we find

$$Y = \frac{C_L}{C_G} = \frac{50}{0.035} = 1429 \qquad \text{and} \qquad F = Y^{0.2} = 4.3 \qquad (10.5)$$

Device sizes for the first stage are those typical of the on-chip logic. Each following stage in the output buffer has devices F times wider than the preceding stage.

Many VLSI components have multiple outputs, ranging from 8 outputs for PROMs and EPROMs up to 32 or more for advanced microprocessors. Serious problems can arise when many outputs switch simultaneously due to transient voltage drops in the inductances of ground and power pins on the chip packages. These are the inductances that are shown in Fig. 10.13. Often the power and ground pins are located at the furthest corners of dual in-line packages. (This standard was adopted around 1965 to simplify the layout of one- and two-layer printed wiring boards.) The inductance of the conductor between the chip bonding pad and the external pin ranges from 5 to 15 nH, depending on the specific dimensions.

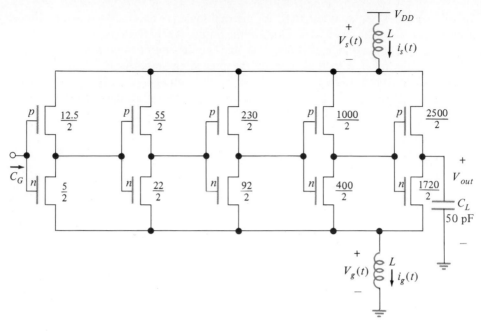

FIGURE 10.13
Five-stage CMOS output buffer.

A buffer designed as suggested above will conduct large currents when switching typical off-chip loads. The ground line current and voltage as a function of time for a buffer switching from high to low is shown in Fig. 10.14. Assuming a package lead inductance L in the ground conductor, the inductive voltage drop is given by

$$V_g(t) = L\frac{di_g(t)}{dt} \tag{10.6}$$

Typical numbers for one output buffer are a rise or fall time of 5 ns with a peak current of 50 mA reached halfway through the switching time. Then with $L = 10$

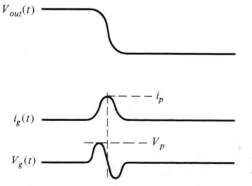

FIGURE 10.14
Transient signals for CMOS output buffer.

nH, the peak value of $V_g(t)$ is

$$V_P = 10 \text{ nH} \times \frac{50 \text{ mA}}{2.5 \text{ ns}} = 0.2 \text{ V} \qquad (10.7)$$

Although this is not enough to cause problems, consider a component with 16 outputs. A bad situation arises when one output is low and remains low while 15 outputs simultaneously switch from high to low. The peak ground line voltage would be $15 \times 0.2 = 3$ V if the currents remained as assumed above. The sixteenth output that was supposed to remain low will momentarily produce a high logic level, quite possibly introducing a logic error.

The serious practical problem of noise errors caused by transient voltages across package lead inductances can be reduced by several means, including assigning the lowest inductance package pins (usually the center pins) for power and ground, using additional power and ground pins in parallel, reducing the off-chip capacitance loads, and designing output buffers for longer rise and fall times.

10.6 SUMMARY

This chapter describes various design techniques aimed at increasing the complexity (gate count) and performance of large-scale integrated digital circuits. Simple circuits employing minimum numbers of devices, reduced voltage swings, and low power consumption are desired for LSI. Array structures are desirable for high circuit density, moderate design costs, and reduced power consumption. Specific techniques described are the following:

• Gate arrays in CMOS and bipolar technology offer quick, economical fabrication of complex digital chips.
• Standard cells drawn from a predesigned library offer more design flexibility than gate arrays but require more unique steps in their fabrication.
• Programmable logic arrays (PLAs) are ROM-based structures that can efficiently replace randomly interconnected gates for many control logic applications. PLAs are available in several forms as standard components and are also useful as elements of custom-designed LSI and VLSI chips.
• Microprocessors and microcomputers are popular standard digital components. They should be used whenever their speed and functional capabilities can satisfy specific requirements. Other books should be consulted for guidance in microprocessor applications.
• The internal circuit design of custom LSI and VLSI components involves important departures from the ideas followed in the design of SSI and MSI components. Special techniques are used to reduce power consumption, increase circuit speed, and reduce the number of transistors required for a particular function. Specific examples relating to array design and output buffers are described.

REFERENCES

1. "Semi-Custom LSI—Gate Arrays," 1980 WESCON Professional Program, Session 30, Electronic Conventions, Inc., El Segundo, Calif.
2. R. Walker, "Logic Arrays, Technologies and Design," MIDCON 1981 Professional Program Session Record, Electronic Conventions, Inc., El Segundo, Calif.
3. D. G. Schweikert, "CAD and Gate Arrays: The Keys to Fast Turnaround IC Design," MIDCON 1981 Professional Program Session Record, Electronic Conventions, Inc., El Segundo, Calif.
4. W. D. Burkard, "Semi-Custom LSI at Storage Technology Corp.," *VLSI Design II*, no. 3, third quarter 1981, pp. 14–18.
5. J. Lohstroh, "Devices and Circuits for Bipolar (V)LSI," *Proceedings of the IEEE*, vol. 69, no. 7, July 1981, pp. 812–826.
6. J. R. Tobias, "LSI/VLSI Building Blocks," *IEEE Computer*, August 1981, pp. 83–101.
7. M. I. Elmasry (ed.), *Digital MOS Integrated Circuits*, IEEE Press, New York, 1981.
8. P. M. Solomon, "A Comparison of Semiconductor Devices for High-Speed Logic," *Proceedings of the IEEE*, vol. 70, no. 5, May 1982, pp. 489–509.
9. C. Mead and L. Conway, *Introduction to VLSI Systems*, Addison-Wesley, Reading, Mass., 1980.
10. L. A. Glasser and D. W. Dobberpuhl, *The Design and Analysis of VLSI Circuits*, Addison-Wesley, Reading, Mass., 1985.
11. N. H. E. Weste and K. Eshraghian, *Principles of CMOS VLSI Design*, Addison-Wesley, Reading, Mass., 1985.
12. *IEEE Journal of Solid-State Circuits*, special issues on semiconductor memory and logic, October of each year since 1970.

PROBLEMS

P10.1. Draw the full transistor-by-transistor schematic for the JK flip-flop layout shown in Fig. 10.1c.

P10.2. Draw the schematic diagram for a single CMOS gate to realize the logic function $\overline{A + (B \cdot C)}$. On a copy of Fig. 10.1a, draw in with colored pencil the metal connections needed to realize this logic function.

P10.3. Find the voltage transfer characteristics, noise margins, and average power consumption for the five gates shown in Fig. 10.3. The voltage swings may be so small that simple piecewise-linear models are not adequate. Assume the following device parameters:

$$npn: \beta = 100 \qquad I_{ES} = 10^{-16} \text{ A}$$

$$pnp: \beta = 5 \qquad I_{ES} = 10^{-13} \text{ A}$$

Diode SD 1 and all transistor clamps: $I_s = 10^{-13}$ A

Diodes SD 2 and SD 3: $I_s = 10^{-10}$ A

P10.4. Calculate the value of W/L for $M_3 = M_4$ in Fig. 10.6b such that the average load charging current $I_{LH(av)} = 38.7 \ \mu\text{A}$, as in the circuit of Example 3.5. Find the average power consumption for the circuit of Fig. 10.6b, assuming a duty cycle of 50%, and compare this with the power consumed by the circuit of Example 3.5. For this and following problems, refer to Examples 3.4 and 3.5 for device and circuit data.

P10.5. For the circuit of Fig. 10.7, make a rough calculation of the time required to charge and discharge $C_1 = 0.5$ pF. All transistors have $W/L = 1.0$; $V_{DD} = 5$ V.

P10.6. Consider the design of the NMOS superbuffer shown in Fig. 10.10a. Assume this circuit must drive a capacitive load of 50 pF.

(a) Specify the W/L ratios for M_3 and M_4 so that $t_{PHL} = t_{PLH} = 50$ ns. For this hand calculation, assume M_3 an M_4 are driven by ideal (zero transition time) complementary 5-V steps. Use the device data from Example 3.4.

(b) Suppose that the W/L ratios for M_2 and M_1 are 20 and 5, respectively, and that M_3 and M_4 have W/L ratios as determined in part (a). Assume that V_{in} makes transitions from 0 to 5 V and back with a 20-ns transition time. Use SPICE to simulate the overall propagation delay of the superbuffer with a 50-pF capacitive load. In calculation of transistor areas for SPICE input, assume the layout follows the rules shown for Examples 3.4 and 3.5.

P10.7. Calculate and sketch V_{in}, V_2, V_3, and V_{out} as a function of I_x for the threshold detector of Fig. 10.10b. Use SPICE to simulate the transfer characteristics, and compare with the results of your hand analysis.

P10.8. Calculate and sketch V_{in}, V_2, and V_{out} as a function of I_x for the threshold detector of Fig. 10.10c. Compare the results of hand analysis with SPICE simulation.

P10.9. Calculate the total number of transistors required for a 1-of-256 row decoder of the type shown in Fig. 10.11, and compare this with the count for a single-stage NOR decoder that uses conventional CMOS logic gates.

P10.10. Design the schematic for a three-stage CMOS decoder for selection of 1 of 512 rows in a 256K ROM. (No unique answer!) Compare the transistor count for your design with that for an expanded version of the two-stage decoder from the previous problem.

P10.11. Design the schematic for a column decoder in CMOS for a 1-of-256 selection. A maximum of three series pass transistors is allowed in order to achieve fast access time.

CHAPTER
11

GALLIUM ARSENIDE DIGITAL ICs

11.0 INTRODUCTION

Up until this point our study of digital ICs has considered only silicon semiconductor technology. That is, we have described circuits that employ silicon MOS-FETs or silicon BJTs. In this we have reflected the past and present marketplace where all major digital IC products have been based upon silicon technology. However, since about 1980 there has been a growing effort to develop digital ICs based upon gallium arsenide technology. Active devices made from the semiconductor gallium arsenide (GaAs) have long been used as discrete components in microwave circuits. Many of the advances in high-speed digital communications and high-resolution radar systems at frequencies beyond a few gigahertz (GHz) have been possible only with gallium arsenide devices. As we have seen in silicon technology, ECL circuits demonstrate the shortest propagation delay times, and therefore they are used in digital systems where very high data rates are required. However, at the present time the maximum frequency of operation of commercial silicon ECL circuits is a data rate of about 4 Gb/s. While it is anticipated that the upper speed limit for future ECL circuits might reach 10 Gb/s, this is near fundamental limits for silicon technology. It is at these and higher data rates that gallium arsenide digital integrated circuits may provide performance not available with silicon ICs. Table 11.1 presents a summary of the physical and electrical properties of gallium arsenide and silicon that are important in the realization of integrated circuits.

TABLE 11.1
Principal properties of GaAs and Si at 27°C

Property	Symbol	Units	GaAs	Si
Low-field electron mobility	μ_n	cm²/V·s	4,000–9,000	500–1,200
Maximum electron drift velocity	v_d	cm/s	2×10^7	1×10^7
Critical field for high-field transport	E_c	V/cm	3×10^3	1×10^4
Low-field hole mobility	μ_p	cm²/V·s	400	480
Relative permittivity	ϵ_r		12.9	11.7
Band gap energy	E_g	eV	1.43	1.11
Intrinsic carrier concentration	n_i	cm⁻³	9.0×10^6	1.45×10^{10}
Maximum resistivity		Ω·cm	1×10^9	1×10^5
Schottky-barrier heights	ϕ_B	V	0.6–0.8	0.4–0.6

Here we consider primarily digital circuits employing the most widely used GaAs transistor, the metal-semiconductor (Schottky-gate) field-effect transistor, or MESFET. GaAs MESFETs offer several advantages as devices for use in high-speed digital integrated circuits:

1. Electrons travel faster in n-type GaAs than in silicon. If other factors are equal, logic circuit propagation delay is directly proportional to carrier velocity. At the usual doping levels for MESFETS, electron mobility at low electric fields is about 4,000 to 5,000 cm²/V·s at 300 K, 5 to 10 times greater than in silicon. Above a critical electric field of about 0.3 V/μm in GaAs, electron velocity begins to level off and reaches a limit of about 2×10^7, about twice the corresponding limit for electrons in silicon. Channel length for MESFETS is typically 0.25 to 1 μm, and applied voltages range from 1 to several volts. Thus electrons often travel at the velocity limit. The net result at circuit level is that GaAs MESFET logic circuits may be up to 2 times faster than silicon ECL for the same minimum feature size. In lightly doped GaAs, the low-field electron mobility can reach the range 8,000 to 9,000 cm²/V·s at room temperature. Another type of MESFET is the GaAs high electron mobility transistor (HEMT). It uses a more complex structure to provide light doping in the channel, leading to higher mobility at low electric fields, particularly at 77 K. However, the maximum electron velocity at high electric fields is only slightly higher.

2. Because of the larger band gap and the absence of critical gate oxide or isolating oxides used in Si devices, GaAs devices are claimed to have greater immunity to radiation effects. This is an important consideration for space and military applications.

3. Pure GaAs is "semi-insulating" with a resistivity in the 10^7- to 10^9-Ω ·cm range at room temperature. The resistivity is so high that devices made of doped GaAs sometimes can be adequately isolated for practical circuits simply by separating them by a micrometer or two on the surface of a semi-insulating substrate. (Selective ion implantation may be used to improve the isolation.) Also, parasitic capacitances are determined by considering the entire substrate as an insulator, so parasitic device capacitances to ground are smaller than the corresponding capacitances for silicon ICs.

4. High-quality Schottky barriers can be realized on GaAs with a variety of metals, such as Al, Pt, Ti, and W, leading to high-quality Schottky diodes. Schottky barriers are used to form transistor gates and as diodes for performing logic and level shifting.

The most important limitation of GaAs technology is that the yield for complex circuits is much lower than for silicon ICs, due in part to more defects in the basic semiconductor material. The cost of GaAs ICs is therefore much higher than for silicon ICs of the same functional complexity, and the feasible scale of integration is much lower. Other drawbacks of GaAs technology are that the hole mobility is about the same as in silicon, so complementary circuits are not attractive, and metal-oxide-semiconductor transistors are not possible due to very high levels of surface-state charge (see discussion of Q_{ss} and Q_{ox} in Sec. 2.3). Finally, because new materials and processes are involved, the reliability of GaAs integrated circuits today is 10 to 100 times worse than for silicon integrated circuits.

11.1 FABRICATION

The fabrication of a planar n-channel GaAs MESFET involves many of the same processes as with silicon ICs. Among these are cleaning, oxidizing, masking, etching, implanting, annealing, metalizing, and dicing. Each of these processes is much the same as described in Sec. 1.6, but because of the importance of high-speed operation the methods used are the best possible and the minimum dimensions used are a micrometer (μm) or less. Figure 11.1 shows a cross section of an n-channel GaAs MESFET and a GaAs Schottky diode. As in the fabrication of a silicon IC, the starting material for a GaAs IC is a wafer cut from a bar of crystalline material. Compared to silicon, gallium arsenide wafers and chips are far more fragile, and they must be handled with delicate care. The diameter of the wafers used to process GaAs digital ICs is usually only 2 or 3 in, whereas for silicon, 4- to 6-in diameter wafers are common.

MASK 1. *Channel doping implant.* An implant using silicon or selenium as the dopant establishes a doping level of about 10^{17} cm^{-3} for the conducting channels of the MESFETS. This results in a final sheet resistance of the undepleted channel of 1,000 to 2,000 Ω per square.

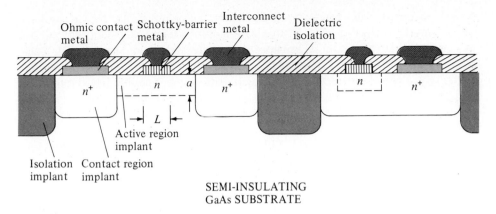

FIGURE 11.1
Cross section of n-channel GaAs MESFET and Schottky diode.

MASK 2. *Source and drain areas.* A sulfur implantation is common for the n^+ source and drain contact regions, producing a sheet resistance of 100 to 200 Ω per square. A controlled thermal anneal cycle at 800 to 850°C electrically activates the first two implants.

MASK 3. *Ohmic contact areas.* The areas where metal will make contact to the source and drain are defined. Ohmic contact is made to the heavily doped regions with a Au/Ge/Ni (gold-germanium-nickel) alloy.

MASK 4. *Isolation areas.* Isolation of devices is achieved by implanting ions of hydrogen or boron in the regions between devices. This increases the resistivity in the isolation regions.

MASK 5. *Areas for metal removal.* The Schottky-barrier gates and source-drain connections are made with a Ti/Pt/Au (titanium-platinum-gold) thin-film sandwich that is initially deposited over the entire wafer. The Ti makes good Schottky barriers and contacts and has excellent mechanical adhesion properties. The Au has excellent conductivity and corrosion resistance. The Pt is needed to prevent an unacceptable reaction between Ti and Au. Mask 5 removes all metal except where it connects to sources, gates, and drains.

MASK 6. *Contact vias.* The wafer is covered with an insulating layer of SiO_2 or Si_3N_4. Windows are opened over all desired connection points.

MASK 7. *Metal interconnections.* A second layer of Ti/Pt/Au, or another metal, is deposited over the entire wafer. The final mask patterns this metal to produce the desired circuit connections. After wafer probing and dicing, the devices are ready for packaging.

Micrometer and submicrometer geometries are required for the integrated devices in order to fully benefit from the potential speed of GaAs. Hence, present-day integrated GaAs MESFETS have gate lengths ranging between 0.5 and 1.0 μm, with similar minimum dimensions for source and drain contacts. The basic simplicity of the MESFET structure facilitates control of such small geometries, but achievement of satisfactory pattern delineation, reproducible line widths, and adequate alignment accuracy requires sophisticated microlithography techniques such as reduction projection photolithography with direct on-wafer stepping, direct-writing E-beam pattern generation, or x-ray replication.

An advanced form of the MESFET, now the subject of widespread development, is the *high-electron-mobility transistor (HEMT)*. The HEMT also goes by the name of *MODFET*, for *modulation-doped field-effect transistor*. Particularly at reduced temperatures, HEMT circuits offer a significant speed improvement over MESFET circuits. However, the fabrication process for the HEMT is much more complex, involving *molecular beam epitaxy (MBE)* and *metal-organic chemical vapor deposition (MOCVD)* to form alternating layers of GaAs and AlGaAs with greatly different doping concentrations. A cross-sectional view of a HEMT is shown in Fig. 11.2. In these structures the key is to place a heavily doped AlGaAs layer adjacent to an undoped GaAs channel layer. Because electron potential energy is lower in the GaAs, electrons "fall" into the lightly doped GaAs channel. Here, electron mobility is higher due to the lower doping level. At room temperature, the mobility advantage for the HEMT over the MESFET is a factor of 2 at most. But for HEMTs at 77 K (the temperature of liquid nitrogen), electron mobilities approaching 100,000 cm^2/V·s have been reported, with switching times of less than 15 ps. This represents a significant improvement compared to the corresponding results for MESFETs at 77 K.

11.2   GALLIUM ARSENIDE (GaAs) TRANSISTOR

Before analyzing the performance of GaAs digital circuits we must understand the basic electrical characteristics of the *n*-channel GaAs MESFET. The phys-

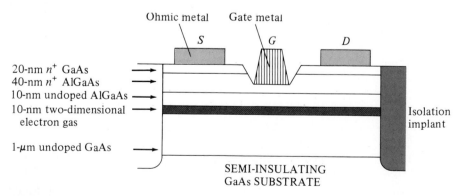

FIGURE 11.2
Cross section of depletion-mode high-electron-mobility transistor.

ical principles of operation and the current-voltage relationships for the GaAs MESFET are similar to those of the silicon junction FET and MOSFET. Figure 11.3 shows a simplified cross-sectional view of an *n*-channel depletion-mode MESFET. Conduction takes place in a surface channel of thickness *a*. The gate forms a Schottky barrier with the channel. With zero bias applied, the built-in voltage of the barrier depletes a portion of the channel of mobile carriers, as seen in Fig. 11.3. When the gate is biased positively with respect to the channel, the depletion layer is narrowed and the channel is more conductive. However, at around 0.7 V a significant forward current begins to flow in the Schottky diode and no further increase in channel conductance is obtained. For negative gate bias, the thickness of the depletion layer increases until it extends through the entire thickness *a* of the channel, and the channel conductance goes to zero. In MESFET parlance, the voltage at which this occurs is known as the *pinch-off* voltage V_P. For rough hand calculations, the same equations as for the MOSFET [Eqs. (2.11) and (2.13) in Sec. 2.4] can be used, with V_P replacing V_T. However, when carrier velocity saturation is present (almost always), MESFET drain current saturates at a drain-source voltage significantly lower than $V_{GS} - V_T$.* A better fit to MESFET characteristics is obtained by using the following equation for both saturation and linear regions:

$$I_D = \beta(V_{GS} - V_T)^2(1 + \lambda V_{DS}) \tanh(\alpha V_{DS}) \qquad V_{GS} \geq V_T \qquad (11.1)$$

The threshold voltage V_T ranges from -0.3 to -3 V for depletion MESFETs. It is 0 to $+0.3$ V for enhancement-mode devices. The channel-length modulation parameter λ is the same as for the MOSFET. The empirical tanh factor must be determined by fitting to data. It produces a good approximation to MESFET characteristics in both the linear and the saturated operating regions. The tanh x function is monotonic; for $x = 0, 0.5, 1.0, 2.0, \infty$, $\tanh x = 0, 0.46, 0.76, 0.96, 1.0$. Coefficient α ranges from 0.3 for a long-channel ($L \approx 20\ \mu m$) device to about 2

---

*Note that the word "saturation" commonly is used in several ways. Carrier velocity saturation and drain current saturation are different matters, both relating to MOSFETS and MESFETS. In this book, the word "saturation" alone in a FET context means drain current saturation. In a bipolar transistor context, there is still another meaning.

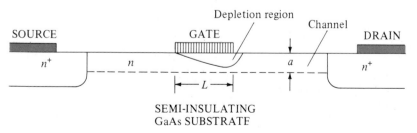

FIGURE 11.3
Detailed cross section of GaAs MESFET illustrating the depletion region under the gate.

for a MESFET with $L = 1$ μm. When the product of α and V_{DS} approaches 2, the drain current ceases to increase with increasing V_{DS} (except for the effect of λV_{DS}), even though the pinch-off point may not have been reached.

For MESFETs, β customarily denotes the device transconductance parameter in units of A/V^2 per millimeter width. This β is unrelated to the β of a bipolar transistor. It conforms to common usage by MESFET specialists and to the notation of the PSPICE model described below. Because MESFET circuits always aim for highest possible speed, channel length L of all transistors typically is held at the minimum allowed value. Therefore the designer is allowed to specify only the device width W. Then we can relate the device transconductance parameter β and the process transconductance parameter β' as follows:

$$\beta = \beta'W \tag{11.2}$$

The current-voltage (I-V) characteristics for a typical MESFET with $L = 1$ μm, $W = 0.02$ mm are shown in Fig. 11.4. These characteristics were obtained by using Eq. (11.2) and the "Example" device parameters listed in Table 11.2 for V_{GS} ranging from -1.2 to $+0.4$ V.

An important parameter in the fabrication of the MESFET is the variation of the threshold voltage from the mean value across the die. For an E-MESFET the standard deviation should be no larger than about 10 mV, while for a D-

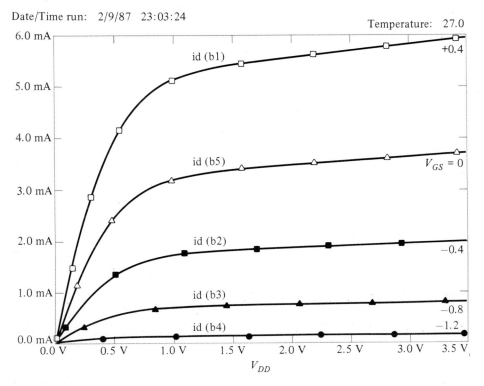

FIGURE 11.4
GaAs MESFET *I-V* characteristics.

TABLE 11.2
PSPICE MESFET model parameters

Symbol	Name	Parameter	Units	Default	Example	Area
V_T	VTO	Threshold voltage	V	-2.5	-1.5	
β	BETA	Transconductance parameter	A/V^2	0.1	0.07	*
α	ALPHA	Saturation voltage parameter	V^{-1}	2	2	
λ	LAMBDA	Channel-length modulation parameter	V^{-1}	0	0.05	
R_D	RD	Drain ohmic resistance	Ω	0	1	*
R_S	RS	Source ohmic resistance	Ω	0	1	*
C_{GS}	CGS	Zero-bias gate-source junction capacitance	F	0	0.5 pF	*
C_{GD}	CGD	Zero-bias gate-drain junction capacitance	F	0	0.5 pF	*
V_{BI}	VBI	Gate junction potential	V	1	0.8	
I_s	IS	Gate SBD saturation current	A	1E–14	1E–14	*
m	FC	Coefficient for forward-bias depletion capacitance formula		0.5	0.5	

* Denotes parameters that scale with width. Example parameters are typical for a channel width of 1 mm, channel length of 1 μm.

MESFET it can go to 50 mV. These variations place important constraints on circuit design.

11.3 SPICE MESFET MODEL

No MESFET model is provided in SPICE2. However, PSPICE,* a popular commercial version for use on personal computers, does include a simple MESFET model employing the equations presented in the previous section. SPICE3 includes a more complex MESFET model. In the PSPICE model, dc characteristics are defined by the parameters VTO and BETA, which determine the variation of drain current with gate voltage, ALPHA, which determines saturation voltage, and LAMBDA, which determines the output conductance. Two ohmic resistances, RD and RS, are included. Actual resistances are inversely proportional to channel width W. In practice, these resistances can cause serious degradation of circuit performance, and they cannot be ignored! Charge storage is modeled by total gate charge as a function of gate-drain and gate-source voltages and is defined by the parameters CGS, CGD, VBI, and FC. A schematic representation of this model is shown in Fig. 11.5. MESFET model parameters are listed in Table 11.2.

11.4 GaAs CIRCUIT FAMILIES

There are at present three different popular circuit configurations for GaAs digital ICs, each with certain advantages and disadvantages.

---

*PSPICE is available from MicroSim Corporation, Laguna Hills, Calif. The authors have no connection with MicroSim Corporation.

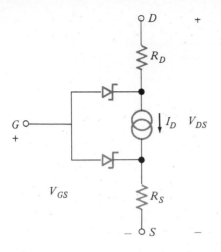

$$I_D = \beta(V_{GS} - V_T)^2 (1 + \lambda V_{DS}) \tanh (\alpha V_{DS})$$

FIGURE 11.5
PSPICE model for GaAs MESFET.

11.4.1 Buffered FET Logic

The earliest GaAs MESFET digital IC, and the one with the greatest maturity, is *buffered FET logic (BFL)*, shown in Fig. 11.6. In this circuit Q_1 and Q_2 are the switching inverter transistors, Q_3 is the load transistor, Q_4 is the buffer source follower, diodes D_1 and D_2 are used as voltage level shifters, and Q_5 is the current source transistor for the source follower. Each of the transistors is an *n*-channel D-MESFET, with V_T in the range −0.5 to −2.0 V, i.e., with $V_{GS} = 0$ V the transistor is conducting. For an input switching transistor to turn off, its V_{GS} must be made more negative than the threshold voltage. However, the drain voltage must be positive with respect to the source. Hence, this circuit requires two supply voltages. If $V_T = -1.5$ V, supply voltages might be $V_{DD} = +2.5$ V and $V_{SS} = -2.5$ V. The input and output voltages of the gate circuit are made compatible by using level-shifting diodes. These are Schottky diodes with each having $V_{D(on)} \approx 0.7$ V. The voltage transfer characteristic of a representative BFL inverter is included in Fig. 11.7. All channel lengths are 1 μm. Device widths in meters are marked on Fig. 11.6. This and the following simulations were done with a 1000-Ω series source resistance for V_{in}, representing a realistic lower limit for the output resistance of a driving circuit. If the V_{in} source impedance were zero, the output low level would increase for V_{in} above 0.7 V. This is due to forward current in the Schottky-barrier diode gate flowing through the series source resistance. From Fig. 11.7, the logic voltage swing is seen to be about 1.8 V, and the noise margins are found to be $NM_H = 0.45$ V and $NM_L = 0.9$ V. Noise margins could be made more equal by adjustment of relative device widths.

The source follower output driver provides that the output voltage levels are relatively unaffected by fan-out loading and load capacitance. Also, no dc current is required to drive subsequent BFL load gates. In practical cases, the fan-out is limited to about 3 by the time taken to charge and discharge the output

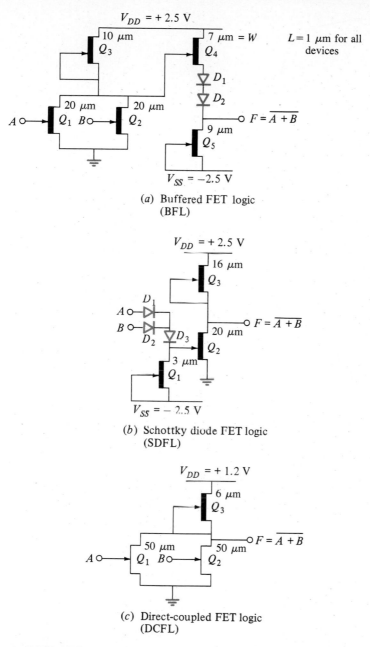

(a) Buffered FET logic
(BFL)

(b) Schottky diode FET logic
(SDFL)

(c) Direct-coupled FET logic
(DCFL)

FIGURE 11.6
Basic GaAs MESFET gate circuits.

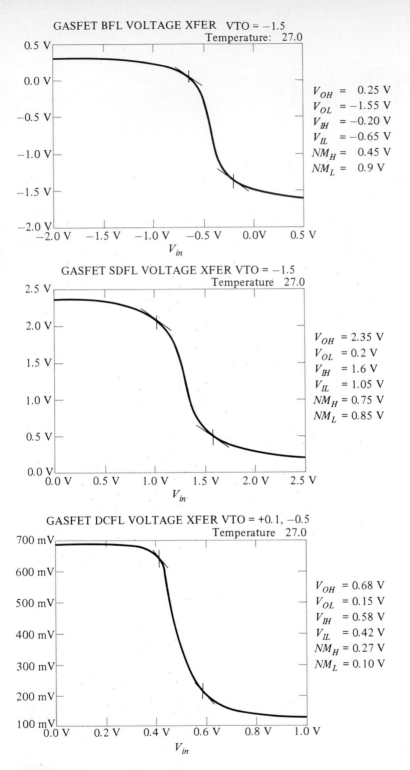

FIGURE 11.7
Voltage transfer characteristics for BFL, SDFL, and DCFL.

load capacitance. With a fan-out of 1, this circuit has a propagation delay of about 100 ps. Power consumption is 10 mW. At this power level the number of gates on a chip is limited to a few hundred. Thus these circuits are useful in SSI, MSI, and LSI circuits but not in VLSI circuits.

11.4.2 Schottky Diode FET Logic

A later development, and an alternative to BFL, is the *Schottky diode FET logic (SDFL)*, also shown in Fig. 11.6. A very similar family is *unbuffered FET logic (UFL)*. In this circuit, Schottky diodes D_1 and D_2 make a simple diode OR circuit as well as acting as voltage level shifters. D_3 is another level shifter, Q_1 is the current source transistor for the level-shifting diodes, Q_2 is the inverting switch transistor, and Q_3 is the load transistor. The transistors are all D-MESFETs so that two voltage supplies are required for this logic family also. Supplies of $+2.5$ and -2.5 V are typical. Voltage transfer characteristics are included in Fig. 11.7.

An advantage of SDFL is that unlike BFL, which requires an extra transistor for each additional input, SDFL only requires an extra diode. The Schottky diodes are very small in area (1 μm $\times$ 2 μm). Thus the packing density of SDFL is superior to that of BFL. Because inputs draw current from the driving circuit, unbuffered SDFL is more susceptible to fan-out loading. Like BFL, in the practical case the fan-out is generally restricted to about 3. A source follower can be added to the output of this circuit if a larger output driving current is required.

The logic voltage swing, noise margins, and propagation delays of SDFL circuits are comparable to those of BFL under the same conditions of power consumption, fan-out, and fan-in. The smaller area of the basic gate permits SDFL to be used in LSI circuits with up to a few thousand gates, but with some sacrifice in speed to stay within the bounds of a maximum power dissipation of about 2 W.

11.4.3 Direct-Coupled FET Logic

The *direct-coupled FET logic (DCFL)* circuit shown in Fig. 11.6 makes use of both enhancement-mode and depletion-mode MESFETs or HEMTs. Q_1 and Q_2 are the switching inverter E-MESFETs, and Q_3 is the D-MESFET load transistor. Voltage transfer characteristics are included in Fig. 11.7. The circuit is similar to the E-D NMOS inverter as described in Sec. 3.2.3. DCFL offers the advantage of extremely simple circuit design, small area, and low power consumption. Only one voltage supply is required. But since the input Schottky diode clamps the high input level at about 0.7 V, the logic swing cannot exceed this magnitude and the noise margins are in the range 0.1 to 0.3 V. Circuit speed is increased by operating with a supply voltage of about 1.2 V. Since device conductance β is about the same as for depletion MESFETS, the maximum current per unit device width is several times smaller for enhancement MESFETS. Thus their ability to drive capacitance loads is poor. The low noise margins in DCFL are a disadvantage. Another limitation is the close control that is required of the threshold voltage of the E-MESFET. Nevertheless, due to its simplicity, small area, and low power consumption ($\approx$ 0.25 mW for the circuit shown), DCFL is the circuit that is presently receiving the most attention in the development of LSI MESFET circuits.

Additional MESFET logic families are being studied. Recent results on *source-coupled FET logic (SCFL)* have been encouraging. It has high tolerance to threshold voltage variations and can be made with fairly low power dissipation (0.5 mW per gate) (Ref. 4).

11.4.4 Gate Circuits

The logic function of each of the MESFET circuits presented in Fig. 11.6 is that of a two-input NOR gate. The number of inputs to each circuit, i.e., the fan-in, may be extended but due to speed requirements is usually limited to four. The NAND logic function may be performed with a series connection of the input transistors, as is illustrated in Fig. 11.8a for the BFL circuit. However, due to the small logic swings and narrow noise margins, the number of series gating transistors is limited to two. Figure 11.8b shows how two levels of logic, as in the AND-OR-INVERT function, may be implemented in BFL. In each of these BFL circuits the output of the logic transistors would connect to a buffer source follower, Q_4 in Fig. 11.7.

The design of two levels of logic in SDFL is presented in Fig. 11.8c, which shows an OR-AND-INVERT function. The series connection of two transistors is implied in the dual-gate inverting switch transistor. In DCFL, the design of more complex logic functions than the simple NOR gate is difficult due to the critical control required for the threshold voltage of the enhancement-mode switching inverter.

Ring oscillator circuits are widely used to obtain a measurement of the propagation delay time and power consumption of logic inverters or gate circuits. Table 11.3 summarizes typical performances of the three circuits just described. Propagation delay (second column) is for fan-out = 1. The increases in propagation delay as a function of additional fan-out and interconnect capacitance are given in the next two columns. The BFL and SDFL circuits employ $V_T = -1.5$ V, with a nominal logic swing of about 2 V. The DCFL circuits employ $V_{TD} = -0.5$ V and $V_{TE} = +0.1$ V, with a nominal logic swing of about 0.5 V. The data given in this table are nominal results. Variations of $\pm 50\%$ in speed and power consumption can be expected due to lot-to-lot process variations.

A comparison of BFL, DCFL, and silicon TTL for an arithmetic carry-lookahead generator, a representative MSI gate logic function, is presented in Table 11.4. The logic design follows that of the commercial 2902 component. Signals propagate through nine cascaded gates within this circuit. This implies an average gate delay of 1.1 ns/9 = 120 ps for the BFL version, consistent with the values cited above. The BFL and DCFL circuits were made by the same MESFET process. It should be noted that a silicon ECL version of the 2902 is entirely feasible. Its propagation delay would likely be 2 ns or less.

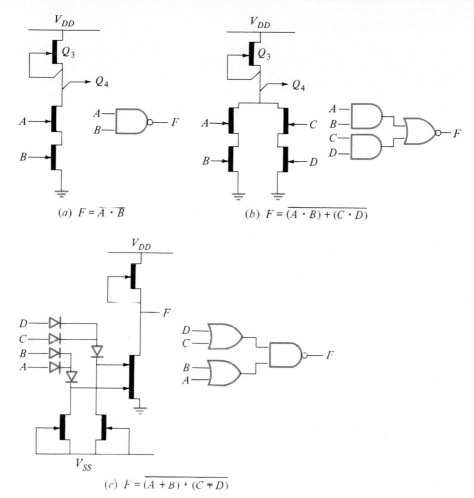

(a) $F = \overline{A \cdot B}$

(b) $F = \overline{(A \cdot B) + (C \cdot D)}$

(c) $F = \overline{(A + B) \cdot (C + D)}$

FIGURE 11.8
GaAs MESFET gate circuits.

TABLE 11.3
Typical logic family performance for different technologies

Circuit family	Propagation delay, ps	Fan-out sensitivity, ps/fan-out	Capacitance sensitivity, ps/fF	Power consumption, mW/gate
BFL (1 μm)	90	20	0.67	10.
BFL (0.5 μm)	54	12	0.67	10.
SDFL/UFL (1 μm)	146	35	2.8	2.5
DCFL (1 μm)	54	35	1.84	0.25
DCFL (0.5 μm HEMT, 77 K)	11	7	0.32	1.3

Source: Ref. 5.

11.4.5   Sequential Circuits

Although basic power-delay characteristics of digital ICs are usually inferred from ring oscillator measurements, a more realistic index of performance is provided by flip-flop circuit operation, since it includes important features such as in-circuit noise immunity and actual interconnection capacitances. Thus the maximum toggle frequency $f_{c(max)}$ of flip-flop circuits when operated as binary frequency dividers is a figure of merit widely used to estimate the maximum clock frequency achievable in complex digital systems.

The most commonly used GaAs sequential circuits are the edge-triggered D flip-flop (DFF) and the master-slave T (toggle) flip-flop (single-clocked SCT or dual-clocked DCT), shown in Fig. 11.9a, b, and c, respectively. The theoretical maximum toggle frequency of these divider circuits depends on the number of logic gates which must stabilize before the output reaches its correct state. This factor ranges from $2t_d$ for a dual-clocked T flip-flop to $5t_d$ for a D flip-flop. Thus equal gate delays will produce clock frequencies a factor of 2.5 times higher in the former circuit than in the latter.

A comparison of the speed and power performance of these various designs, as has been reported in the literature, is presented in Table 11.5. The table clearly demonstrates the capability of each of the GaAs circuits to perform logic at gigahertz frequencies. The 3.8 GHz at 1.2 mW per gate for the DCFL circuit is impressive.

11.5   GaAs LSI CIRCUITS

The characteristics required of a basic gate circuit in GaAs LSI circuits are very high speed, low power consumption, and small die area. To meet the third requirement it helps if the basic circuit is a simple one using only one supply voltage. It has already been noted that while for any given feature size and technology BFL and SDFL digital circuits tend to be faster than DCFL circuits, they also consume more power per gate. Additionally, their die area per gate is also larger. As a consequence, while BFL and SDFL are well-suited for very high speed logic in SSI and MSI circuits, only DCFL is a viable circuit in LSI and VLSI applications.

Several digital LSI circuits will be described to give some idea of the

TABLE 11.4
Carry-lookahead circuit comparison

Technology	Total chip power dissipation, mW	Overall propagation delay, ns	Power-delay product, nJ	Supply voltages, V	Logic swing, V	Noise margin, V
BFL	520	1.1	0.57	+3.5, −3.0	2.0	0.75
DCFL	34	1.6	0.055	+1.2	0.8	0.21
Si TTL	260	4.5	1.2	+5.0	3.0	0.9

Source: Ref. 6.

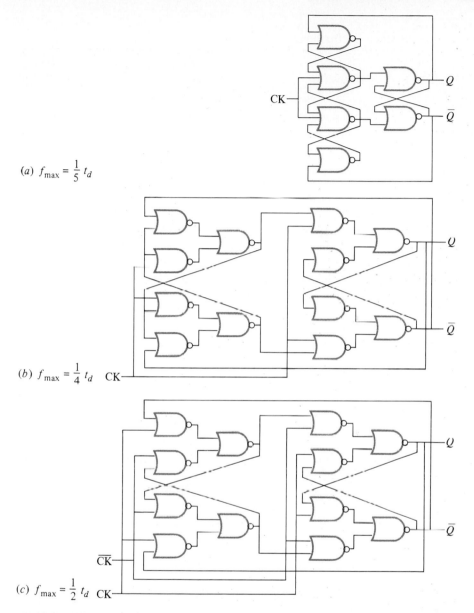

(a) $f_{max} = \dfrac{1}{5} t_d$

(b) $f_{max} = \dfrac{1}{4} t_d$ CK

(c) $f_{max} = \dfrac{1}{2} t_d$ CK

FIGURE 11.9
Sequential circuits. (a) Edge-triggered D flip-flop (DFF). (b) Single-clocked master-slave T flip-flop (SCT). (c) Dual-clocked master-slave T flip-flop (DCT).

TABLE 11.5
Frequency divider performance

Circuit type	Logic function	Geometry L, μm	$f_{c(max)}$, GHz	Power, mW/gate
BFL	Divide by 2 (DFF)	0.8	2.2	5.3
	4-bit shift register (DFF)	0.8	1.3	6.2
SDFL	Divide by 2 (DDF)	1.0	1.9	2.5
DCFL	Divide by 8 (SCT)	0.6	3.8	1.2
	Divide by 2 (DCT)	1.2	2.4	3.9

Source: Ref. 7.

scope of GaAs LSI circuits up to the present time (1987). They are random-access memories, a parallel multiplier, and a gate array. All the circuits make use of enhancement- and depletion-mode n-channel MESFETs in a basic DCFL circuit. DCFL is chosen because only its power consumption is low enough to be acceptable in LSI applications.

Many 1K-bit and 4K-bit RAMs have been described in the literature, and a 16K-bit RAM has been reported recently. They all make use of a basic six-transistor memory cell similar to the static NMOS memory cell illustrated in Fig. 9.12 and described in Sec. 9.2.1. Operating voltage is typically in the range 0.7 to 1.5 V.

An early example of a 1K-bit SRAM is the NTT device (Ref. 8). This 3.3 mm × 3.3 mm chip contains 7,084 FETs with 1 μm gate length. The basic 6-T cell is 69 μm × 56 μm and contains four E-mode and two D-mode MESFETs. The mean value of the threshold voltage for the E-MESFETs was +145 mV with a standard deviation of 73 mV. For the D-MESFETs the threshold voltage was −227 mV and the standard deviation was 88 mV. The address access time of the SRAM was 2.0 ns with a power dissipation of 459 mW at a supply voltage of 0.7 V.

A later development is a 4K-bit SRAM, again from NTT (Ref. 9). The organization is 4,096 words × 1 bit, and more than 26,600 MESFETs are contained on the 5.6 mm × 6.0 mm chip. Again, this is an E-D DCFL design, with the basic 6-T cell measuring 62 μm × 65 μm. As earlier noted, a concern of GaAs LSI fabrication is the uniformity of the threshold voltage, especially for the E-MESFET. In this development the uniformity of the FET characteristics was measured wafer to wafer and chip to chip. Some chip size areas were found to have sufficient uniformity for a 4K-bit RAM operation, but no wafer-size areas demonstrated the necessary uniformity. In each wafer there were nine chips, each containing a 4K-bit SRAM, monitor FETs, and other test patterns. The minimum address access time was measured to be 2.8 ns with a power dissipation of 1.2 W. Two supply voltages were used, 1.3 V for the memory cell and 1.8 V for the peripheral circuits.

Also of interest is a HEMT 1K-bit SRAM developed by Fujitsu (Ref. 10). This 3.7 mm × 2.9 mm chip integrates 7,244 FETs in a 1,024 words × 1 bit organization. A 6-T memory cell is used, with the mean threshold voltage and standard deviation for the E-MESFETs being 131 and 19 mV, and for the D-MESFETS −419 and 56 mV, respectively. Of particular interest is the address access time that was measured at 300 K to be 3.4 ns with a power consumption of 290 mW, but at 77 K was reduced to 0.9 ns, though the power increased to 360 mW.

Most recent is a 16K-bit SRAM reported by Mitsubishi (Ref. 11). The same 6-T cell circuit is used, with a cell size of 31.5 μm × 24 μm. Overall chip size is 5.79 mm × 4.73 mm. Two levels of metal were used for interconnections. Within a single chip access time ranged from 5 to 11 ns, with an average of 7 ns. This variation is due primarily to a 30-mV variation of threshold voltage across the chip. To reduce the access time variation about the mean to 10% (as is achieved with silicon bipolar RAMs), threshold voltage variation would have to be reduced to 5 mV. Power consumption was 1.0 W at $V_{DD} = 1.0$ V.

A parallel multiplier presents at the output terminals the product of two binary words presented at the input terminals. Thus an 8 × 8 multiplier forms the binary product of two 8-bit input words. An example is the 8 × 8 bit multiplier from Sony (Ref. 12). This 3.7 mm × 4.2 mm chip contains approximately 1,800 NOR gates and employs a basic E-R DCFL circuit. The D-MESFET load transistor is replaced by a resistor, typically about 3 kΩ. The multiply time was 6 ns with a power dissipation of 876 mW operating at a supply voltage of 1.46 V. This corresponds to a propagation delay time of 182 ps per gate with a power dissipation of 0.48 mW per gate.

Also of note is a 16 × 16 bit multiplier from Fujitsu (Ref. 13). This is a DCFL circuit using a tungsten silicide self-aligned structure. The 4 mm × 4 mm chip contains 10,624 FETs with 2 μm gate length. There are 3,168 E-D gates in the multiplier designed into 224 full adders, 16 half adders, 256 NOR gates, and 96 input-output buffers. The mean value of the threshold voltage for the E-MESFETs was +2 mV, and for the D-MESFETs was −779 mV. Multiplication was performed in 10.5 ns with a power dissipation of 952 mW at a supply voltage of 1.6 V. This performance corresponds to a propagation delay of 162 ps per gate and a power dissipation of 0.30 mW per gate.

Finally we describe a 1,000 gate array from Toshiba (Ref. 14). The chip size is 3.7 mm × 3.7 mm, and the basic cell is an E-D DCFL three-input NOR gate. The mean value and the standard deviation of the threshold voltage for the E-MESFETS was 100 and 53 mV, respectively, while for the D-MESFETS it was −700 and 110 mV. Operating at a power dissipation of 0.2 mW per gate, the propagation delay time was 100 ps per gate with no load and 350 ps per gate with a fan-out of 3. The gate array was configured as a 6 × 6 bit parallel multiplier, and a 10.6-ns multiplication time was measured at a 380-mW power consumption.

11.6 SUMMARY

The principal appeal of gallium arsenide digital circuits lies in the potential for operation at speeds up to 2 times faster than silicon-based digital ICs. Operation at higher speeds is possible because the mobility of electrons in GaAs is higher than in silicon. Today the most widely used GaAs transistor is the *n*-channel MESFET. There is active work on *n*-channel HEMTs for still faster operation. MESFETs and HEMTs are similar in structure to junction field-effect transistors but employ a Schottky-barrier diode as the gate. Both depletion-mode and enhancement-mode transistors are feasible.

Several circuit families are used. Buffered FET logic and Schottky diode FET logic are characterized by high speed and high power consumption; heat removal problems limit them to a few thousand gates per chip at most. Direct-coupled FET logic has the simplest configuration and the lowest power consumption, so it is attractive for LSI and VLSI usage. However, since it is limited to operation with voltage swings less than 0.8 V, speed is sacrificed particularly when driving off-chip. Acceptance of GaAs circuits would be speeded by the adoption of standard supply voltages and logic levels so that chips from different vendors could be interconnected readily.

The maximum feasible scale of integration for GaAs ICs is two or three orders of magnitude smaller than for silicon ICs because the defect density for finished wafers is much higher for GaAs. Consequently GaAs circuits today cost at least 10 times more per circuit function than silicon ICs. Although this situation will improve with time, it is unlikely that digital GaAs circuits will ever compete in cost with silicon ICs.

REFERENCES

1. W. R. Curtice, "A MESFET Model for Use in the Design of GaAs Integrated Circuits," *IEEE Transactions on Microwave Theory and Techniques,* vol. MTT-28, no. 5, May 1980, pp. 448–455.
2. K. Lehovec and R. Zuleeg, "Analysis of GaAs FET's for Integrated Logic," *IEEE Transactions on Electron Devices,* vol. ED-27, June 1980, pp. 1074–1091.
3. S. I. Long et al., "High Speed GaAs Integrated Circuits," *Proceedings of the IEEE,* vol. 70, January 1982, pp. 35–45.
4. K. Hasegawa et al., "Low Dissipation Current GaAs Prescaler IC," *Electronics Letters,* vol. 22, Feb. 27, 1986, pp. 251–252.
5. L. E. Larson, J. F. Jensen, and P. T. Greiling, "GaAs High-Speed Digital IC Technology: An Overview," *IEEE Computer,* vol. 19, no. 10, October 1986, pp. 21–27.
6. R. N. Sato et al., "Performance of Carry Lookahead Generator Circuit Fabricated with Gallium Arsenide," *IEEE Journal of Solid-State Circuits,* vol. SC-22, February 1987, pp. 121–124.
7. G. Nuzillat et al., "GaAs MESFET IC's for Gigabit Logic Applications," *IEEE Journal of Solid-State Circuits,* vol. SC-17, June 1982, pp. 569–582.
8. K. Asai et al., "1Kb Static RAM Using Self-Aligned FET Technology," *IEEE International Solid-State Circuits Conference, Digest of Technical Papers,* February 1983, pp. 46–47.
9. M. Hirayama et al., "A GaAs 4 Kb SRAM with Direct Coupled FET Logic," *IEEE Journal of Solid-State Circuits,* vol. SC-19, October 1984, pp. 716–720.
10. K. Nishiuchi et al., "A Subnanosecond HEMT 1 Kb Static RAM," *IEEE Journal of Solid-State Circuits,* vol. SC-21, October 1986, pp. 869–874.
11. S. Takano et al., "A 16K GaAs SRAM," *IEEE International Solid-State Circuits Conference, Digest of Technical Papers,* February 1987, pp. 140–141, 370.

12. K. Gonoi et al., "A GaAs 8 × 8-bit Multiplier/Accumulator Using JFET DCFL," *IEEE Journal of Solid-State Circuits,* vol. SC-21, August 1986, pp. 523–529.
13. Y. Nakayama et al., "A GaAs 16 × 16b Parallel Multiplier Using Self-Alignment Technology," *IEEE International Solid-State Circuits Conference, Digest of Technical Papers,* February 1983, pp. 48–49.
14. Y. Ikawa et al., "A 1K-Gate GaAs Gate Array," *IEEE Journal of Solid-State Circuits,* vol. SC-19, October 1984, pp. 721–728.

APPENDIX

A

BASIC
LOGIC
DESIGN

A.0   INTRODUCTION

In a digital system information is represented solely in discrete (or *quantized* or *digitized*) form. Most commonly a binary form is used, which means only two discrete states are allowed, normally denoted as 0 and 1.

Logic design with binary quantities has some peculiarities, but it also presents some useful opportunities. For instance, the answer to a question can be only *yes* or *no,* it never can be *maybe*! A special algebra applicable to the binary system was invented by George Boole (1815–1864). This form of algebra can be useful to a logic and/or circuit designer, and some familiarity with it is essential in the analysis and design of digital integrated circuits.

A.1   LOGIC FUNCTIONS

The three basic operations performed with Boolean algebra are included in Table A.1. The $\cdot$ symbol denotes the logic AND operation, although as the table shows, the $\cdot$ between the variables is usually omitted. The $+$ symbol indicates a logic OR. A bar over the variable indicates the NOT operation, or logic inversion. That is, since only two states of the variable are permitted, if $A = 0$ then $\overline{A} = 1$.

Consider first the AND function of the two binary variables A and B. Shown in Fig. A.1a are two ways of expressing the AND function. A *truth table* is simply a systematic listing of the values of the dependent variable (F) in terms of all the possible values of the independent variables (A and B). Since we are

TABLE A.1
Basic Boolean operations

Operation	Boolean expression
AND	$F = A \cdot B = AB$
OR	$F = A + B$
NOT	$F = \overline{A}$
NAND	$F = \overline{A \cdot B} = \overline{AB}$
NOR	$F = \overline{A + B}$

working with a binary system there are 2^N combinations, where N is the number of independent variables being considered. Note the AND statement is true ($= 1$) only if $A = 1$ *and* $B = 1$. The standard logic symbol for a two-input AND gate is also illustrated in Fig. A.1a. The requirement here is that both the inputs (A and B) be at a 1 for the output (F) to be at a 1.

The OR function, more properly designated the inclusive-OR function, is illustrated in Fig. A.1b. Notice from the truth table that the OR statement is true if *either* $A = 1$ *or* $B = 1$, but in addition to the either/or condition, there is an *or both* condition. It is this *or both* condition that leads to the name inclusive-OR. Also shown in Fig. A.1b is the standard logic symbol for a two-input inclusive-OR gate.

The truth table and logic symbol of the NOT function is illustrated in Fig. A.1c. The small circle at the output of the symbol indicates logic inversion.

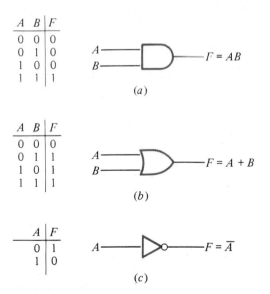

A	B	F
0	0	0
0	1	0
1	0	0
1	1	1

$F = AB$

(a)

A	B	F
0	0	0
0	1	1
1	0	1
1	1	1

$F = A + B$

(b)

A	F
0	1
1	0

$F = \overline{A}$

(c)

FIGURE A.1
Truth table and standard symbol for the three basic logic operations. (a) AND function, $F = AB$. (b) OR function, $F = A + B$. (c) NOT function, $F = \overline{A}$.

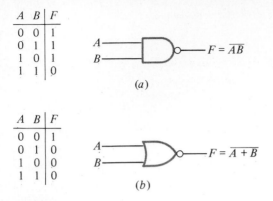

A	B	F
0	0	1
0	1	1
1	0	1
1	1	0

$F = \overline{AB}$

(a)

A	B	F
0	0	1
0	1	0
1	0	0
1	1	0

$F = \overline{A + B}$

(b)

FIGURE A.2
Truth table and standard logic symbol. (a) NAND function, $F = \overline{AB}$. (b) NOR function, $F = \overline{A + B}$.

From these three basic logic operations, two other common logic functions may be derived. The inverter can be combined with the AND gate to form the NOT-AND, or NAND, function, illustrated in Fig. A.2a. The inverter may also be combined with an OR gate to form the NOT-OR, or NOR, function, shown in Fig. A.2b.

Two other useful logic functions are illustrated in Fig. A.3. The exclusive-OR (X-OR) function excludes the *or both* condition of the inclusive-OR. Notice from the truth table that the statement is true if either $A = 1$ or $B = 1$, but not with both at a 1. Combined with an inverter, the combination yields an exclusive-NOR (X-NOR) function.

The basic logic functions have been described here with just 2 input

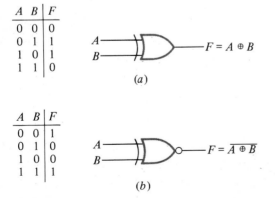

A	B	F
0	0	0
0	1	1
1	0	1
1	1	0

$F = A \oplus B$

(a)

A	B	F
0	0	1
0	1	0
1	0	0
1	1	1

$F = \overline{A \oplus B}$

(b)

FIGURE A.3
Truth table and standard logic symbol. (a) Exclusive-OR function, $F = A \oplus B$. (b) Exclusive-NOR function, $F = \overline{A \oplus B}$.

variables. Gates with more than 2 inputs are also available, though in practice the limit is generally 4 or 8.

BOOLEAN IDENTITIES. With the three basic operations (AND, OR, and NOT) it is possible to deduce a set of Boolean identities. These are listed in Table A.2. These identities are useful in simplifying a complex logic expression for a clearer understanding of the logic to be performed. There also may be an economic benefit in the saving in silicon area when implementing the logic as an integrated circuit (IC).

All the operations described in Table A.2 may be proven with a truth table, but most of them are obvious. From the AND gate in Fig. A.1a we see that with $B = 0$, always $F = 0$. Hence $0 \cdot A = 0$. From the OR gate in Fig. A.1 we note that with $B = 0, F = A$, and $0 + A = A$ follows.

The last two identities in Table A.2 are the DeMorgan theorems, extremely useful in reducing complex Boolean expressions to simple and workable proportions. The proof of these two theorems, through the use of truth tables, is illustrated in Fig. A.4. Similarly, the other identities presented in Table A.2 can be proven.

The use of these identities is demonstrated in the following two examples.

Example A.1. Simplify

$$WXY\overline{Z} + WX\overline{Y}Z + W\overline{X}YZ + W\overline{X}\,\overline{Y}Z$$

Solution

$$WXY\overline{Z} + WX\overline{Y}Z + W\overline{X}Y\overline{Z} + W\overline{X}\,\overline{Y}Z = WX\overline{Z}(Y + \overline{Y})$$

$+ \ W\overline{X}Z(Y + \overline{Y})$	from identity *14*
$= \ WX\overline{Z} + W\overline{X}Z$	from identity *8*
$= \ W\overline{Z}(X + \overline{X})$	from identity *14*
$= \ W\overline{Z}$	from identity *8*

TABLE A.2
Boolean identities

1. $0 \cdot A = 0$	11. $A + B = B + A$
2. $0 + A = A$	12. $A(BC) = (AB)C$
3. $1 \cdot A = A$	13. $A + (B + C) = (A + B) + C$
4. $1 + A = 1$	14. $A(B + C) = AB + AC$
5. $A \cdot A = A$	15. $(A + B)(A + C) = A + BC$
6. $A + A = A$	16. $A + AB = A$
7. $A \cdot \overline{A} = 0$	17. $A + \overline{A}B = A + B$
8. $A + \overline{A} = 1$	18. $\overline{A + B} = \overline{A}\,\overline{B}$
9. $A = A$	19. $\overline{AB} = \overline{A} + \overline{B}$
10. $AB = BA$	

A B	A + B	$\overline{A+B}$	$\overline{A}$	$\overline{B}$	$\overline{A}\,\overline{B}$
0 0	0	1	1	1	1
0 1	1	0	1	0	0
1 0	1	0	0	1	0
1 1	1	0	0	0	0

(a)

A B	AB	$\overline{AB}$	$\overline{A}$	$\overline{B}$	$\overline{A}+\overline{B}$
0 0	0	1	1	1	1
0 1	0	1	1	0	1
1 0	0	1	0	1	1
1 1	1	0	0	0	0

(b)

FIGURE A.4
Proof of DeMorgan's theorem. (a) $\overline{A + B} = \overline{A}\,\overline{B}$. (b) $\overline{AB} = \overline{A} + \overline{B}$.

The initial expression contains four terms and six variables. The result has been simplified to only one term and two variables.

By invoking DeMorgan's theorem we can show that the NAND gate is equivalent to an OR gate with inverted inputs. This is illustrated in Fig. A.5. From identity 19 in Table A.1 we have

$$F = \overline{A B} = \overline{A} + \overline{B}$$

Thus the NAND gate may be described as an AND gate with an inverted output, as in Fig. A.5a, or as an OR gate with inverted inputs, as in Fig. A.5b. Of course, by simply connecting the two inputs together the NAND gate performs the NOT operation. The ability of this simple logic gate to perform all the basic logic operations is extremely useful when implementing logic expression as a hardware design.

Similarly, from identity 18 in Table A.2 we have

$$F = \overline{A + B} = \overline{A}\,\overline{B}$$

This allows us to represent the NOR gate as an OR gate with an inverted output, as in Fig. A.6a, or, as seen in Fig. A.6b, as an AND gate with inverted inputs.

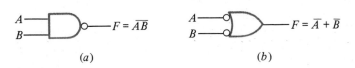

(a)   (b)

FIGURE A.5
Equivalent representation of NAND gate. (a) AND gate with inverted output. (b) OR gate with inverted input.

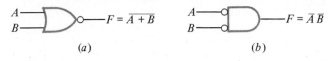

FIGURE A.6
Equivalent representation of NOR gate. (*a*) OR gate with inverted output. (*b*) AND gate with inverted input.

A.2   KARNAUGH MAPS

The algebraic reduction of a Boolean expression is not always easy and generally requires a certain amount of intuition or luck. Many techniques have been developed to aid in this reduction. A most helpful technique is the Karnaugh map. This is a matrix array of all the possible combinations of the independent variable. As such, the Karnaugh map (or K map) provides exactly the same information as the truth table, but in a different form.

   The truth table and K map of a two-variable function are shown in Fig. A.7. Notice there are four entries to the truth table and four squares to the K map. In the map, the columns contain the two states of A ($\overline{A}$ and A, or 0 and 1), while the rows contain the two states of B. From the truth table we have

$$F = \overline{A}B + A\overline{B} + AB$$

   This simple example will prove the usefulness of the K map. In the second column of the map ($A = 1$), we have

$$A\overline{B} + AB = A(\overline{B} + B) \qquad \text{from identity 14}$$
$$= A \qquad \text{from identity 8}$$

Similarly, in the second row ($B = 1$) we have

$$\overline{A}B + AB = B$$

We now see that the original expression

$$F = \overline{A}B + A\overline{B} + AB$$

can be reduced or simplified to

$$F = A + B$$

The original expression containing three terms and four variables has been reduced

A	B	F
0	0	0
0	1	1
1	0	1
1	1	1

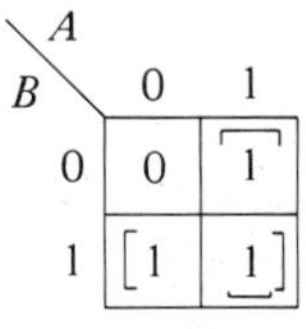

	A	0	1
B			
0		0	1
1		1	1

FIGURE A.7
Truth table and Karnaugh map of a two-variable function: $F = \overline{A}B + A\overline{B} + AB$.

to one containing only two terms and two variables. In the K map the grouping of two adjacent squares in any column or row will indicate a redundant variable, due to the Boolean identities 14 and 8.

The truth table and K map of a three-variable function are shown in Fig. A.8. There are eight entries (since $2^3 = 8$) to the truth table and eight squares to the K map. For ease of reduction the order of the listing of the variables is important. There must be only a one-variable change between any two adjacent squares. The listing has a special cyclic symmetry. That is, the columns are listed as 00 01 11 10. The leftmost digit refers to the leftmost variable. The cyclic order of listing is continuous in both the horizontal and vertical directions. From the truth table we see that

$$F = \overline{A}\,\overline{B}C + \overline{A}B\overline{C} + \overline{A}BC + ABC$$

To simplify this expression we group adjacent squares to eliminate a redundant variable. By grouping the two 1s in the second column we eliminate the variable C. By a horizontal grouping of the two 1s in columns 1 and 2 we eliminate the variable B, and by a further horizontal grouping of the two 1s in columns 2 and 3 we eliminate the variable A. We therefore have a simplified form

$$F = \overline{A}B + \overline{A}C + BC$$

Of course for this problem an algebraic reduction is also possible:

$$F = \overline{A}\,\overline{B}C + \overline{A}B\overline{C} + \overline{A}BC + ABC$$
$$= \overline{A}B(\overline{C} + C) + \overline{A}C(\overline{B} + B) + BC(\overline{A} + A)$$
$$= \overline{A}B + \overline{A}C + BC$$

A four-variable map, containing 16 squares ($2^4 = 16$) is shown in Fig. A.9. Again notice the listing of the variables. This is important to readily identify the redundant variable. Notice also that the cyclic order of the listing is continued by repeating the pattern to the right or left, up or down. From the truth table

$$F = \overline{A}\,\overline{B}\,\overline{C}D + \overline{A}\,\overline{B}CD + \overline{A}B\overline{C}\,\overline{D} + \overline{A}B\overline{C}D + \overline{A}BC\overline{D} + \overline{A}BCD$$

$$+ AB\overline{C}D + ABCD$$

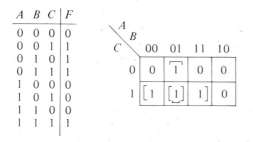

A	B	C	F
0	0	0	0
0	0	1	1
0	1	0	1
0	1	1	1
1	0	0	0
1	0	1	0
1	1	0	0
1	1	1	1

FIGURE A.8
Truth table and Karnaugh map of a three-variable function: $F = \overline{A}\,\overline{B}C + \overline{A}B\overline{C} + \overline{A}BC + ABC$.

A	B	C	D	F
0	0	0	0	0
0	0	0	1	1
0	0	1	0	0
0	0	1	1	1
0	1	0	0	1
0	1	0	1	1
0	1	1	0	1
0	1	1	1	1
1	0	0	0	0
1	0	0	1	0
1	0	1	0	0
1	0	1	1	0
1	1	0	0	0
1	1	0	1	1
1	1	1	0	0
1	1	1	1	1

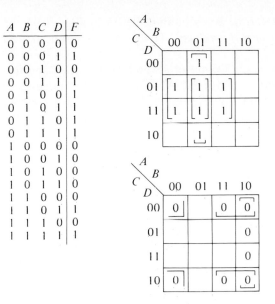

FIGURE A.9

Truth table and Karnaugh map of a four-variable function: $F = \overline{A}\,\overline{B}\,\overline{C}D + \overline{A}\,\overline{B}CD + \overline{A}B\overline{C}\,\overline{D} + \overline{A}B\overline{C}D + \overline{A}BC\overline{D} + \overline{A}BCD + AB\overline{C}D + ABCD$.

Generally the squares of the K map are marked with a 1, indicating them to be true. It is implied that open squares are marked with a 0, or false. However the marking of the 0s is not really necessary. This is the case in Fig. A.9a.

We have seen that the grouping of two adjacent squares eliminates one variable. It logically follows, and it can be proven with the Boolean identities that the grouping of four adjacent squares should eliminate two variables. This is done in Fig. A.9a, with the simplified result

$$F = \overline{A}B + \overline{A}D + BD$$

Sometimes it is more convenient in the logic design to group the 0s in a NOT statement, as in Fig. A.9b. Then a solution is developed using DeMorgan's theorem. From Fig. A.9b,

$$F = NOT\,(A\overline{B} + A\overline{D} + \overline{B}\,\overline{D})$$

The last term of this equation is from a grouping of the four corner squares and is due to the cyclic listing of the columns and rows in the map.

$$F = \overline{(A\overline{B} + A\overline{D} + \overline{B}\,\overline{D})}$$
$$= \overline{(A\overline{B})}\,\overline{(A\overline{D})}\,\overline{(\overline{B}\,\overline{D})}$$
$$= (\overline{A} + B)(\overline{A} + D)(B + D)$$

STANDARD FORMS. The task of converting the required logic into the actual hardware of the digital system is generally simplified if the logic expression is arranged in one of two standard forms. By grouping the 1s in Fig. A.9a the expression has been arranged as a *sum of products*. That is, F is the sum of $\overline{A}B$, $\overline{A}D$, and BD.

The sum of products results in the ORing of several AND terms, and this solution may be implemented with 3 two-input AND gates and 1 three-input OR gate, as shown in Fig. A.10a. In Fig. A.10b we have inverted the output of each AND gate and each input of the OR gate. Due to identity 9 in Table A.2, this double inversion has no effect on the overall logic expression. Now from Fig. A.5, we again note the equivalent representation for a NAND gate. Thus, converting the output gate to an AND gate with an inverted output, as in Fig. A.10c, we see that the sum-of-products form leads to an all-NAND gate configuration.

By grouping the 0s in Fig. A.9b, the resulting expression is in the form of a *product of sums*. Here, F is the product of three summations $(\overline{A} + B)$, $(\overline{A} + D)$, and $(B + D)$. The product of sums results in the ANDing of several OR terms. How the product-of-sums solution leads to an all-NOR gate configuration is illustrated graphically in Fig. A.11.

Whether a logic expression is best implemented with NAND gates, NOR gates, or AND and OR gates is subject to other decisions, which are part of the total design criteria.

We conclude this introduction to logic design with one final example.

Example A.2. A car will not start when the ignition switch is turned on if either:

1. The doors are closed but the seat belts are not buckled.
2. The seat belts are buckled but the parking brake is on.
3. The parking brake is off but the doors are not closed.

Write a minimal expression allowing the car to start when the ignition switch is turned on as:
(*a*) A sum of products, and implement as an all-NAND solution.
(*b*) A product of sums, and implement as an all-NOR solution.

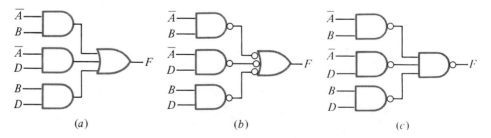

FIGURE A.10
The development of a sum-of-products logic expression $(F = \overline{A}B + \overline{A}D + BD)$ to an all-NAND gate configuration.

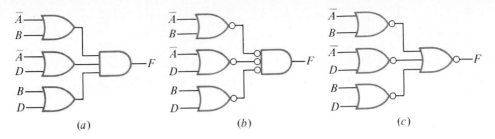

FIGURE A.11

The development of a product-of-sums logic expression $F = (\overline{A} + B)(\overline{A} + D)(B + D)$ to an all-NOR gate configuration.

Solution

For the doors closed, let $D = 1$
For the seat belts buckled, let $S = 1$
For the parking brake on, let $P = 1$
The car will start with $C = 1$

We next complete the truth table, as shown below. With three independent variables there are eight entries to the truth table. For each row of the truth table, the dependent variable (C) is checked against the state of the independent variables to see whether it is true ($= 1$) or false ($- 0$).

D	S	P	C
0	0	0	0
0	0	1	1
0	1	0	0
0	1	1	0
1	0	0	0
1	0	1	0
1	1	0	1
1	1	1	0

(a) From the truth table, notice C is true ($=1$) in only two of the rows. That is,

$$C = \overline{D}\,\overline{S}P + DS\overline{P} \qquad \text{a sum-of-products solution}$$

From the K map, in Fig. A.12a, we see that no further reduction of this expression is possible. The all-NAND solution to this problem is also given in Fig. A.12a.

A sum-of-products and all-NAND solution always result from identifying those rows of the truth table where the dependent variable is true.

(b) Also from the truth table, we note that C is false ($=0$) in six of the rows. That is,

$$\overline{C} = \overline{D}\,\overline{S}\,\overline{P} + \overline{D}S\overline{P} + \overline{D}SP + D\overline{S}\,\overline{P} + D\overline{S}P + DSP$$

Grouping the 0s in the K map of Fig. A.12b we have

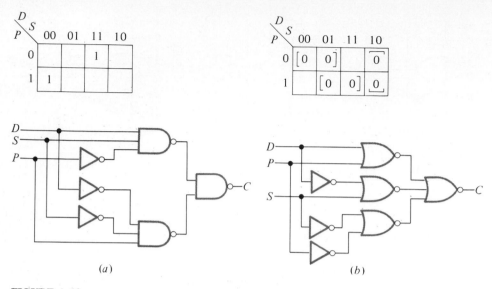

FIGURE A.12
(a) Sum-of-products solution: $C = \overline{D}\,\overline{S}P + DS\overline{P}$. (b) Product-of-sums solution: $C = (D + P)(\overline{S} + P)(\overline{D} + S)$.

$$\overline{C} = \overline{D}\,\overline{P} + SP + D\overline{S}$$

We will now show that a reduction of the $\overline{C}$ term leads to a product-of-sums and an all-NOR-gate solution. Inverting both sides of this equation,

$$C = \overline{\overline{D}\,\overline{P} + SP + D\overline{S}}$$

$$= (\overline{\overline{D}\,\overline{P}})\,(\overline{SP})\,(\overline{D\overline{S}}) \qquad \text{from identity } 18$$

$$= (D + P)(\overline{S} + \overline{P})(\overline{D} + S) \qquad \text{from identity } 19$$

This is a product-of-sums solution, and the all-NOR realization to this problem is shown in Fig. A.12b.

Exercise A.1. Simplify the following functions and express each (a) as a sum of products and (b) as a product of sums:

$$\overline{R}S\overline{T} + R\overline{S}T + RST$$

$$W\overline{Z}R + T(WR + \overline{Z}) + \overline{T}Z$$

Exercise A.2. Use a truth table and map to write a minimal expression for fighting, knowing that:
(a) Al will start a fight with anyone but only if someone else is watching.
(b) Bill will only start a fight with Chuck but only if no one else is around.
(c) Chuck will start a fight with anyone but not if Don is around.
(d) Don never starts a fight.

A.3   DIODE LOGIC

The first transistorized digital circuits, used in computers in the late 1950s and early 1960s, made use of individually packaged ("discrete") diodes and transistors in a circuit form known as *diode logic*. An understanding of the basic logic blocks used in digital systems can be obtained by studying OR and AND functions implemented with just diodes and resistors.

DIODE OR GATE. A diode logic gate formed with just two diodes and a resistor is illustrated in Fig. A.13*a*. The function of the diodes is to isolate the inputs and assure directivity of information flow. The simple current-voltage (*I-V*) relationship shown in Fig. A.13*b* is assumed for each diode. That is, with $V < 0.7$ V the diode is not conducting and $I = 0$. But with $V = 0.7$ V the diode current is unlim-

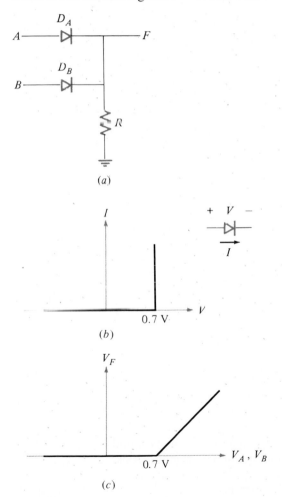

FIGURE A.13
Diode OR gate.

ited and $I \rightarrow \infty$, as seen in the plotted I-V characteristic. Such a characteristic is described by writing $V_{D(on)} = 0.7$ V.

The voltage levels at the input of the gate are chosen such that:

A voltage of 0 V represents a logic 0 (0) and is indicated as the low level (L).

A voltage of 4 V represents a logic 1 (1) and is indicated as the high level (H).

With a low at both A and B inputs of the gate, the diodes are nonconducting, there is no current in the resistor, and the voltage at the output F is 0 V.

Now with a high at input A and a low at input B, the diode D_A is forward-biased, and current now flows in the resistor. The voltage at F is now 3.3 V, and diode D_B is reversed-biased by 3.3 V. Of course, with a high at input B and a low at A, the output voltage is again 3.3 V, but now diode D_A is reversed-biased.

With a high at both A and B, both diodes are conducting, but F is still 3.3 V. The current in the resistor is shared equally by each diode. We conclude that the output F is at a high, or 1, with either A or B, or both, at a high, or 1. This, then, is a two-input inclusive-OR gate:

$$F = A + B$$

The voltage transfer characteristic (VTC) for this OR circuit is plotted in Fig. A.13c. The VTC is a common, useful means for describing important aspects of circuit performance, especially the nominal logic levels and noise margins. In the case of the diode OR circuit, the VTC is found by noting that

With $V_A = V_B < 0.7$ V: $\qquad V_F = 0$ V

With V_A or $V_B > 0.7$ V: $\qquad V_F = V_A$ or $V_B - 0.7$ V

A disadvantage of this simple OR circuit is the loss of 0.7 V in passing through the logic gate. This could be disastrous in a series string of such gates. It is very desirable to have the voltage levels at the output be the same as at the input.

A simple solution to the problem is shown in Fig. A.14a. Diode D_F acts as a voltage *level shifter*. Provided this diode is always conducting, the voltage drop across the input diodes is canceled by the voltage drop across diode D_F. Assuming $V^+ > 4$ V and $V^- < 0$ V, the values of the resistors R_1 and R_2 are chosen so that there is current in the level-shifting diode over the range of input voltages V_A and V_B. Now with a low of 0 V at the inputs A and B, there will be a low of 0 V at the output F. With a high of 4 V at either A or B, or both, there will be a high of 4 V at F.

The VTC for the modified OR gate is shown in Fig. A.14b. The positive excursion of V_F is limited by the current in D_F going to 0, while the negative swing is limited by the current in the input diodes going to 0.

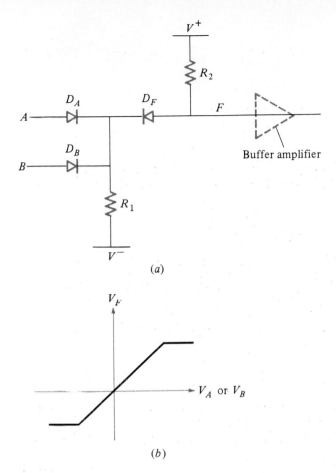

FIGURE A.14
Modified diode OR gate.

However, there is still a problem. Loading of similar gates at F causes V_F to decrease. For any reasonable *fan-out* (number of load gates) this could be troublesome. To overcome this undesired attenuation of logic levels, and to provide the capability of driving other inputs, a *buffer amplifier* is added to the output circuit, as shown dashed in Fig. A.14a. The task of the buffer amplifier is to maintain the output logic levels, in this case 4 V and 0 V, when the output is loaded by a reasonable number of load gates. Typically, this is 5 to 10.

DIODE AND GATE. With the same components of three diodes and two resistors a two-input AND gate can be constructed; this is shown in Fig. A.15a. Again, diode D_F acts as a level-shifting diode and should be conducting at all times. Then, with a low ($=0$ V) at both inputs, current is flowing in all diodes, and

$$V_F = V_A = V_B = 0 \text{ V}$$

The voltage at the common anodes would be 0.7 V.

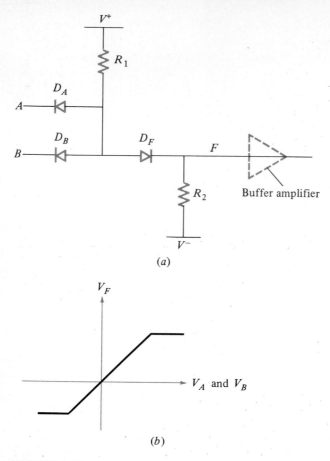

FIGURE A.15
Modified diode AND gate.

Now, with a high ($=4$ V) at A and a low at B, current continues to flow through D_B, D_F is still conducting, and

$$V_F = V_B = 0 \text{ V}$$

Diode D_A is reverse-biased by 3.3 V.

However, with a high at both A and B inputs, the voltage at the common anodes is increased to 4.7 V, the current in D_F is increased, and

$$V_F = V_A = V_B = 4 \text{ V}$$

This then is a diode AND gate, in that F only responds to a high, or 1, at both A and B.

$$F = AB$$

The VTC of this modified diode AND gate is similar to that of the modified OR gate and is shown in Fig. A.15b. However, for the AND gate the positive swing of V_F is limited by the current in the input diodes D_A and D_B going to 0, and the negative swing is limited by the current in D_F going to 0.

With the addition of a buffer amplifier at the output (shown dashed in Fig. A.15a), this circuit is similar to the latest development in transistor-transistor logic (TTL).

Exercise A.3. In Fig. A.14a, $V^- = -4$ V, $V^+ = +4$ V, $R_1 = 1$ kΩ, and $R_2 = 2$ kΩ. With $V_A = V_B$ sketch the VTC as V_A varies from -4 to $+4$ V. Assume a diode characteristic as in Fig. A.13b.

Exercise A.4. Repeat Exercise A.3 for the circuit shown in Fig. A.15a.

REFERENCES

1. V. H. Grinich and H. G. Jackson, *Introduction to Integrated Circuits*, McGraw-Hill, New York, 1975.
2. H. Taub and D. Schilling, *Digital Integrated Electronics*, McGraw-Hill, New York, 1977.
3. M. M. Mano, *Computer Logic Design*, Prentice-Hall, Englewood Cliffs, N.J., 1972.

ANSWERS TO EXERCISES

Chapter 2

E2.1. $V_T = -0.24 - 0.70 - 1.39 - 0.18 = -2.51$ V

$\gamma = 1.67$ V$^{1/2}$

For $V_T = -1$, $N_I = 3.3 \times 10^{11}$ cm^{-2}

For $V_T = +3$, $N_I = 1.2 \times 10^{12}$ cm^{-2}

E2.2. In saturation, $I_D = 123$ μA, so transistor is not saturated. Solving Eq. (2.11b), get $V_{DS} = 2.0$ V.

E2.3. $C_{ox} = 4.9 \times 10^{-8}$ F/cm^2

$C_{J0} = 6.95 \times 10^{-9}$ F/cm^2

$C_g = 0.05$ pF; $C_{sb} = C_{db} = 0.010$ pF

E2.4. $C = 2.07$ pF

E2.5. $I_{(av)} = 6.7$ mA

Chapter 3

E3.1. $V_{OH} = 3.5$ V; $V_{OL} = 1.4$ V

$V_{IH} = 2.2$ V; $V_{IL} = 1.0$ V

$NM_L = -0.4$ V (negative—no margin!); $NM_H = 1.3$ V

E3.2. $K_R = 18.6$.

E3.3. $I_D = 38.2, 34.7, 30.4, 19.1$ μA

Chapter 4

E4.1. $\phi_0 = +816$ mV

$\phi_n = +348$ mV

$\phi_p = -467$ mV

E4.2. (a) $Q_j = (3.4)(10^{-12})$ C

$E_o = (1.3)(10^5)$ V/cm

$$x_n = (8.6)(10^{-5}) \text{ cm}$$
$$x_p = (8.6)(10^{-7}) \text{ cm}$$
$$C_j = (0.30)(10^{-12}) \text{ F}$$
(b) $Q_j = (8.0)(10^{-13}) \text{ C}$
$$E_o = (3.1)(10^4) \text{ V/cm}$$
$$x_n = (2.0)(10^{-5}) \text{ cm}$$
$$x_p = (2.0)(10^{-7}) \text{ cm}$$
$$C_j = (1.3)(10^{-12}) \text{ F}$$
(c) $C_{eq} = (0.48)(10^{-12}) \text{ F}$

E4.3. $I_s = (1.7)(10^{-15}) \text{ A}$
$V = 0.70 \text{ V}$

E4.4. $\tau_s = 6.2 \text{ ns}$

E4.5.   (i) Open collector: short base
(ii) Open emitter: short base
(iii) BC short: short base
(iv) BE short: short base
(v) CE short: short base

E4.6. *pn*: $V = 780 \text{ mV}$ Al: $V = 480 \text{ mV}$
Mo: $V = 380 \text{ mV}$ Pd: $V = 540 \text{ mV}$
Ti: $V = 380 \text{ mV}$ Pt: $V = 660 \text{ mV}$

Chapter 5

E5.1. $I_E = I_{DE} - \alpha_R I_{DC} = I_{ES}(e^{V_{EB}/V_T} - 1) - \alpha_R I_{CS}(e^{V_{CB}/V_T} - 1)$
$I_C = \alpha_F I_{DE} - I_{DC} = \alpha_F I_{ES}(e^{V_{EB}/V_T} - 1) - I_{CS}(e^{V_{CB}/V_I} - 1)$

E5.2. $\beta_{sat} \ll \beta_F$ and $\beta_{sat} \ll \beta_R$
Alternatively, $V_{BE(sat)} - V_{BC(sat)} \ll I_{C(EOS)} r_c$

E5.3. $I_E = (4.98)(10^{-4}) \text{ A}$ $C_{BE} = 4.73 \text{ pF}$
$I_C = (4.93)(10^{-4}) \text{ A}$ $C_{BC} = 0.19 \text{ pF}$

Chapter 6

E6.1. $N = 22.3 \rightarrow 22$

E6.2. $\tau_F = 4.65 \text{ ns}$; $\tau_{BF} = 200 \text{ ns}$; $\beta_F = 43$

E6.3. $Q_F = (4.98)(10^{-14}) \text{ C}$; $Q_R = (-1.0)(10^{-23}) \text{ C}$

E6.4. First notice:

$$\beta_F + \beta_R + 1 = \frac{\alpha_F}{1 - \alpha_F} + \frac{\alpha_R}{1 - \alpha_R} + 1$$

$$= \frac{\alpha_F(1 - \alpha_R) + \alpha_R(1 - \alpha_F) + (1 - \alpha_F)(1 - \alpha_R)}{(1 - \alpha_F)(1 - \alpha_R)}$$

$$= \frac{1 - \alpha_F \alpha_R}{(1 - \alpha_F)(1 - \alpha_R)}$$

Now,

$$\tau_s = \frac{\beta_F \tau_F(\beta_R + 1)}{\beta_F + \beta_R + 1} + \frac{\beta_F \beta_R \tau_R}{\beta_F + \beta_R + 1}$$

$$T_s = \frac{\alpha_F T_F}{1 - \alpha_F \alpha_R} + \frac{\alpha_F \alpha_R T_R}{1 - \alpha_F \alpha_R}$$

$$= \frac{\alpha_F (T_F + \alpha_R T_R)}{1 - \alpha_F \alpha_R}$$

E6.5. (*a*) $V_{OH} = 5.0$ V; $V_{IH} = 1.5$ V; $NM_H = 3.5$ V
   $V_{OL} = 0.4$; $V_{IL} = 0.7$ V; $NM_L = 0.3$ V
   (*b*) $t_d = 0.8$ ns; $t_f = 4.4$ ns; $t_s = 0$ ns; $t_r = 15$ ns
   $t_{PHL} = 2.9$ ns; $t_{PLH} = 7.7$ ns

Chapter 7

E7.1. (*a*) $V_{OH} = 1.01$ V: $V_{OL} = 0.15$ V; therefore, $LS = 0.86$ V
   $V_{IH} = 0.73$ V; therefore, $NM_H = 0.28$ V
   $V_{IL} = 0.65$ V; therefore, $NM_L = 0.50$ V
   (*b*) $N = 2.87 \rightarrow 2.0$
E7.2. (*a*) BP 1: $V_{in} = V_{IL} = 1.4$ V; $V_{out} = V_{OH} = 4.0$ V
   BP 2: $V_{in} = V_{IH} = 1.5$ V; $V_{out} = V_{OL} = 0.1$ V
   (*b*) $N = 6.99 \rightarrow 6.0$
E7.3. (*a*) $N_H = 177$; $N_L = 77$; therefore, $N = 77$
   (*b*) $P_D = 10.6$ mW
E7.4. (*a*) $N_H = \infty$; $N_L = 112$; therefore, $N = 112$
   (*b*) $P_D = 2.2$ mW
E7.5. (*a*) $V_R = -1.18$; $V_{OH} = -0.75$ V; $V_{OL} = -1.58$ V
   (*b*) $N = 84$
   (*c*) $NM_H = 0.33$ V; $NM_L = 0.30$ V
E7.6. (*a*) Hand calculations:
   BP 1: $V_{in} = V_{IL} = 590$ mV ($I_C = 0.01 I_{C(max)}$); $V_{out} = V_{OH} = 715$ mV
   BP 2: $V_{in} = V_{IH} = 715$ mV ($I_C = 0.99 I_{C(max)}$); $V_{out} = V_{OL} = 19$ mV
   $NM_H = 0$ mV; $NM_L = 570$ mV
   (*b*) From SPICE:
   BP 1: $V_{in} = V_{IL} = 586$ mV ($I_C = 0.01 I_{C(max)}$); $V_{out} = V_{OH} = 725$ mV
   BP 2: $V_{in} = V_{IH} = 715$ mV ($I_C = 0.99 I_{C(max)}$); $V_{out} = V_{OL} = 19$ mV
   $NM_H = 10$ mV; $NM_L = 566$ mV
E7.7. $(W/L)_N = 145$ $(W/L)_P = 4.7$

Chapter 8

E8.1. See Fig. E8.1
E8.2. See Fig. E8.2
E8.3. See Fig. E8.3
E8.4. $V_{GS1} = 2.49$ V
E8.5. $D = J\bar{Q} + \bar{K}Q$ (see also Fig. E8.5)

   $= \overline{J\bar{Q}} + \overline{(K + \bar{Q})}$

   $= \overline{(\overline{J\bar{Q}})(K + \bar{Q})}$

E8.6. $R_1/R_E = 2.91$, $R_2/R_E = 6.32$

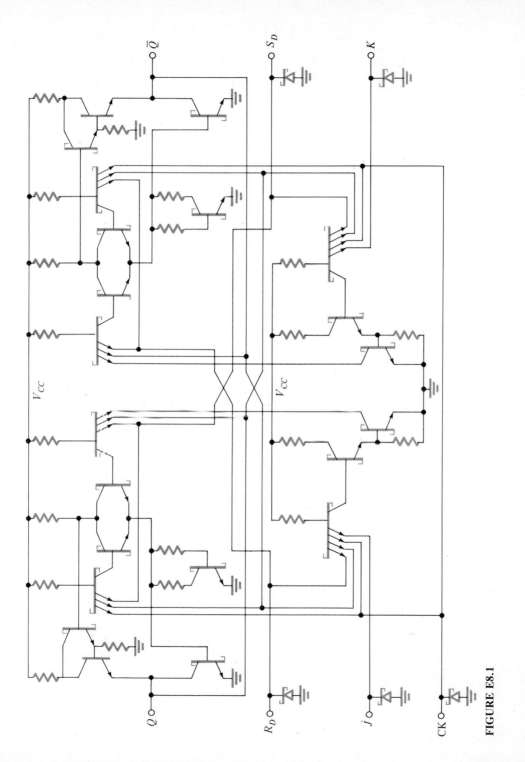

FIGURE E8.1

453

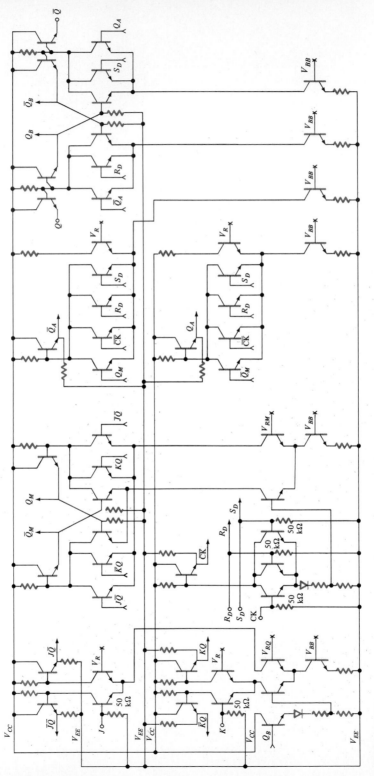

FIGURE E8.2

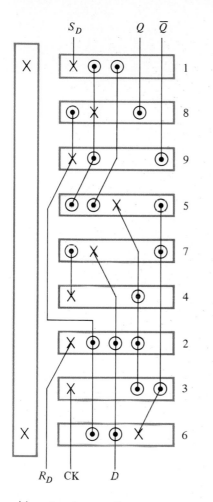

S_D Q $\overline{Q}$

R_D CK D

X   - Input connection

⊙   - Output connection

——   - Metal connection

FIGURE E8.3

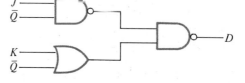

J
$\overline{Q}$

K
$\overline{Q}$

D

FIGURE E8.5

E8.7. (a) PW $= 6.9$ μs
  (b) $V_{Th} = 4$ V, PW $= 5.1$ μs, that is, -26%
    $V_{Th} = 6$ V, PW $= 9.2$ μs, that is, $+33\%$

E8.8. $V_X(t) = A + Be^{-t/\tau}$, where $\tau = R_X C_X$
    $t_0 \rightarrow t_1$: $V_X(t_1) = V_T = (V_T + V_{DD})e^{-t_1/\tau}$
    $t_1 = \tau \ln \dfrac{V_T + V_{DD}}{V_T}$
    But $V_T - V_{DD}/2$. Therefore,
    $t_1 = \tau \ln 3 = 1.1\tau$

$$t_1 \to t_2: V_X(t_2) = V_T = V_{DD} + (V_T - 2V_{DD})e^{-t_2/\tau}$$

$$t_2 = \tau \ln \frac{V_T - 2V_{DD}}{V_T - V_{DD}}$$

$$t_2 = \tau \ln 3 = 1.1\tau$$

Now, $t_p = t_1 + t_2 = 2.2\tau$

Hence, $f = \dfrac{1}{t_p} = \dfrac{1}{2.2\tau}$

E8.9. See Fig. E8.9

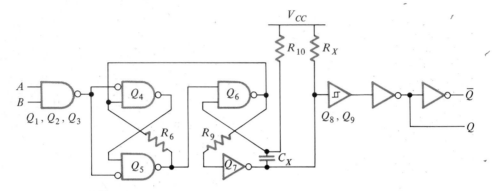

FIGURE E8.9

Appendix A

EA.1. i. (a) $F = RT + \bar{R}S\bar{T}$

(b) $F = (R + S)(R + \bar{T})(\bar{R} + T)$

ii. (a) $F = RW + T\bar{Z} + \bar{T}Z$

(b) $F = (R + T + Z)(R + \bar{T} + \bar{Z})(\bar{T} + W + \bar{Z})(T + W + Z)$

EA.2. $F = AC + ABD + BC\bar{D}$

EA.3. See Fig. EA.3

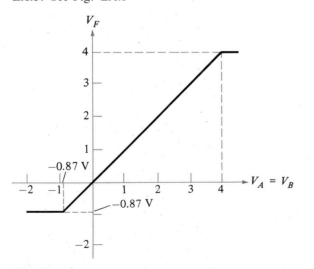

FIGURE EA.3

EA.4. See Fig. EA.4

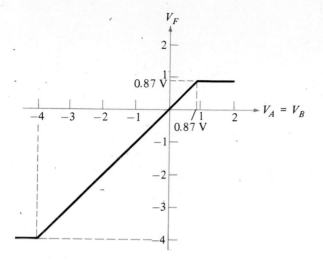

FIGURE EA.4

INDEX

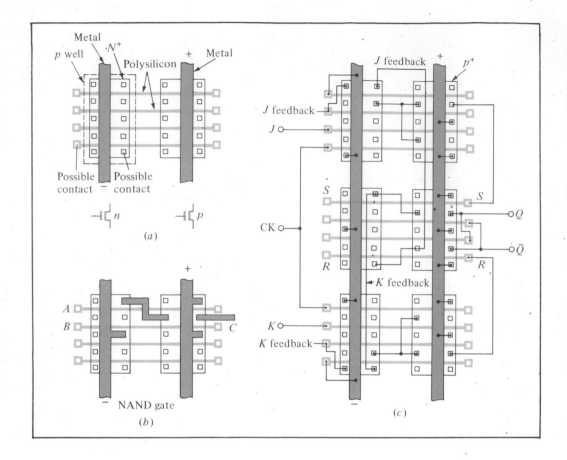

CMOS gate arrays.